1983

MATHEMATICS
concrete behavioral
foundations

MATHEMATICS
concrete behavioral
foundations

Joseph M. Scandura

Mathematics Education Research Group
University of Pennsylvania

with the assistance of
John Durnin
George Lowerre

Harper & Row, Publishers
New York, Evanston, San Francisco, London

To
Jeanne, Janie,
Joey, Julie

Contents

* indicates optional section

The instructor who is entirely satisfied with the text he is presently using is not likely to adopt this one. If you fall into this category, you might just as well close the book now and save yourself the trouble of reading further. On the other hand, if you have been concerned that students who have studied your present text still don't understand what mathematics is all about (although they may know a fair amount of detail and/or new vocabulary), or cannot relate the mathematics they have learned to the real world, or have little feeling for the wide variety of behavior which knowing mathematics makes possible, then, under these conditions, it may pay you to look further.

This book has been designed as an intermediate level text on mathematics for elementary and middle school teachers, and may also be used in terminal courses for liberal arts students or as a supplement in courses for secondary school teachers. It would be appropriate as a first course in some colleges and universities, as it has been used at Pennsylvania, or as a first *year* course, or as a second semester (year) course in other colleges. Our experience with preliminary versions of the book shows that it is especially well suited for use by experienced teachers, and the book could very profitably be used either as a graduate text or for in-service teacher training courses.

The emphasis throughout the text is on building a cohesive and concrete mathematical edifice. Even where specific topics are being covered, a concerted attempt is made to show how these topics fit into the overall conception. Several techniques have been used to help accomplish this. First, basic mathematical ideas recur in new guises throughout the text. Sets, relations, operations, systems, and so on, are not only treated as important concepts in their own right, but are used throughout the text to help unify the treatment. Second, where possible, as in most of Part III, the chapters have been constructed in parallel fashion. It has been our experience that this makes it easier for the student to see the similarities as well as the differences which exist among number systems. In addition, where parallelisms seem to break down (or become cumbersome), I have tried to indicate why. Third, explicit attention has been given throughout to the relationship between mathematical ideas and reality. In fact, the book is unique among full-length texts in the degree of emphasis given to concrete embodiments of mathematical ideas. We hear much about relevance today, and students and teachers alike are no longer as enamored by "modern mathematics" as they were even a few years ago. If we are to keep them from "throwing out the baby with the bath water," we as mathematics educators are going to have to show them why modern mathematics is important. This text is designed to help make this possible. Fourth, although the emphasis is on mathematical content, the book is not limited to just that. Chapter 1 contains a unique analysis of the processes (e.g., detecting regularities, making inferences, etc.) which, although not normally considered to be part of mathematical content, are extremely crucial in *doing* mathematics. In many ways, this chapter sets the tone for much of the rest of the book. It illustrates the advantages, especially for teachers, of thinking about mathematics in terms of the kinds of behavior which knowing mathematics makes possible. This point of view recurs throughout the book and helps provide further insight into both the mathematical ideas themselves and how they might be taught to students. To take full advantage of these insights, methodological questions are included in most of the exercises.

The book requires very little in the way of prerequisite mathematics, but it does require that the student be able to read precise college level prose. Mathematical notation is used sparingly and only where necessary for clarity and precision of meaning. Many examples are given and the reader is asked questions at many points in the text. The latter are designed to elicit the student's involvement and to help ensure that he thinks about what he is reading and does not gloss over important details.

The text is divided into three parts. Part I (Chapter 1) on objectives and mathematical processes is the first concerted attempt to deal systematically at an elementary level with those basic mathematical competencies (processes) which are invariably involved in performing mathematical tasks but which are rarely made an object of explicit study.* This chapter has been singled out by many of our students as one which they have found most useful in their teaching. It has also been used with secondary school mathematics teachers with equal success. In addition, the point of view may be useful to the liberal arts student by providing him with a broader perspective on mathematical thinking.

Part II (Chapters 2–4) is designed to introduce the reader to many of the major ideas in contemporary mathematics. Chapter 2 deals with those general mathematical ideas which are basic to essentially all of mathematics (i.e., sets, relations, and operations) and emphasizes the similarities and differences both within and among them. The treatment is more complete than in most texts at this level. Chapter 3 deals with logic and deductive reasoning and shows informally how the set operations (i.e., Boolean algebra) provide a model for the statement logic. Chapter 4 introduces mathematical (algebraic) systems and embodiments, and theories, as the basic objects of mathematical study. The emphasis throughout this part is on the concrete reality of the mathematics in the lives of both the reader (elementary school teacher) and her students. The

---

* Among the other treatments presently available, the monograph *Mathematics in the Primary School* (Melbourne: Macmillan, 1966) by Z. P. Dienes is particularly recommended.

treatment has a number of other features which are not common at this level. These features include (1) the central role played by sets, relations, and operations in organizing content, (2) relationships between logical reasoning and the real world, (3) an introduction to relationships which may exist among different mathematical systems, and (4) a sharp distinction between semantics (systems) and syntax (properties of systems, or theories).

Part III deals systematically with the more traditional topics on number systems, beginning with the natural numbers and building up through the positive rationals and integers to the rationals, reals, and further extensions. Although the main approach is constructive, the number line is introduced as an alternative in discussing the positive rationals and integers, and is used heavily in discussing the reals in Chapter 9. In addition to an emphasis on the concrete foundations of the subject, the various chapters (particularly Chapters 7 and 8) are explicitly constructed so as to parallel one another. This procedure provides increased insight into the relationships that exist both within and between the different number systems, and at the same time makes possible a more sophisticated approach than would otherwise be possible. In line with the difference between denoted entities and descriptions (introduced in Chapter 1 and continued in Part II), a sharp distinction is made throughout between numbers and number systems, on the one hand, and numerals and systems of numeration, on the other. Chapter 9 ends by describing a basic relationship between algebra (i.e., number systems) and geometry, and in the process introduces the idea of a limit as a prelude to analysis.

Although there is little geometry in the text itself, it is too important to leave out entirely, and Appendix A provides a short, self-instructional introduction to geometry which parallels a treatment developed by the School Mathematics Study Group for primary school teachers. Rather than using a text format, the material has been organized by listing the specific geometric tasks, which the reader is expected to perform, and then describing a rule for solving such tasks, together with examples.

Appendix B will also be useful to teachers and prospective teachers, because it is based on an analysis of ten elementary school textbook series in mathematics. This research was conducted under my direction by members of the Mathematics Education Research Group (MERG) under contract with Research for Better Schools, U.S. Regional Educational Laboratory, Philadelphia, Pennsylvania.

In order to facilitate use of the text, an answer book with many worked-out solutions is available on request from the publisher. (Answers to the odd-numbered exercises, together with many detailed solutions, are included at the end of this book.) In addition, a workbook has been designed to directly parallel the text. In this workbook, many of the key ideas have been identified and formulated as specific tasks to be performed. The reader may learn to solve these tasks either by reading (and interpreting) explicit rule statements for solving each kind of task or by inducing the underlying rule from the examples provided. Numerous exercises are also included for each kind of task. The approach used is based on a theory of mathematical knowledge,* which goes beyond the usual behavioral objectives point of view, and makes it possible to build transfer potential directly into the workbook. In particular, the workbook includes tasks which may be solved by combining in predetermined ways rules that have been learned earlier. This workbook has been evaluated empirically as part of a doctoral dissertation by Walter Ehrenpreis and the results unequivocally support the "higher order" form of analysis proposed in the aforementioned theory.

The selection of exercises has also been facilitated by keying them as follows: *Section* exercises are prefaced by an *S* and are based directly on the material in the particular section involved. *Extra* exercises are prefaced by an *E* and require information from other chapters or elsewhere. (This information may or may not be readily available to the student.) *Methods* exercises are prefaced by an *M* and deal with methodological issues involved in teaching the content in question. Exercises marked with an *asterisk* are more difficult than the others and require more ingenuity on the part of the student. *Brackets* indicate that the question deals with one or more of the ideas introduced in Chapter 1. Frequently these ideas do not affect the working of the problem, and the brackets are included merely to help make the student more aware of how the fundamental processes identified in Chapter 1 enter into *doing* mathematics. Finally, most sections have an additional exercise or two for *further thought and/or study*. In writing many of the exercises, a conscious effort has been made to supplement, as well as simply reflect, the text. I hope that the instructor will find these exercises, and the solutions provided, a valuable aid in promoting student learning as well as in testing.

The text may be used in a variety of ways depending on the instructor's interests and the needs of the students. In particular, the three parts of the book have been written so as to be largely independent of one another and the text can be entered equally well at Chapter 1, Chapter 2, or Chapter 5. In addition, Chapters 2 and 3, 3 and 4, and 7 and 8, respectively, may be interchanged (except that 2 should come before 4). It would be quite feasible, for example, to start with Chapter 5 of Part III, and then alternate between chapters in Parts I and II (Chapters 1–4) and those in Part III. Changes should not be made indiscriminately, however, and without good reason.

The book also lends itself to individualization in the sense that the development tends to parallel the order in which various topics are introduced in the elementary school. Thus, preschool and primary school teachers (N, K–2) may be allowed to concentrate on Chapter 5 and the initial parts of Chapters 6–8 (on the nature of numbers and numerals), together with the relevant parts of Parts I and II. Teachers in the upper elementary grades will correspondingly want to spend relatively more time on the algorithms (Chapter 6) and the body of Chapters 7, 8, and 9.

In teaching undergraduate elementary school teachers in training, we have

---

* J. M. Scandura, *A Theory of Mathematical Knowledge: Can Rules Account for Creative Behavior?* Mathematics Education Research Group, Report No. 52, Structural Learning Series, University of Pennsylvania, Philadelphia, Pa. 19104 (also in *Journal for Research in Mathematics Education*, in press).

found that it makes an important difference whether the material is presented before, during, or after practice teaching. If presented before, or particularly after, the indicated order is just as good as any other. For students who have never taught or observed, however, it is sometimes useful to postpone the introduction of Chapter 1 until after they have covered Part II (or Part III). If the course is taken at the same time as practice teaching, it is frequently useful to get into Part III, particularly Chapters 5–7, as soon as possible. The students are frequently fearful and concerned more with what they perceive as immediate matters—for example, the more traditional topics on arithmetic which they see in the children's texts—than with topics which the instructor may deem more crucial and of long-range benefit. Under these circumstances, it is better to let the students get their feet wet in practice teaching, and thereby get a feeling for what is important in the school setting, before getting into Parts I and II. Nonetheless, the ideas inherent in these chapters can be profitable and should be introduced implicitly in discussing Part III. For example, the question "How would you teach this?" immediately raises the question of whether to use examples (i.e., teach by discovery) or exposition. Students who have had to face up to this kind of question find the discussion in Chapter 1 particularly relevant and useful.

In graduate courses for experienced teachers, just the reverse has been true. Teachers take very naturally to the ideas discussed in Parts I and II. This, in turn, tends to provide them with a useful background for increasing their depth of understanding of the more traditional topics discussed in Part III. Parts of Chapters 3 and 4 are somewhat more advanced than the rest (e.g., the discussion of homomorphisms) and can be made optional. However, many teachers who do succeed in mastering these ideas have been most anxious to try them out with their pupils. (We have been pleased with the many success stories we have heard from teachers who have tried this.)

Chapters 5 and 6 tend to go rather smoothly. Special care should be taken in introducing Chapters 7 and 8, as the approach is generally unfamiliar to most teachers. The basic arithmetical operations in these chapters are defined first and then their concrete realizations are discussed. Although contrary to the order preferred by many instructors, this tended to facilitate the discussion as well as make it more precise and efficient. Furthermore, there is some research by Suppes and Dienes which suggests that this order of presentation may in the long run be the most efficient. The instructor, of course, is free to use whichever order he deems best for his classes.

It is also worth noting that the parallel construction of Chapter 7 and Chapter 8 tends to make things get progressively easier for the student as he goes along, even where he may have been confused at the beginning. The student should be made aware of this and provided with encouragement where needed. Chapter 9 is more compact than the others and is included largely as icing on the cake for the better student. The instructor should feel free to pick and choose from this chapter as he sees fit. The introduction and summary sections are particularly recommended, as these provide the student with a useful overview of the different number systems. The treatment of the rationals, and particularly the reals, is more thorough than in other books written for the elementary school teacher.

We have found that both pre- and in-service elementary school teachers are well able to study Appendixes A and B independently. With preservice teachers, however, it will generally be useful to discuss Appendix B in class (in order to supplement the views expressed there with your own).

In a terminal course in mathematics for liberal arts and junior college students, the text may be used as is, although the instructor may wish to post-pone the introduction of Chapter 1 until after Part II has been completed. In addition, of course, he may want to ignore the material pertaining to method-ology. This can be accomplished by simply eliminating those exercises marked with an M. It might be possible to cover most of the book in one semester in some colleges and universities, but in most, and certainly in junior colleges, there is more than enough material for a full year. Parts I and II, together with sections from Part III, constitute a particularly appropriate curriculum for a one semester course for liberal arts students.

With secondary school teachers, the text is probably best used as either a supplement or as one of two texts. Chapter 1 should be covered thoroughly and the students required to make up examples of their own for each of the basic processing skills identified. The remainder of the book can be covered more rapidly than with elementary school teachers, but we have been continually appalled by how much of the material even our better students do not know. Coverage of this content is particularly important for the junior high school teacher (who frequently has to be an expert in remedial mathematics). Mathe-matics majors can generally cover most of the text (including Chapter 9) in about half a semester, leaving the remainder for other purposes.

Although the text is entirely self-contained, it has been our experience that the major ideas in the text can be made available to even the weakest student by carefully coordinating the use of the text with the workbook (description above) which has been prepared for this purpose. This workbook has been designed to parallel the text closely and can be used either to introduce the key topics be-fore reading the text or as highly directed practice which can be accomplished after the student has read the text, possibly in lieu of the text exercises. The instructor may want to suggest both alternatives to the students and let them select the way which seems to best fit their own learning style.

It is an open secret that no book is the product of simply one man. This book is no exception. In writing the book, I have had more than a little assist-ance from a number of members of the Mathematics Education Research Group at the University of Pennsylvania. Joanna Burris, Walter Ehrenpreis, and Judy Gera assisted me when I introduced many of the ideas for the first time in a class for elementary school teachers. Later during a sabbatical year, John Durnin assisted me with the book in any number of ways. His help ranged from pulling together needed materials, to administrative assistant, to proofreader par excellence. Furthermore, I incorporated his comments, criticisms, and sug-gestions on more than one occasion and the book is a far better product for it. This help came at a most appropriate time, as I was then also engaged in writing my research monograph on *Mathematics and Structural Learning* (Englewood Cliffs, N.J.: Prentice-Hall) and a number of research articles. Upon my return I received the ready assistance of a number of my students in fine-tuning the text and in writing exercises and answers. George Lowerre, Louis Ackler, Julia Gatter, Sister Jeannine Grammick, Julia Hirsch, George Luger, and Christopher Toy all played important roles in this process. I must single out George Lowerre, however, for the special role he played in collating, criticizing, and editing both the exercises and the comments on the text itself. Both Durnin and Lowerre also assisted me later in going over the edited manuscript and in reading galleys and page proofs. It is for these reasons that I have singled them out by acknowledging their help on the title page: "with the assistance of." In addition, Walter Ehrenpreis provided substantial assistance with Appendix A, Gerald Satlow with Appendix B, and Linda Hunsicker with the index.

In spite of all the help I have received, however, I must bear full responsibility for the contents. I planned and wrote the book, I decided what kinds of exercises to include, and I decided what changes if any were to be made and how. I readily acknowledge that whatever limitations still remain in the text may well be due to the many suggestions or comments made that I chose to ignore.

In addition, I want to thank those of the Harper & Row staff who have been most helpful during the final phases of the work. Blake Vance, the editor, and Tony Asch, his field assistant, played no small part in my decision to publish with Harper & Row. I only hope that their trust in me and the purposes of this book prove justified. Karen Judd and Susan Emry of the Harper production staff, who labored so long over the manuscript itself, made my job much easier than it would otherwise have been.

Credit is also due the typists who worked so long and so hard. Mrs. Mary Tye did such a fine job for me in California that I still frequently send her things to do all the way from Philadelphia. Katherine Whipple and Lee Carvalho did an equally competent job in making the final corrections.

Last, but not least, I would like to express my deep gratitude to my parents and to my wife, Alice, and my children for their support while I was writing the book. During this period, as well as throughout much of my professional career, my wife has borne the burden of most of the day-to-day problems of raising a family of four children and has left for me most of the joys.

I shall feel more than repaid for the effort that went into this book if mathematics teaching in some small way may benefit as a result. Since my children would be among the beneficiaries of such improvement, it is to them that I dedicate this book.

JOSEPH M. SCANDURA

The book you are about to study is different from most college mathematics texts. To be sure, it has its so-called formulas and symbols but it has fewer of them than is common in most books which cover the same content and in the same depth.

Although much of the book is written in ordinary prose, you must not be deceived by this. As with any mathematics book it is not possible to skip sentences, phrases, or even words without taking the chance that you will overlook something important.

Even more important than reading the material carefully is that you *think* about what you read. After you read through a paragraph or short section, or even a particularly meaning-filled sentence, ask yourself what it means before going on. Try to visualize the idea. Ask yourself questions about the material. Or, rephrase the material in your own words. These are all useful techniques for making sure that you know exactly what is being said. To help you in this regard, numerous examples are included and questions are interspersed throughout the text. It is important that you check these examples and answer the questions.

In studying the text, you should also have a pencil and paper in hand so that you can work things out on paper where necessary. Remember, even professional mathematicians do much of their hard thinking on paper. Don't try to do it all in your head. It is simply not possible. (Of course, many questions will be easy enough to answer in this way—and, if so, you will know it.)

One final point on the text: The exercises are *not* optional. Full understanding is rarely achieved simply through reading. It is absolutely necessary to also work enough of the exercises to get a feeling for how the various ideas apply, how they interact, and when to use them and why.

If you have the workbook as well as the text, there are several ways in which you might use them. Try different alternatives until you find out which one best fits your own learning style. If you are are a good reader, for example, you might want to first read the text somewhat more rapidly than you otherwise would. Then, go through the workbook and exercises. Finally, you might pick out and work those text exercises that still seem novel to you. (You can check the answers to the odd-numbered exercises at the end of the text. Many solutions are worked out in detail for your benefit.) If you are an average reader and feel your mathematics background to be particularly weak, on the other hand, you might want to go through each section in the workbook to firm up the key ideas before going to the text to see how they all fit together. As in the other technique, you would then work those text exercises which still appear novel. (In using the workbook, some students find it best after reading each task to go directly to the examples rather than to first read the corresponding rule.)

Before you begin, you may want to browse through the table of contents to get an overview of the course. For more details, you might even want to peek at the preface for your instructor.

Good luck on your journey. I hope you will enjoy studying this book. Even more, I hope you will profit from it.

JOSEPH M. SCANDURA

One
Objectives
and Process Abilities
in Mathematics

1

Objectives
and Process Abilities
in Mathematics

OBJECTIVES

During the past 10 years, we have witnessed a large-scale revision of the mathematics curriculum at all levels from the preschool through the graduate level. The general tendency has been to organize mathematical content about general notions and to rid the curriculum of obsolete material. With these changes in content have come changes in objectives. Many who have been most active in curriculum reform are quick to point out that their primary aim is not simply mathematical content per se, but to teach children to think creatively—as stated by some, "to think like a mathematician." This objective has been given particular emphasis at the elementary school level.

Few thoughtful individuals, of course, would hold that mathematical thinking is sufficient for solving *all* of life's problems, but almost all would agree that mathematical thinking is desirable, and perhaps indispensable, in a large number of everyday situations—situations that involve much more than simple arithmetic. Mathematical and logical thought are involved in all kinds of planning, in detecting regularities and structural similarities in seemingly disparate situations, in organizing information, in argumentation, and in such other process abilities or skills as technical reading and interpreting maps and graphs. The difficulty with such an objective, however, is that no one has been able to give a satisfactory definition of just what it means to think mathematically. Without such a definition, only "those in the know" would be able to tell when a student is thinking like a mathematician and when he is merely thinking.

This does not mean that we should abandon such general objectives. On the contrary, such broad educational objectives are central to all curriculum planning. Still, the question of objectives has led to a good deal of controversy. There seem to be two fairly well defined, but opposed, points of view. On the one hand, many educational psychologists hold that stating objectives is worthless insofar as educational planning is concerned unless the objectives are stated unambiguously and in operational terms—that is, in terms that make it possible to determine, by testing the learner, whether or not an objective has been attained. Only then, it is argued, can objectives serve any real purpose. Proponents of this view have organized curricula about such easily tested objectives as: "counts orally from one to ten"; "supplies the number which is one more, one less, or in between any two given numbers less than 200"; "adds two numbers with sums up to 20 using expanded notation when required"; "writes any given date in terms of the month, day, and year"; "uses a ruler in measuring real objects or pictures to the nearest inch."

The other group, composed primarily of mathematics educators, has been just as firm. It is useless to specify objectives, they would argue, because only the most trivial objectives can actually be specified. A typical comment might be:

> The ability to make intelligent guesses, the ability to think mathematically, etc.; these are the important objectives of mathematical education but we are, as yet, unable to define them operationally. We think it is important for people to appreciate mathematics, to develop intellectual independence, to develop effective habits of thinking, to appreciate the importance of deductive thought, and furthermore, we feel that we can make good judgments as to which kinds of instructional situations may be helpful in achieving these ends. Do not try to pin us down prematurely as to just what is involved.

The present view is somewhere in between. On the one hand, it is felt that complete reliance on operationally defined objectives has led some to fragmented curricula, curricula based on discrete bits of knowledge with little or no attention to general processing skills. On the other hand, the nonobjectivists have not gone as far as possible in pinning down the vague and nonoperational aims of mathematics education which they propose.

It appears that more attention should be given to identifying general processing skills *and* making them operational. Unfortunately, however, this is easier said than done. To get some appreciation for the difficulties involved, it may be helpful to consider some earlier attempts to identify educational objectives.

There are many statements of such objectives, both of education in general and of mathematics education in particular. One of those that has been widely referred to is known as the Report of the Harvard Committee.[1] According to this report, the aims of general education are: (1) to contribute to the preparation for life's needs, not only those which the student realizes but also those he must be taught to realize; (2) to establish basic relevance between knowledge and everyday experience; (3) to provide a nonspecialized type of training characterized by wide application, universal value, and great intellectual appeal; and (4) to lay the foundation of basic information essential to later intelligent pursuit of individual interests and special aptitudes.

[1] The Harvard Committee, "General Education in a Free Society," Cambridge, Mass., Harvard University Press, 1945, p. 4.

The Harvard Committee also proposed a list of those competencies that a program designed to meet such aims should instill in students. This list included such competencies as the ability to: (1) think effectively, (2) communicate thought, (3) make relevant judgments, (4) improve and maintain health, (5) understand his physical environment, etc. There were 13 competencies listed in all. Not many people, of course, would disagree with the content of these competencies. The difficulty with the competencies is that they do not adequately indicate how one would go about teaching them or, in fact, how to determine whether they have been achieved. In short, the competencies listed are not operational. Furthermore, such lists as these typically overlap in uncertain ways, so it is difficult, if not impossible, to know how they relate to one another—or, put another way, to know the role each plays in defining the educated man.

The efforts of prestigious groups such as the Harvard Committee, however, are not to be belittled. The committee took on a very difficult job, a job we are clearly not prepared to go into here in depth. Nonetheless, to provide a perspective within which to view the general process abilities for mathematics education described below, we shall propose an admittedly tentative, but potentially operational, definition of the educated man. According to this definition, *the central aim of the schools is to maximally increase the students' capacity to deal effectively with those situations that are most apt to occur during his lifetime in a way that is both desirable to him and not incompatible with the aims of the society in which he lives.* This definition is extremely general; it includes everything from the competencies required of a physicist working on a theoretical problem to those required of a stunt man who finds himself in a situation in which his parachute has not opened. Still, it is operational in the sense that one could determine in a more or less objective manner the extent to which a person has or has not been "educated." For example, a school that graduates a habitual thief, no matter how clever, would not have achieved the stated objective.

Restricting this definition to mathematics education, we might arrive at a statement not incompatible with that expressed in 1927 by the National Committee on Mathematical Requirements:

> The primary purposes of the teaching of mathematics should be to develop those powers of understanding and of analyzing relations of quantity and of space which are necessary to an insight into and a control over our environment and to an appreciation of the progress of civilization in its various aspects, and to develop those habits of thought and of action which will make these powers effective in the life of the individual.

Stated more operationally, *the person educated in mathematics should be able to deal effectively with situations in which mathematical knowledge and/or the methods of thought used in mathematics are applicable.*

SIX BASIC PROCESS ABILITIES IN MATHEMATICS

Although these statements provide an umbrella under which to work, they are certainly not very specific concerning the kinds of competencies that might be involved. Furthermore, although the last statement is suggestive, none of these prescriptions clearly distinguishes between the *general processing skills* involved in *doing* mathematics and those abilities more aptly associated with *mathematical content.* This is understandable—it is often difficult to make this distinction.

To circumvent this problem in our own research,[2] we have identified six basic kinds of processing skills, from which a whole host of others may be derived. Competencies that do not fall into one of these six categories, or cannot be derived from them as indicated in Section 7, are simply not considered to be processing skills—*by definition.*

The six basic kinds of processing skills are:

1. the ability to *detect* mathematical regularities and
2. the reverse ability, to *particularize*, which involves constructing instances (examples) of given regularities;
3. the ability to *interpret* descriptions of mathematical ideas, which includes the ability to learn by exposition as a special case, and
4. the reverse ability, to *describe* (to others), mathematical ideas which have already been learned;
5. the ability to *make logical inferences* and
6. the reverse ability, to *axiomatize*—to identify basic ideas (e.g., axioms) in a set of ideas (e.g., properties of a mathematical theory) from which all the other ideas may be logically derived.

In the following sections, we will first consider each of the six kinds of abilities in turn. Then we will examine several kinds of more complex processing skills that involve combinations of these six basic kinds of ability. As will be seen, mastery of such skills may greatly increase one's ability to solve problems and, more generally, to engage in effective mathematical thinking of all types.

As a result of studying this chapter, you should gain a better feeling for the kinds of processing skills associated with knowing mathematics and with the interrelationships among these skills. This should help make it possible for you to classify tasks according to the types of processing skills required for their solution and also to construct tasks that will require your students to use such processing skills in order to solve them.

1. DISCOVERY—THE ABILITY TO DETECT REGULARITIES

1.1 What Is Involved in Detecting Regularities?

One of the major aims of mathematics instruction is to improve a person's ability to detect regularities in displays of various kinds. Roughly speaking, detecting a regularity involves perceiving a pattern or drawing out or abstracting that which is common to a number of examples. This abstracting process is often called *reasoning by induction.*

The presence of a regularity in a display implies the existence of some common underlying rule or rules, which, once discovered, make it possible for the

---

[2] This research is being conducted by members of the Mathematics Education Research Group (MERG) of which the author is director. Headquarters of the group is at the Graduate School of Education, University of Pennsylvania.

learner to behave in ways not previously possible. More specifically, we can determine whether or not a particular regularity has been detected by testing the learner on new examples. If he responds to the new examples according to the underlying rule, he is then said to have detected the regularity. If he cannot, then he has not detected the regularity. (We could, of course, ask him to *describe* the regularity or the corresponding underlying rule but, as we shall see in Sections 3 and 4, this requires additional processing skills which the learner may or may not have.) Some examples will help make clear what is involved.

1.2 Some Examples

One of the techniques most commonly used by elementary school teachers to promote discovery involves the sequential presentation of examples. From these examples, the student is required to discover the regularity in question.

(1) Consider the regularity evident in the display

$$
\begin{aligned}
1 &= 1 \\
1 + 3 &= 4 \\
1 + 3 + 5 &= 9 \\
1 + 3 + 5 + 7 &= 16 \\
1 + 3 + 5 + 7 + 9 &= 25
\end{aligned}
$$

Having examined these examples, can you predict the sum of the number series $1 + 3 + 5 + 7 + 9 + \cdots + 21$? If so, you have probably detected the relationship between the various number series and their respective sums. If not, consider the number of terms in each number series and its relationship to the sum. What happens when you multiply the number of terms by itself?

Clearly, to obtain the sum of a series of consecutive odd numbers beginning with 1, all one needs to do is square the number of terms in the series.

Is this rule applicable to $1 + 3 + 5 + \cdots + 21$? What is the sum?

What about $2 + 4 + 6 + \cdots + 14$? If the rule above does not apply, can you discover one which does? *Hint:* You may find it helpful to construct other series of this type and to order them as above—that is, consider

$$
\begin{aligned}
2 &= 2 \\
2 + 4 &= 6 \\
2 + 4 + 6 &= 12
\end{aligned}
$$

(2) Another example is provided by a display that classifies numbers according to divisibility by 3. For the time being, ignore the numbers in the two left-hand columns (in parentheses).

((7))	(16)	97 — is *not* (divisible by 3)
((3))	(12)	5412 — is
((3))	(12)	2433 — is
((4))	(13)	319 — is *not*
((7))	(7)	322 — is *not*
((5))	(14)	12335 — is *not*
((6))	(15)	456 — is

Can you predict, without dividing, whether 392016 is divisible by 3? If not, examine the sum of the digits in the respective numerals above. You may find it helpful to actually write the sums to the left of the numerals (as shown above in single parentheses). If you still cannot tell whether 392016 is divisible by 3 without dividing, write the sums of the digits of the numerals in parentheses further to the left (as shown above in double parentheses). Are these sums divisible by 3? What is the relationship between the numbers in parentheses and the numbers being tested for divisibility by 3?

Does this rule work with 29742? What about 461? By now you have probably discovered that if the sum of the digits in a base ten numeral is divisible by 3, then so is the original number—but not otherwise.

(3) Next, consider the geometric display

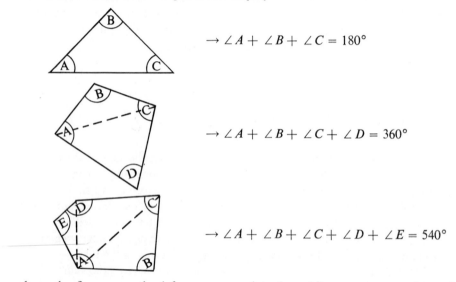

$$\rightarrow \angle A + \angle B + \angle C = 180°$$

$$\rightarrow \angle A + \angle B + \angle C + \angle D = 360°$$

$$\rightarrow \angle A + \angle B + \angle C + \angle D + \angle E = 540°$$

where the figures on the left are assumed to be arbitrary representatives of three-, four-, and five-sided polygons (i.e., triangles, quadrilaterals, and pentagons). The question in this case is, What is the relationship (if any) between the number of sides in a figure and the sum of the "interior" angles. Once detected, of course, this relationship should make it possible to give the sum of the angles of any polygon (n-sided figure). On the basis of the three examples in the display, what would you expect the sum of the angles of a hexagon (six sides) to be? An octagon (eight sides)? Draw an arbitrary hexagon (or octagon) and measure the angles with a protractor. Is the sum what you expected?

The dotted lines may help you to understand why this relationship holds. What is the sum of the interior angles of a triangle (as above)? Can the polygons above be broken down into triangles? How many? What about a hexagon? An octagon? An arbitrary n-sided figure? As a check, we note that the sum for a *12*-sided polygon is $12 - 2 = 10$ (the number of triangles) times 180°, or 1800°.

Regularities may also be displayed in various kinds of patterns and arrays.

(4) Consider, for example, the sequence

$$0 \ 0 \ \square \ 0 \ 0 \ \square \ 0 \ 0 \ \dots$$

Can you continue the pattern? One way in which this might be done is

0 0 □ | 0 0 □ | 0 0 ▨ | 0 0 ▨ | ...

(The vertical lines indicate the repeating cycle.)

Although this is the pattern which most people would probably detect, it is not the only one possible. Can you find any others? One other way to continue the pattern is

0 0 □ 0 | 0 □ 0 0 | | 0 0 ▨ 0 | 0 ▨ 0 0 | | ...

Among other things, this example illustrates that reasoning by induction does *not* necessarily lead to a single regularity. In general, there may be any number of alternatives.

(5) The sequence

□ 0 □ 0 0 □ 0 0 0 □ ...

may also be continued in more than one way. For example, consider

(a) □ 0 | □ 0 0 | □ 0 0 0 | □ 0 0 0 0 | ▨ 0 0 0 0 0 | ...

(b) □ 0 □ 0 0 □ 0 0 0 | □ 0 ▨ 0 0 ▨ 0 0 0 | ▨ 0 ...

Notice that the regularity depicted in (a) involves successively increasing the number of 0's by one. Young children just becoming familiar with natural numbers typically do not detect this regularity without guidance. A number of 5-year-olds, however, who were otherwise unable to detect this pattern, were able to do so after they were presented with the pattern in the form

□ 0
□ 0 0
□ 0 0 0
□ 0 0 0 0

(As an added inducement to discovery, children might be asked to count the number of 0's in the successive rows.)

(6) The numbers 0, 1, 2, 3, 4, 5, 6, 7, 8, 9, . . . are represented by the numerals 0, 1, 10, 11, 100, 101, 110, 111, 1000, 1001, . . . in the base 2 numeration system. Even if you know nothing about base 2 numerals, you should be able to see some regularity in this sequence. The next three numerals are 1010, 1011, 1100. Can you continue the pattern?

(7) Can you continue the sequence

...

Perhaps the most natural way to do this is

...

Multidimensional arrays often present an even richer source of regularities.

(8) Consider, for example, the tabular array

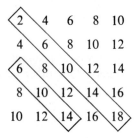

The following regularities (among others) are present in the display.

1. In each row, the numbers jump up by two as you go from left to right.
2. Every number in this table is even.
3. Each upper left to lower right diagonal sums to a multiple of 10 (e.g., $6 + 10 + 14 = 30, 2 + 6 + 10 + 14 + 18 = 50$).

What other regularities can you detect? Compare the elements in the rows, columns, and diagonals. Also consider such things as the sums of the elements in the rows, etc.

(9) The display

6	7	2
1	5	9
8	3	4

is called a magic square. Observe that

1. The sum of the numbers in each row is 15.
2. All of the numbers in the corners are even.

What other regularities are evident? Examine the rows, columns, and diagonals. Now compare the *sums* of the numbers in each row, column, and diagonal. What do you find? (This regularity is the defining characteristic of all magic squares: the sum of the numbers in every row, column, and diagonal is the same.)

EXERCISES 1–1.2

S–1. Find the next three elements in each of the following sequences. Describe the pattern verbally.

    a.   1, 4, 9, 16, 25, ...
    b.   ↑ → ↓ ← ↑ → ...
    c.   2, 4, 8, 16, 32, ...
    d.   △, □, ◇, ◇, ...
    e.   1, 1, 2, 3, 5, 8, 13, 21, ...
    f.   101, 1001, 10001, 100001, ...

S–2. Find a rule that accounts for each of the following displays. Then answer the specific questions asked.

a. 25 is divisible by 5   110 is divisible by 5
    27 is not divisible by 5   111 is not divisible by 5
    45 is divisible by 5   2740 is divisible by 5
    70 is divisible by 5   4768 is not divisible by 5
    58 is not divisible by 5   3009 is not divisible by 5

Give 3 numbers larger than 100 which are divisible by 5 and 3 which are not.

b. $\dfrac{1}{1\cdot 2} = \dfrac{1}{2}$

$\dfrac{1}{1\cdot 2} + \dfrac{1}{2\cdot 3} = \dfrac{2}{3}$

$\dfrac{1}{1\cdot 2} + \dfrac{1}{2\cdot 3} + \dfrac{1}{3\cdot 4} = \dfrac{3}{4}$

$\dfrac{1}{1\cdot 2} + \dfrac{1}{2\cdot 3} + \dfrac{1}{3\cdot 4} + \dfrac{1}{4\cdot 5} = \dfrac{4}{5}$

$\dfrac{1}{1\cdot 2} + \dfrac{1}{2\cdot 3} + \dfrac{1}{3\cdot 4} + \dfrac{1}{4\cdot 5} + \dfrac{1}{5\cdot 6} = \dfrac{5}{6}$

Find $\dfrac{1}{1\cdot 2} + \dfrac{1}{2\cdot 3} + \dfrac{1}{3\cdot 4} + \cdots + \dfrac{1}{9\cdot 10}$ without actually summing the series.

Find $\dfrac{1}{1\cdot 2} + \dfrac{1}{2\cdot 3} + \dfrac{1}{3\cdot 4} + \dfrac{1}{4\cdot 5} + \cdots + \dfrac{1}{20\cdot 21}$.

c. $11 \times 9 = 99$
    $21 \times 19 = 399$
    $31 \times 29 = 899$
    $41 \times 39 = 1599$
    $51 \times 49 = 2499$
Find 91×89.

d. 2 points determine at most 1 line segment.

3 points determine at most 3 line segments.

4 points determine at most 6 line segments.

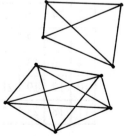

5 points determine at most 10 line segments.

6 points determine at most 15 line segments.

7 points determine at most how many line segments?
8 points determine at most how many line segments?

e. 1^3 $= 1$
    $1^3 + 2^3$ $= 9$
    $1^3 + 2^3 + 3^3$ $= 36$
    $1^3 + 2^3 + 3^3 + 4^3$ $= 100$
    $1^3 + 2^3 + 3^3 + 4^3 + 5^3 = 225$

Find $1^3 + 2^3 + 3^3 + 4^3 + 5^3 + 6^3$ without adding.

S–3. What prior knowledge is necessary to discover each of the regularities in Problem S–2?

S–4. Describe the regularity (pattern) present in each of the following:

a. Days of the week.
b. Traffic lights. Is the pattern changed if there is a four-way "walk" light?
c. Courses in a meal.
d. Getting dressed.

S–5. Look for regularities in the following display of counting numbers, even numbers, and odd numbers.

$\underline{n}$ 1 2 3 4 5 6 7 8 9 ...

$\underline{e}$ 2 4 6 8 10 12 14 16 18 ...

$\underline{o}$ 1 3 5 7 9 11 13 15 17 ...

State as many regularities as you can find.

S–6. List all of the regularities you can find in the following display. Consider rows, columns, and diagonals, sums, odd and even numbers, etc. (This is another example of what is known as a magic square.)

17	24	1	8	15
23	5	7	14	16
4	6	13	20	22
10	12	19	21	3
11	18	25	2	9

S–7. Arrange the natural numbers into five columns as indicated.

1	2	3	4	5
6	7	8	9	10
11	12	13	14	15
16	17	18	19	20
21	22	23	24	25
26	27	28	29	30

a. Can you discover a regularity with respect to the above display in each of the following groups of sums?

i. $1 + 2 = 3$
 $6 + 2 = 8$
 $1 + 7 = 8$
 $11 + 2 = 13$
 $6 + 12 = 18$
 $11 + 17 = 28$

ii. $2 + 4 = 6$
 $7 + 4 = 11$
 $2 + 19 = 21$
 $12 + 9 = 21$
 $7 + 9 = 16$
 $12 + 14 = 26$

iii. $3 + 4 = 7$
 $3 + 9 = 12$
 $13 + 4 = 17$
 $8 + 14 = 22$
 $13 + 14 = 27$
 $18 + 9 = 27$

b. If any number in column 5 is added to any number in column 1, where is the resulting sum?
 If any number in column 5 is added to any number in columns 2, 3, 4, or 5, where is the resulting sum?
 Can you state this regularity verbally?

c. Can you find similar regularities involving multiplication?

d. If you arrange the natural numbers into 6 columns, can you detect regularities similar to those in parts a, b, and c?

E–8. Sometimes we infer a general rule that turns out to be wrong. Consider the problem of determining the maximum number of regions into which n planes divide ordinary 3-space. What regularity can be detected from the display?

NUMBER OF PLANES	NUMBER OF REGIONS
0	1
1	2
2	4
3	8

You will see later whether the general rule underlying the regularity you detected is correct when there are more than 3 planes (see Section 7).

M–9. What are some of the regularities children can detect in the ordinary addition and multiplication tables? Consider regularities across rows, down columns, and along diagonals.

M–10. How would you help elementary school children to detect the regularity known as the commutative law for addition: $A + B = B + A$.

1.3 Further Observations

Any given regularity may be inherent in any number of displays—even in displays that have a completely different character.

Consider, for example, the regularity "Repeat 00□" of Example (4). Have you recently noticed this same pattern in some sequence of concrete objects or events? It might be found, for example, in a porch railing:

on in a two-way lamp: on-on-off-on-on-off-on-on-off. You might even hear, rather than see, the pattern in a series of drum beats.

Similarly the regularity exhibited by the symbolic display of Example (6) (Section 1.2) can be embodied in somewhat disguised form in a display consisting of strings of switches, where each switch may be either "on" (corresponding to 1 and indicated ↑) or "off" (corresponding to 0 and indicated ↓):

$$↓, ↑, ↑↓, ↑↑, ↑↓↓, ↑↓↑, ↑↑↓, ↑↑↑, ↑↓↓↓ \cdots$$

A concrete situation that displays the regularity of Example (7) is immediate. In this case, we simply represent each icon (e.g., □) with a piece of paper with a mark in the corresponding corner. Thus the iconic display

□ □ □ □

and the corresponding row of marked papers have the same regularity. Using symbols, such as the letters of the alphabet, to do the job requires a little more ingenuity. Here we might let "*a*" correspond to □ and "*b*," "*c*," "*d*," and "*e*" to a *dot* in the upper left, upper right, lower right, and lower left, respectively. This would result in the sequence

$$a\,b, a\,c, a\,d, a\,e\,/\,a\,b, a\,c, a\,d, a\,e\,/\,\ldots ^3$$

In general, two or more displays may have any number of regularities in common. This happens particularly often in mathematics. For example, compare the display of Example (8) with

1	3	5	7	⑨
3	5	7	9	11
5	7	9	11	13
7	9	11	13	15
9	11	13	15	17

---

[3] Of course, if we did not want to distinguish between the □ and the *dot*, we could just as well let, say, *p* correspond to □, and *q*, *r*, *s* to the other icons, respectively. In this case, we could represent the sequence more simply as *p*, *q*, *r*, *s*/*p*, *q*, *r*, *s*/ . . .

Of the three regularities listed in Example (8), the first (i.e., the numbers in each row go up by twos) is present in both displays, but the second is not. The third regularity, that each upper left to lower right diagonal sums to a multiple of 10, also does not hold in the above display. But a generalization of this third regularity does hold in both displays. In each display, compare the sums of the diagonals with the number in the upper right-hand corner. This is 10 in Example (8) and 9 in the above example. What do you find?

The common regularity may be expressed by saying that the numbers in each upper left to lower right diagonal sum to a multiple of the *number in the upper right corner.*

In Example (8), you may have noticed that the numbers in each lower left to upper right diagonal are identical. Is this true of the above display?

There is no need to stop here. With a little perseverance, you may be able to extend the list of regularities common to both.

Research evidence indicates that it is easier to detect regularities when examples are presented simultaneously in a systematic fashion, as we have done in this chapter. It is important to recognize, however, that regularities are not normally presented this way in the real world, and people should be provided with experiences in detecting regularities in the sort of displays they are apt to meet in everyday life. Thus, for example, the examples in a display might be presented individually and in random order over an extended period of time.

EXERCISES 1–1.3

S–1.   Give a real-life example that involves the same pattern as each of the following regularities:

    a.   "repeat □ △ ○" (i.e., □ △ ○ □ △ ○ □ △ ○ ...)
    b.   "repeat each symbol and add one more" (i.e., *a*, *ab*, *abc*, *abcd*, ...)

S–2.   Find as many regularities as possible that are common to *all three* of the following displays:

a.					b.				c.				
1	2	3	4		2	5	8		1	4	7	10	13
2	3	4	5		5	8	11		4	7	10	13	16
3	4	5	6		8	11	14		7	10	13	16	19
4	5	6	7						10	13	16	19	22
									13	16	19	22	25

1.4 Summary and Specific Techniques

In this section, we attempt to convey by illustration what is involved in detecting regularities and the diversity of forms this detection can take. Its essential nature, we have seen, involves going from a set of observables to a rule or rules that characterize certain aspects of these observables. Once detected, a regularity (or, more accurately, the rule associated with a regularity) makes it possible to respond appropriately to new examples of the regularity, examples that the learner may never have seen before. In addition, we noted that a given display may have any number of regularities and that essentially the same regularity may appear in many different guises. More examples of such displays are given in Chapter 4. (There they are called *embodiments.*)

So far we have said very little about specific techniques that might be helpful in detecting regularities. There are two reasons for this, reasons which apply equally, if not more so, to the other processing skills discussed below. First, very little research effort has gone into identifying such techniques in an explicit way and, second, even where such techniques can be identified, it would probably be unwise for us to go into the details here, because this would detract from the basic framework we are trying to describe.

It will suffice to mention one or two general techniques that may be helpful in detecting regularities and to encourage the reader to try to modify them and, indeed, to invent new ones of his own. Perhaps the most useful advice one might give for making discoveries is to *organize* the given display. Although most of the illustrative displays given above were already organized (to facilitate discovery), this need not always be the case. Thus the examples in any of these displays could just as well have been displayed in random fashion. For example, the display of Example (1) might have been presented as

$$1 + 3 + 5 + 7 + 9 = 25$$
$$1 + 3 + 5 \qquad\quad = 9$$
$$1 \qquad\qquad\qquad = 1$$
$$1 + 3 + 5 + 7 \quad = 16$$
$$1 + 3 \qquad\qquad = 4$$

It seems reasonably clear that it would be harder to detect the regularity of Example (1) from this display than from the one given earlier.[4]

Another technique that is sometimes useful is to change the form of the display. Thus it is frequently easier to detect regularities in a symbolic display than in a corresponding concrete one—simply because the latter type is frequently quite cumbersome to work with. Consider, for example, the display (0 0 □ 0 0 □ 0 0 □) of Example (4) and the corresponding display composed of rungs in a porch railing (in Section 1.3).

Although such techniques are notoriously hard to pin down in detail, they are clearly important to learn and use. They have been shown to be helpful in a wide variety of discovery situations, and more attention should be given to them in mathematics education. *Nevertheless, detecting most regularities depends, at least in part, on the prior acquisition of other information that relates specifically to the regularity (or regularities) in question.* A child who does not have a reasonable mastery of multiplication, for example, could hardly be expected to detect the regularity displayed in Example (1). Similarly, the ability to divide (by 3) is clearly a prerequisite to discovering the regularity of Example (2). Multiplication and certain logical operations appear to be involved in Example (3). Example (4), on the other hand, seems to involve only relatively primitive perceptual

---

[4] These comments in no way conflict with what was said in the last paragraph of Section 1.3. There we noted that it is important to have experiences in detecting regularities in "random" displays. Here we are suggesting that one technique for dealing with random displays is to organize them.

abilities. The display in Example (5) could not be continued very far by a child who could not count. Example (6) would seem to require still more advanced competencies such as the idea of "place value." What prior learning would be necessary in detecting the regularities of Example (7)? Example (8)?

EXERCISES 1–1.4

S–1.   Organize each of the following displays. Then find the rule that accounts for the display and answer the specific questions asked.

a. $2 + 4 + 6 + 8 \qquad = 20 = 5 \cdot 4$
     $2 \qquad\qquad\qquad\quad = 2 = 2 \cdot 1$
     $2 + 4 + 6 + 8 + 10 = 30 = 6 \cdot 5$
     $2 + 4 + 6 \qquad\qquad = 12 = 4 \cdot 3$
     $2 + 4 \qquad\qquad\quad = 6 = 3 \cdot 2$
     Sum $2 + 4 + 6 + 8 + \cdots + 20$.
     Sum $2 + 4 + 6 + 8 + \cdots + 100$.

b. $3 \rightarrow 6$
     $5 \rightarrow 13$
     $1 \rightarrow 3$
     $4 \rightarrow 9$
     $7 \rightarrow 24$
     $2 \rightarrow 4$
     $6 \rightarrow 18$
     $8 \rightarrow\ ?$
     $9 \rightarrow\ ?$

c.

X	Y
9	8
3	24
8	9
2	36
4	18
6	12

     Find Y for $X = 1$, $X = 12$, $X = 18$.

d. $3 \rightarrow 3$
     $5 \rightarrow 6$
     $9 \rightarrow 9$
     $7 \rightarrow 9$
     $1 \rightarrow 3$
     $8 \rightarrow 9$
     $10 \rightarrow 12$
     $6 \rightarrow 6$
     $4 \rightarrow 6$
     $2 \rightarrow 3$
     $13 \rightarrow\ ?$
     $21 \rightarrow\ ?$
     $35 \rightarrow\ ?$

E–2.   Another useful technique to aid in detecting regularities involves adding more elements to the display by additional computations. Can you find a regularity in the display

$$\frac{3}{4} = \frac{3}{4}$$

$$\frac{3}{4} \times \frac{8}{9} = \frac{2}{3}$$

$$\frac{3}{4} \times \frac{8}{9} \times \frac{15}{16} = \frac{5}{8}$$

If you detect a regularity, check it by computing $\frac{3}{4} \times \frac{8}{9} \times \frac{15}{16} \times \frac{24}{25}$ and

$\frac{3}{4} \times \frac{8}{9} \times \frac{15}{16} \times \frac{24}{25} \times \frac{35}{36}$.

If you cannot detect a regularity, compute $\frac{3}{4} \times \frac{8}{9} \times \frac{15}{16} \times \frac{24}{25}$ and $\frac{3}{4} \times \frac{8}{9} \times \frac{15}{16} \times$

$\frac{24}{25} \times \frac{35}{36}$ and add them to the display. Now can you detect a regularity?

If not, compute $\frac{3}{4} \times \frac{8}{9} \times \frac{15}{16} \times \frac{24}{25} \times \frac{35}{36} \times \frac{48}{49}$. Use the regularity you have de-

tected to evaluate $\frac{3}{4} \times \frac{8}{9} \times \frac{15}{16} \times \frac{24}{25} \times \frac{35}{36} \times \frac{48}{49} \times \frac{63}{64}$ and $\frac{3}{4} \times \frac{8}{9} \times \frac{15}{16} \times \frac{24}{25} \times$

$\frac{35}{36} \times \frac{48}{49} \times \frac{63}{64} \times \frac{80}{81} \times \frac{99}{100}$.

E–3.   Often the addition of a line (or lines) to a geometric figure will help. Can you detect the regularity in the following input-output pairs?

INPUT   OUTPUT

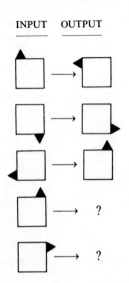

Does an auxiliary line help?

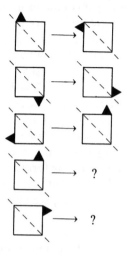

2. PARTICULARIZATION—THE ABILITY TO CONSTRUCT EXAMPLES

In order to determine whether a student has learned a particular idea, rule, or pattern, we frequently ask him to construct an example. We are asking this, for example, when we pose such problems as:

1. "Give an example of a prime number."
2. "Find a (new) number series to which the summation rule, n^2, applies."
3. "Construct a (new) magic square."
4. "Draw a square (i.e., a rectangle with four equal sides)."

Clearly, in order to answer such questions, the student must know, for example, (1) what a prime number is, or (2) what "n^2" means. What does he need to know in order to answer (3)? (4)?

In an important sense, then, constructing examples requires a kind of processing skill that is the reverse of that required in detecting regularities. In detecting regularities, the learner is required to "get the idea" or discover the underlying rule(s) from examples. Constructing examples, on the other hand, requires one to construct new examples given the idea or rule(s). (The idea or rule(s) itself, of course, may be acquired either by discovery or by some other means—see Sections 3 and 5.)

People should learn to construct examples for many reasons. One reason is to determine whether certain ideas can be realized. For example, it is relatively easy to think of ideas to solve some of our social problems, but it is not quite so easy to find ideas that will actually work in the reality in which we live. This frequently becomes clear to the inventor (or potential user) of such an idea when he tries to determine some of its implications by constructing examples—that is, by considering what would happen in particular situations if the idea were applied.

In advanced mathematics, the major technique for determining whether a given *conjecture* (i.e., educated guess) is false is to see if one can construct what is called a *counterexample*. (The major technique for showing that a conjecture is true is to prove the conjecture—see Sections 5 and 6.) By a counterexample is meant an example that illustrates the *premises* of the conjecture but not the *conclusion*. For example, consider Fermat's famous conjecture that the equation $x^n + y^n = z^n$ has natural number (i.e., non-zero whole number) solutions for x, y, and z only when $n = 1$ or 2. Thus, when $n = 1$, $x^1 + y^1 = z^1$ is just $x + y = z$ and this has many solutions. Among them are: $1 + 2 = 3$, $2 + 2 = 4$, etc. When $n = 2$, $x^2 + y^2 = z^2$ also has many solutions. Here, $3^2 + 4^2 = 5^2$ $(9 + 16 = 25)$, $5^2 + 12^2 = 13^2$ $(25 + 144 = 169)$, and $6^2 + 8^2 = 10^2$ $(36 + 64 = 100)$ but, for example, $1^2 + 2^2 \neq 3^2$ $(1 + 4 \neq 9)$. Now, no one has ever found three natural numbers x, y, and z such that $x^n + y^n = z^n$ for any $n \geq 3$. But, then, no one has ever proved that no such numbers are to be found. In the former case, of course, we would have a counterexample of Fermat's conjecture and hence would know it to be false. (For further discussion of related ideas, see also Sections 5 and 6 and Chapters 3 and 4.)

Another reason for constructing new examples is that a much better intuitive grasp of the underlying idea, rule, or pattern can often be obtained by so doing. This may help the learner to extend the scope and richness of an idea or rule already learned, particularly if he is encouraged to devise as many different examples as possible. To some extent, this is what the child does when he is asked to make up new stories to fit a particular kind of verbal problem in arithmetic or when, having learned what "triangle" means, he is asked to find as many different kinds of triangular objects as he can.

Additional instances are sometimes easy to find or construct, once an idea, rule, or pattern has been identified. Think, for example, of the regularity of Example (2) of Section 1.2, where all one needs to do is to test additional numbers to see which are divisible by 3. Where additional examples cannot be found so easily, known examples may provide a starting point for the construction of new ones. This can frequently be accomplished, for example, by modifying a known example slightly so that the result will still illustrate the underlying idea or rule. Thus, in Example (4), we might replace the circles with triangles to obtain the sequence

$$\triangle \ \triangle \ \square \quad \triangle \ \triangle \ \square \quad \triangle \ \triangle \ \cdots$$

which exhibits the same regularity as Example (4).

At a more sophisticated level, suppose that we want a new display that exhibits the common regularities of the tables in Example (8) and on page 14. In this case, for example, we can construct a table of four rows and four columns by starting with a new number, say, 6, in the upper left corner. In order to preserve regularity 1, we increase each successive number by two as we move right and by two as we move down. This gives

6	8	10	12
8	10	12	14
10	12	14	16
12	14	16	18

Did we succeed? Check to see whether the other regularities mentioned on page 10 are characteristic of this table.

Can you construct new examples of other regularities exhibited in Section 1.2? Try to be inventive!

EXERCISES 1–2

S–1. Give a *concrete* example of a sequence that exhibits the same regularity as the sequence of

    a. Example (4)
    b. Example (5)
    c. Example (6)

S–2. a. Find a new number series of the type given in Example (1) whose sum is the square of the number of terms.

    b. Find a number series whose sum is 5 times the square of the number of terms. (*Hint:* How could you modify the series of Example (1) to achieve this?)

S–3. a. Positive examples can lend plausibility to an assertion that a rule or regularity is true. A single negative instance or counterexample is sufficient to prove an assertion false. "Test" the following assertions by finding several examples of each.

    i. If n is an odd number, then $n^2 + n$ is odd.
    ii. Given any 3 consecutive whole numbers, one of them is divisible by 3.
    iii. If n is any whole number, then $n^2 - n + 41$ has no divisors except 1 and itself ($n^2 - n + 41$).
    iv. Every map can be colored with only 4 colors (i.e., so that no two contiguous countries have the same color).
    v. A divisibility rule for 11: Add alternate digits of the numeral. Add the remaining alternate digits. Subtract one sum from the other. If the result is divisible by 11, so is the original number.
    b. Which of the assertions in (a) do you know are false; which do you think are true?

S–4. Construct an example of each of the following rules:

    a. The sequence starts with a square and the nth square is followed by n circles.
    b. The sequence starts with a single square and then alternates pairs of circles and (pairs of) squares.
    c. To add two fractions, express each in terms of a common denominator, and add the numerators.

M–5. Making up stories whose solutions can be found by solving a given mathematical problem is a good exercise in particularizing. What sorts of stories would you expect a child to invent to fit the problem $(5 \times 4) + 2 = ?$

For further thought and study: Which method of instruction do you believe is more effective: (1) presenting many examples and leading students to arrive at a rule (discovery), or (2) stating a rule followed by several particular examples (exposition) and encouraging the students to produce more examples? Investigate some of the current research on this question, such as that reported in W. G. Roughead and J. M. Scandura, " 'What Is Learned' in Mathematical Discovery," *Journal of Educational Psychology*, **59**, 1968, pp. 283–289.

3. THE ABILITY TO INTERPRET MATHEMATICAL DESCRIPTIONS

In spite of the current emphasis on discovery learning in mathematics, much of what is learned in schools takes place by exposition.

If students are to progress in their study of mathematics, they must necessarily learn to read and interpret statements, graphs, and other forms of verbal and iconic descriptions of mathematical ideas (including oral presentations). Thus the student who can read and interpret precise statements will be in a far better position to learn new mathematics than one who cannot. The importance of such skills in solving so-called verbal problems, for example, is so obvious as to eliminate the need for further discussion.

3.1 What Is Involved in Interpretation?

Stated generally, the ability to interpret involves determining the *meaning* of descriptions, irrespective of whether the descriptions are given verbally (i.e., orally) or in written form, or whether they involve mathematical symbols or words, are of a pictorial (*iconic*) nature, or otherwise. By the meaning of a description, we mean the idea, rule, or rules that the description denotes. Our concern here, of course, is with mathematical ideas and mathematical rules.

Two basic ways exist for determining whether an intended meaning has been obtained. The first is to see if the learner can *paraphrase* the description—that is, describe it in his own words. For example, a young child might demonstrate his knowledge of the commutative law ($a + b = b + a$) by saying something like, "They're the same both ways." Similarly, a young child might rephrase a teacher's explanation of the subtraction algorithm as, "Subtract the numbers, except when you can't. Then, you borrow." Although neither paraphrase reflects the precise meaning intended, such statements suggest that the child knows the main ideas. When the underlying idea is a rule(s), of course, a learner may also demonstrate his mastery of the meaning by showing that he can apply the rule(s). Hence a child can show that he understands a description of *how* to do something (e.g., add) by showing that he *can* do it.

Note, however, that interpreting the meaning of a verbal problem—that is, knowing what the goal is and what information is given in the statement of the problem—is not the same as solving the problem. In order to solve a problem, one must know not only what it is (i.e., means) but how to go about solving it.

Some further examples will make clear what is intended. Mathematical descriptions may be given orally, as well as visually, but we must limit ourselves for obvious reasons to written and graphic descriptions.

3.2 Some Examples

(1) We begin by considering verbal, iconic, and symbolic descriptions of the rule underlying the display of Example (6).

Verbal—The next string (of 0's and 1's) is constructed from the preceding string by: (a) changing the first symbol on the right to a 1 if it is a 0; (b) if it is 1, changing it to a 0 and the next symbol to a 1, unless the next symbol is already 1, in which case it is changed to 0 and the next (third) symbol is changed to 1, unless the third is 1, in which case it is changed to 0, and so on until one reaches the last digit in the preceding string. If all of the digits in the preceding string are 1's, each is changed to a 0 and another 1 is appended to the left.[5]

Ionic— $1 \boxed{0/1} \ldots \boxed{0/1}\boxed{0/1} 0 \quad \rightarrow 1 \boxed{0/1} \ldots \boxed{0/1}\boxed{0/1} 1$

$1 \boxed{0/1} \ldots \boxed{0/1} 0\ 1 \quad \rightarrow 1 \boxed{0/1} \ldots \boxed{0/1} 1\ 0$

$1 \boxed{0/1} \ldots \boxed{0/1} 0\ \overbrace{1 \ldots 1}^{n} \rightarrow 1 \boxed{0/1} \ldots \boxed{0/1} 1\ \overbrace{0 \ldots 0}^{n}$

$1\ 0\ 1\ \ldots\ 1 \qquad \rightarrow \quad 1\ 1\ 0 \ldots 0$

$1\ 1\ 1\ \ldots\ 1 \qquad \rightarrow 1\ 0\ 0\ 0 \ldots 0,$

where $\boxed{0/1}$ means that the digit may be *either* 0 or 1.

Symbolic— Let $a_n\, a_{n-1} \ldots a_1$ be an arbitrary string of 0's and 1's, where $a_i = 0$ or 1 for $i = 1, \ldots, n-1$; and $a_n = 1$.

Then,

1. If $a_1 = 0$, the successor element is $a_n\, a_{n-1} \ldots a_2\, 1$.
2. If $a_i = 1$, $i = 1, \ldots, m < n-1$, the successor element is $a_n\, a_{n-1} \ldots$ $1_{(m+1)}0_{(m)} \ldots 0$
   (where $1_{(m+1)}$ means there is a 1 in the $(m+1)$ position, and $0_{(m)}$ means a 0 in the (m) position)
3. If $a_i = 1$, $i = 1, \ldots, n$, the successor element is $1\, 0_{(n)}\, 0 \ldots 0$.

It is not surprising that any of these descriptions might be difficult to understand. We so constructed them in order to demonstrate a very important point. There are many mathematical ideas and rules which are *easier* to discover from examples than to learn by exposition—and numeration systems seem to be one of them. It is particularly important to keep this fact in mind in teaching mathematics to children in the elementary and middle schools—although many teachers will attest to its relevance for high school and college teaching as well.

One must not make the mistake of thinking that it will always be more difficult to learn by exposition, however. On the contrary, the major advantage

[5] A more precise and succinct verbal description is:
1. Read the last digit of the given numeral.
2. If the digit is a 1, change the 1 to 0, write it down (and go to 4).
3. If the digit is a 0 change 0 to 1, write the new numeral (and stop).
4. If there is another digit to the left, read it (and go to rule 2).
5. Write "1" in the next position to the left of the last "0" written (and stop).

man has over other animals is his ability to learn and communicate by verbal means. Man's knowledge has reached the fantastic point it has today for precisely that reason: The next generation does not need to discover for itself everything known to the previous generation. Consider, for example, some of the regularities of Examples (8) and (9). Do you think it would be easier to interpret, "Each upper left to lower right diagonal is a multiple of 10," or to discover the same fact from the display of Example (8)? What about learning that the sum of the elements in each row, column, and diagonal of the magic square of Example (9) is 15?

There are also differences, of course, in the ease with which different kinds of descriptions of the same rule(s) may be interpreted. In the case of ordinary English, for example, one can reduce the number of new (technical) words to a minimum but the descriptions often become so long and cumbersome as to be impractical to use. Using mathematical symbolism, on the other hand, is more efficient and precise. The use of symbols, however, frightens some people even when the basic ideas are already familiar. It should be noted in this regard that recent research indicates that, although symbolic descriptions of mathematical rules can be memorized far more quickly than verbal descriptions, they are more difficult to interpret[6] (see section 3.4 for more details).

It is also worth noting that an iconic description like that above can often be used successfully with young children to impart rather complex directions— even where verbal and symbolic descriptions fail.[7] Why this happens is not entirely clear, but perhaps it is related to the close connection between icons (pictures) and the things they represent. (See Section 3.3 for a detailed discussion of the manner in which symbols and icons denote meanings.)

(2) Perhaps the most common type of iconic description is the *graph*. Thus, for example, the circle graph (Figure 1–1) depicts the world production of raw iron shortly after World War II.

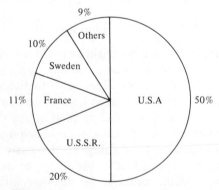

Figure 1–1

Can you read this graph? About what proportion of the world's production of raw iron was the United States responsible for? the U.S.S.R.? France?

[6] J. M. Scandura, "Learning Verbal and Symbolic Statements of Mathematical Rules," *Journal of Educational Psychology*, **58,** 1967, pp. 356–364.

[7] Paul·Rosenbloom (personal communication).

(3) Can you interpret

$$\left(\sum_{k=1}^{2} k \left(\sum_{i=1}^{5} i + \sum_{j=1}^{3} j\right)\right) - 45$$

What does it equal?

Perhaps it would help to know that

$$\sum_{r=1}^{n} r = 1 + 2 + 3 + \cdots + n \text{ (for any letter } r)$$

Can you figure it out now? If not, remember that in simplifying expressions involving parentheses, you work "from the inside out." Can you do it now? *Hint*:

$$\sum_{i=1}^{5} i = 1 + 2 + 3 + 4 + 5 = 15$$

EXERCISES 1–3.2

S–1.   A child must learn to interpret many standard symbols in addition to numerals and letters. Give the standard interpretation of each of the following symbols:

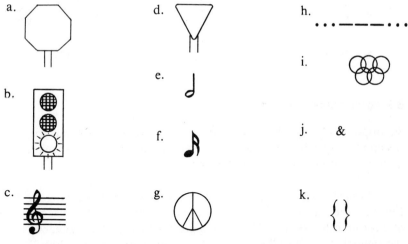

a.

b.

c.

d.

e.

f.

g.

h.

i.

j.   &

k.   {}

S–2.   Compile a list, which you consider complete, of standard symbols other than numerals used in elementary school mathematics.

S–3.   Suggest a reason for regarding > and < (as in 5 > 4 and 1 < 3) as icons rather than symbols.

E–4.   Mathematicians have agreed on the interpretation (meaning) of the following symbols:

$|x|$ stands for the magnitude (size) of the number x whether x is positive or negative, e.g., $|{}^{+}3| = |{}^{-}3| = 3$

$[x]$ stands for the greatest integer less than or equal to the number x, e.g., $[4] = [4.6] = [4.9] = 4$.

Using these standard interpretations, evaluate the following:

    a. $|^-7.6|$; $|7.6|$
    b. $|3.12|$; $[3.12]$
    c. $|3| + |^-4|$; $|^-3| + |^-4|$; $|3 + {^-4}|$
    d. $[2.4] + [1.9]$; $[2.4 + 1.9]$
    e. $[^-2.3]$ (be careful)
    f. $[|^-3.4|]$; $|[^-3.4]|$ (be careful)

E–5.   Mathematicians have agreed on a standard interpretation (meaning) of the following symbol:

$$\sum_{i=1}^{n} f(i) \text{ stands for } f(1) + f(2) + \cdots + f(n) \text{ where } f(i) \text{ represents any term defined over the variable } i, \text{ e.g.,}$$

$$\sum_{i=1}^{3} 2^i = 2^1 + 2^2 + 2^3 = 2 + 4 + 8 = 14$$

$$\sum_{i=1}^{2} i^3 = 1^3 + 2^3 = 1 + 8 = 9$$

Using the standard interpretation of the symbol, evaluate the following:

    a. $\displaystyle\sum_{i=1}^{5} i$ c. $\displaystyle\sum_{i=1}^{4} \frac{1}{i}$ e. $\left(\displaystyle\sum_{i=1}^{2} 2^i\right)\left(\displaystyle\sum_{i=1}^{3} i^2\right)$

    b. $\displaystyle\sum_{i=1}^{3} 3^i$ d. $\displaystyle\sum_{i=1}^{5} i^2$ f. $\displaystyle\sum_{i=1}^{3} i^3$

3.3 Further Observations

To appreciate fully what is involved in interpreting visual materials, it is important to understand how icons and symbols denote.[8] In particular, they relate quite differently to the things they denote.

    Symbols denote arbitrarily. They are given their meaning by definition. For example, the symbol "white" means what it means because we say it does and for no other reason. The French prefer "blanc." The same, can be said for the symbols "5," "five," and "3 + 2," or the Roman numeral "V."
    There is, however, an important difference between the symbols "white" and "blanc," and "5" and "3 + 2." The former two refer to a certain property (i.e., color) of *objects*; the latter symbols each refer to the numerosity property of *sets*. The symbols "*n*," "natural number," and "*X*" refer to still higher-order collections (i.e., properties of sets of sets). Symbols (including words) are not bound to mere things (or sets of things). Examples of symbols that denote actions or relationships are equally easy to come by. "*Add* the first to the last" and "*name*

---

    [8] Icons and symbols may be treated as "objects" in their own right. Thus, as in Example 6 of Section 1.1, they may serve as the elements in a sequence from which a regularity may be acquired. Both derive their primary value from their ability to represent things.

the color," for example, both refer to actions (see Section 1 in Chapter 2 for further discussion of things, sets, actions, and relations).

It is the arbitrary nature of the way in which symbols denote that makes it impossible to infer the meaning from the shape of a symbol alone, and at the same time makes it possible to use symbols to refer equally well to all kinds of meanings, no matter what the level of abstraction. Because there is typically no common way in which symbols denote, the meaning of each symbol must be learned separately. Thus, "7" looks no more as if it should refer to {o o o o o o o} than does the symbol "8."[9] On the positive side, symbols may denote highly abstract ideas just as easily and precisely as first order meanings exactly *because* they need have no resemblance to their denotation. It is important to notice, however, that individual symbols may be combined according to particular patterns and, in this case, their meanings can be determined by reference to previously learned symbol meanings. For example, the number denoted by "23" can be ascertained directly from the meanings of "2" and "3" together with the rules of our numeration system.

Icons, on the other hand, are *object-like*. They represent what they *seem* to represent. Thus, a picture of the landscape represents *the* landscape. The drawing, ⬡ , is an icon that might represent a box. It could also represent any number of other objects having the same general form. Certain relationships between real objects (e.g., relative size, proximity, orientation, etc.) are represented in icons by corresponding relationships between those parts of the icon that refer to the objects in question. In precisely this manner, for example, ▢ may represent a piece of paper with a mark in the upper left corner. Icons, however, need not represent all possible relationships—only those that are deemed important for a particular purpose. Thus if relative size is important, the drawing must be to scale. If only shape is important, relative size need not be preserved—only the shapes of the individual objects. When we draw "in perspective," neither size nor shape is preserved. Very young children often tend to preserve even less in their drawings.

Because icons have features in common with the entities they denote, it is often possible to generalize the correspondence between an icon and its meaning. Thus the icon "卌 |" denotes the same referent as does "6" but it does so in a nonarbitrary way. The slashes in "卌 |" can be put into one-to-one correspondence with the elements in any set whose numerosity is specified by "卌 |." More important than the fact that a particular correspondence can be identified is the fact that this correspondence provides a basis for constructing icon-names for any number. Thus "|," "||," "|||," ..., "卌 卌 ||," and so on represent the number of objects in any set with one, two, three, ..., twelve elements, respectively.

This ability of icons to refer in a generalizable way, however, is bought at a price. Because by definition they are referent-like, icons necessarily retain less

---

[9] Actually, "1," "2," "3," "4," "5," and "6" may have originally been adopted from the icons "ı," "=," "≡," " 4ı," "5ı," and "|ᴄ|," respectively.

detail when used to represent relatively abstract ideas. At higher levels of abstraction it may even be impossible to find iconic representations. For example, although the icon "|||" serves admirably to denote the number 3, there is no true icon that represents the variable "X" or "□."[10] In cases such as these, symbols can sometimes be added (to icons) to retain some of the icon-like features of the original icon. Thus, the icon-like sign "||| ... ||" might be obtained from the icons above by adding the ellipsis, "...," to indicate that the number of slashes may vary.

Arrows, →, also have both icon- and symbol-like properties. They are often used to denote direction, presumably because they serve as icons for weather vanes, and weather vanes are used to determine wind direction. But when used, for example, to indicate the path an object takes, arrows are acting as symbols. Because of the tie-in with weather vanes and the like, however, arrows denote direction in a rather natural way and are almost universally used for this purpose. We will later discuss iconic and symbolic representations as they relate to mathematics.

EXERCISES 1–3.3

S–1. Classify each of the following as a symbol or an icon:

    a. the arabic numeral 3
    b. the roman numeral III

    c. $\dfrac{1}{3}$

    d.

    e.

    f. ≅

    g.

S–2. Give examples of three symbols and three icons (not mentioned in the text) used in everyday life.

3.4 Specific Techniques

Clearly, the ability to read is one basic component of the ability to interpret. It is possible, however, to teach some children to read material that they do *not* comprehend (i.e., that they cannot interpret). Such children presumably detect regularities in the way words are pronounced and are able to apply these pronunciation rules to new sequences of words, even words for which they have learned no referent.

Only after a child is familiar with the meaning of each word, symbol, and/or icon contained in a mathematical description can he be expected to interpret the description as a whole. This suggests that one way to help children

---

[10] Here "□" is being used as a symbol, not as an icon. It refers to a variable, not some sort of box.

learn to interpret mathematical descriptions is to teach them the meanings of common mathematical words, such as number, numeral, five, greater than, triangle, percent; symbols, like 5, 9, $\frac{1}{2}$, $+$, $=$, \$, %, $\sum$; and icons, like $||$, $\bigwedge\hspace{-0.5em}\bigvee$, $\bigotimes$ (graph), $>$.

As obvious as it might seem, this suggestion is often overlooked in teaching children how to solve so-called "verbal problems," in which ordinary English words often have special mathematical interpretations. Consider the following problem statement: "Their grandfather gives a dollar to Jeanne and 60 cents to Julie, telling Jeanne to get change for the dollar and to split the difference with Julie. How much will Jeanne have? How much will Julie have?" A child may know the English meaning of each word in this statement, and even be able to imagine all of the events described, without understanding the intended mathematical meaning. Specifically, he may not know the mathematically relevant meanings of some of the words:

"dollar" means 100 cents,
"difference" means the result of subtracting the smaller from the larger,
"split" (in two) means to divide by two (and give one part to each person).

Of course, knowing these meanings does not guarantee solution of the problem, nor even the ability to interpret the problem statement; on the other hand, no one could be expected to solve or interpret the problem without knowing these meanings.

Many other ordinary words and phrases also have specific mathematical referents. For example:

"5 more than" may mean to add 5
"and" sometimes means to add
"of" and "times" often indicate multiplication
"is" may indicate equality

Can you think of more examples?

In teaching such meanings to children, a word of caution is in order. The meaning of a word or symbol in a mathematical description frequently depends upon its context. The symbol "$-$," for example, has a different meaning in $^{+}3 + {}^{-}4$ and $3 - 4$. The intended meaning of each occurrence, however, is clear from the context. Similarly, the term "difference" calls for subtraction in "What is the *difference* between 13 and 8?" but not in "What is the *difference* between a square and a triangle?" or "What is the *difference* between the numeral '5' and the numeral '8'?" (Numerals and numeration systems are discussed in detail in Chapter 6.) Learning to interpret words and symbols in context requires much experience, and the teacher should provide such experiences as often as is convenient.

As in teaching natural languages, children must be taught more than mere specific symbol meanings. Wherever possible, they should be taught specific techniques for interpreting new combinations of symbols. One example involves the use of parentheses. Mathematicians have adopted the convention that in simplifying any arithmetic (or algebraic) expression, for example, $(5 + ((5 \cdot 6) \cdot$

$(5 + 1))) \cdot 2$, one should simplify the expressions within the innermost parentheses first and work "outward." For example,

$$(5 + ((5 \cdot 6) \cdot (5 + 1))) \cdot 2 = (5 + ((30) \cdot (6))) \cdot 2$$
$$= (5 + 180) \cdot 2 = (185) \cdot 2 = 370.$$

Without parentheses and the indicated convention, or other agreements about the order of operations, even a simple expression like $5 + 5 \cdot 6$ is ambiguous ($5 + 5 \cdot 6 = 10 \cdot 6 = 60$ and $5 + 5 \cdot 6 = 5 + 30 = 35$).

The use of parentheses acts much as a grammar and is applicable to a wide variety of situations, irrespective of the particular symbols involved. In a recent experiment,[11] the ability to use parentheses and knowledge of the meanings of the individual symbols used were found to be not only necessary conditions for interpreting a statement of a mathematical rule, but they were essentially sufficient. Nothing else seemed to be needed to make a statement interpretable.

Following training in the use of parentheses with *neutral* materials, the experimental subjects (who were preservice elementary school teachers) were able to apply statements of rules they had never seen before, given only the meanings of the constituent symbols. Given that $\sum_{i=1}^{n} i$ means the same as $1 + 2 + 3 + \cdots + n$, for example, they were able to interpret, and so to apply correctly, the rule statement, $\sum_{z=1}^{t} z \left(\sum_{x=1}^{r} x + \sum_{y=1}^{s} y \right)$. One should be cautious not to overgeneralize, but it seems likely that there are some identifiable interpreting skills that might well be taught to school children.

Unfortunately, the number of such easily identified techniques is quite small. To help ensure that children learn a wide variety of general heuristics by which they can "figure out" the meanings of new statements, they should be given many varied opportunities for doing so. The teacher, of course, may assist the child by providing hints of various kinds.

EXERCISES 1–3.4

S–1. When driving, we continually have to interpret iconic road signs. Interpret each of the following signs:

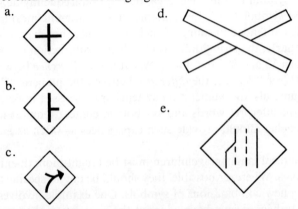

a.

b.

c.

d.

e.

[11] J. M. Scandura, "Learning Verbal and Symbolic Statements of Mathematical Rules," *Journal of Educational Psychology*, **58**, 1967, pp. 356–364.

E–2. Mathematicians have agreed on the convention that one should always simplify the expressions in the innermost parentheses first and work "outward," but that in the absence of parentheses, multiplications and divisions should be done before additions and subtractions. Using this convention, evaluate the following:

a.	$3 \times 2 + 4$	f.	$12 \div (4 + 2)$
b.	$3 \times (2 + 4)$	g.	$20 + 10 \div 2 + 4 \times 2$
c.	$5 + (4 \times 3 + 1) \times 5$	h.	$(20 + 10) \div 2 + 4 \times 2$
d.	$5 + 4 \times 3 + 1 \times 5$	i.	$20 + 10 \div (2 + 4 \times 2)$
e.	$12 \div 4 + 2$	j.	$20 + 10 \div ((2 + 4) \times 2)$
		k.	$(20 + 10) \div ((2 + 4) \times 2)$

M–3. Invent three or more symbols, assign meanings to them, and write a story making use of these symbols which you could use to help children learn to interpret.

M–4. Suggest an activity which would help children learn to interpret (a) maps, (b) graphs.

For further thought and study: Concepts can often be introduced to children via interpretations in the form of activities, games, or stories. Look up some specific examples in books by Z. P. Dienes (e.g., *Mathematics in the Primary School*) or the Minnemast materials for grades 1–3.

4. THE ABILITY TO DESCRIBE MATHEMATICAL IDEAS

Teaching children to communicate has always been a major aim of elementary education in its broadest sense. Although one does not ordinarily think of mathematics teaching as a means toward this end, few elementary school subjects offer as many opportunities to encourage the precise, accurate, and concise use of language as does mathematics. Unfortunately, many teachers do not take advantage of these opportunities. Children can and should be given practice in formulating precise descriptions of things they have already learned or discovered about mathematics.

4.1 What Is Involved in Describing Mathematical Ideas?

Clearly, the ability to describe involves formulating a description of an idea (rule or regularity) in one's mind. Mathematical ideas may be described in *a number of different ways.* Descriptions may be verbal, symbolic, or iconic, or they may involve various combinations of these three modes. They may be presented in visual, auditory, or even tactile form. Furthermore, they may presume much or little knowledge on the part of the reader or the hearer.

In order to tell whether or not a person can describe a mathematical idea (rule), we first need to know whether he has learned the idea (rule). This can be determined by the methods described in Sections 1.1 and 3.1. Even though a person knows how to use an idea, however, we must not mistakenly think that he can necessarily describe it. Every teacher has had the experience of working with children who knew what to do—they could get the right answers—but they could not describe how they did it.

In order to judge the adequacy of a given description, the teacher may keep three basic criteria in mind. Whatever its form, a really good description should:

1. be composed of words, symbols, or icons whose meanings are all known to the hearer (or reader), and which are combined in a grammatical way,
2. completely, precisely, and accurately characterize the idea (object, regularity, or rule) that it is intended to describe,
3. be as concise as is feasible.

Much experience is necessary to acquire the ability to produce descriptions which meet these requirements, however, and in the elementary school we must be willing to accept a good deal less. For example, if asked to describe what a square is, a kindergarten child could hardly be expected to say more than something like, "It's 'square,'" or better, "It has four sides—and they're even." Eventually, of course, we may want to require a more precise description such as, "A square is a rectangle with equal sides."

4.2 Some Examples

The following three examples show how some of the mathematical rules discussed in Section 1.2 may be variously described. Try to decide in each case which mode of description seems most precise, which seems most concise, and which is easiest to understand. First, however, you should go back to the corresponding examples in Section 1.2 and see if you can construct accurate and concise descriptions which might be understood by *your* students. Why not actually try them out?

1. (Example (1) of Section 1.2)

   Verbal—To find the sum of a series of consecutive odd numbers beginning with 1, square the number of terms in the series.

   Ionic—

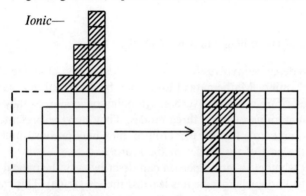

(What specific number series does the icon on the left describe? *Hint:* Count the number of squares in each column.)

Symbolic—To find the sum of an arithmetic series, $\sum_{i=1}^{n} (2i–1)$, compute n^2.

(See Exercises 4.2, Problem E–9 for the meaning of $\sum$.) What is $(2i–1)$ if $i = 1$? If $i = 2$? If $i = 3$? Do you see why $\sum_{i=1}^{n} (2i–1)$ describes the sum of the first n odd numbers? If not, write out $\sum_{i=1}^{5} (2i–1)$.

(*Hint:* $\sum_{i=1}^{3} (2i–1) = 1 + 3 + 5$.)

Which of these three descriptions seems most precise? Which is most concise? Which is easiest to understand? Do all descriptions meet the requirements stated in Section 4.1? Pay particular attention to the iconic description. Does it include all instances of Example (1) of Section 1.2? Do the verbal and symbolic descriptions include all cases?

2. (Example (2) of Section 1.2)

Verbal—To see if a number is divisible by 3, divide the sum of its digits by 3. If the sum is divisible by three (i.e., the remainder is 0), the number is also divisible by 3. Otherwise, it is not.

Iconic—There seems to be no simple iconic description. Can you invent one? (*Hint:* Represent a number in terms of groupings of units, tens, and hundreds. Replace each grouping by (for example) a slash mark. Then see if the slash marks can be grouped by 3's with none left over.)

Symbolic—Let $a_n a_{n-1} \ldots a_2 a_1$ be a whole number where the a_i are digits. Then to see if $a_n a_{n-1} \ldots a_2 a_1$ is divisible by 3, divide $\sum_{i=1}^{n} a_i$ by 3. If there is a whole number m such that $\sum_{i=1}^{n} a_i = 3m$, then $a_n a_{n-1} \ldots a_2 a_1$ is divisible by 3. Otherwise, it is not.

In Example (1), the symbolic description is somewhat more concise than the verbal description. What is the situation in Example (2)?

3. (Example (3) of Section 1.2)

Verbal—To find the sum of the interior angles of a polygon, divide the polygon into triangles by joining the vertices so that no two triangles overlap. Then multiply the number of triangles by 180°.

Iconic –

Again, it is practically impossible not to use symbols in one form or another.

Symbolic—To determine the sum of the interior angles of a polygon, divide the polygon into triangles by joining vertices so that no two triangles overlap. Then compute $(180 \times n)°$ where n is the number of triangles.

Again, the verbal and symbolic descriptions are about equally concise. Why? (*Hint:* Consider the availability of common mathematical symbols to denote the ideas involved.)

4. (Example (4) of Section 1.2)

Given the sequence 0 0 □ 0 0 □ 0

Verbal—The element immediately after a square is a zero; the element after a zero that follows a square is a zero; the element after a zero that follows a zero is a square.

Iconic—... □ → ... □ 0
        ... □ 0 → ... □ 0 0
        ... 0 0 → ... 0 0 □

Symbolic—Let a_i, $i = 1, 2, 3, \ldots$, be the ith element in the sequence.

If $a_i = 0$ and $a_{i-1} = 0$, then $a_{i+1} = \square$.
Otherwise, $a_{i+1} = 0$.

5. Example (5) of Section 1.2 is similar to Example (4). Can you write verbal, iconic, and symbolic descriptions?

6. Example (6) of Section 1.2 was described in three ways in Section 3.2.

7. (Example (7) of Section 1.2)

Given the sequence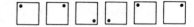

Verbal—The next element is determined by rotating the preceding element by 90° in a clockwise direction.

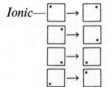

Ionic—

Symbolic—Let a_i, $i = 0, 1, 2, 3$ correspond to ⬒, ⬓, ⬔, and ⬕, respectively. Then the element after a_i is a_{i+1} when $i = 0, 1, 2$ and a_0 when $i \neq 3$. (Or, better, the element after a_i is $a_{i+1 \pmod 4}$.)

8. (Example (8) of Section 1.2)

Verbal—The sum of the numbers in each upper left to lower right diagonal is a multiple of 10.

Iconic—

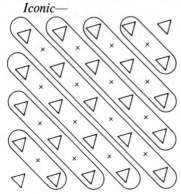

(where each diagonal corresponds to
a multiple of 10)

Figure 1–2

Symbolic—Let a_{ij} represent the number in the ith row and the jth column of the table. Then $a_{1j} + a_{2(j+1)} + \ldots + a_{(6-j)5} = 10(6 - j)$, if $j > 1$ and $a_{i1} + a_{(i+1)2} + \ldots + a_{5(6-i)} = 10(6 - i)$, if $j = 1$.

Compare the verbal and symbolic descriptions. Which one is more concise? Easier to interpret? (Clearly, some mathematical ideas are easier to describe in words than in symbols. Symbols are only necessary where verbal descriptions become imprecise and/or too lengthy.)

9. (Example (9) of Section 1.2)

Verbal—In the three-by-three magic square, the sum of the numbers in each row and column and along each main diagonal is 15.

Iconic—

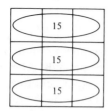 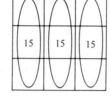 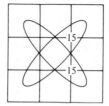

Figure 1–3

Symbolic—If a_{ij} is the entry in the ith row and jth column of the magic square, where $i, j = 1, 2, 3$, then

$$\sum_{j=1}^{3} a_{ij} = \sum_{i=1}^{3} a_{ij} = \sum_{i=1}^{3} a_{ii} = \sum_{i+j=4} a_{ij} = 15$$

These examples make clear that the suitability of a particular kind of description depends on the rule involved. For instance, Example (1) seemed most naturally represented in symbolic form while most of the patterns (i.e., Examples (4), (5), (7)) were readily amenable to iconic description. Whereas verbal and symbolic descriptions could be found in each case, several of the symbolic statements may have appeared particularly difficult to understand, let alone to devise. Thus, while symbolic descriptions are precise and, for this reason, indispensable to the mathematician, it would in general be quite inappropriate to require elementary school children to describe regularities in terms of mathematical symbols. Very young children may find it more natural to devise iconic descriptions such as those given in Examples (4), (5), (7), and (8). We should emphasize, however, that children from about the fourth grade on often become intrigued with mathematical symbols and typically learn to use new ones much more quickly than do adults.

EXERCISES 1–4.2

S–1. How would you use icons to represent the following:

    a. a class of four boys and three girls
    b. a town with three houses, one store, and one church
    c. a right turn of 120°; a left turn of 120°; a rotation through 360°
    d. a sequence of three switches, any one of which might be "on"

S–2. What are the advantages of iconic representations? Are there any disadvantages?

S–3. Each of the concrete objects mentioned below has a distinctive and familiar regularity. Represent this regularity both symbolically and iconically, if possible.

    a. keys on a piano
    b. a waltz
    c. the distress signal for ships at sea
    d. a chess board
    e. molecules of water

E–4. Consider the commands: "attention" (A), "right face" (R), "about face" (T), and "left face" (L). Assume that for each pair of commands given by a scoutmaster, a Boy Scout is to execute the single command equivalent to executing the two commands in order. Make up a table (display) showing the result of each possible pair of commands. (*Hint:* See section 7.1.)

S–5. Give verbal, iconic, and symbolic descriptions (where applicable) of each of the patterns in Problem S–1 of Exercises 1–1.2. Which description seems best for each one? Why?

S–6. Give verbal, iconic (where applicable), and symbolic (where applicable) descriptions of each of the patterns in Problem S–2 of Exercises 1–1.2. Which description seems best for each one? Why?

S–7. Describe any regularity you can detect in the display below, using the type of description that seems most appropriate.

```
                          1
                      1       1
                  1       2       1
              1       3       3       1
          1       4       6       4       1
      1       5      10      10       5       1
  1       6      15      20      15       6       1
```

This well-known display, called Pascal's triangle, has many simple and sophisticated regularities.

E–8. a.   The representation below has both iconic and symbolic elements. Identify the icons and the symbols and tell what they represent.
  b.   What general regularity is represented?
  c.   What features of the original object are retained in the iconic representation? What features are ignored?

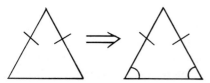

E–9.   Let $f(i)$ represent any term containing the letter i. Then the symbol $\sum_{i=1}^{n} f(i)$ is used to represent $f(1) + f(2) + f(3) + \cdots + f(n)$. That is, successive values 1, 2, 3, ..., n are substituted in the term and the results are added.

For example,

$$\sum_{i=1}^{3} 2^i = 2^1 + 2^2 + 2^3 = 2 + 4 + 8$$

$$\sum_{i=1}^{4} i^2 = 1^2 + 2^2 + 3^2 + 4^2 = 1 + 4 + 9 + 16 = 30$$

Use the $\sum$ notation to describe the following:

a. $1 + 2 + 3 + \cdots + 7$
b. $\dfrac{1}{1} + \dfrac{1}{2} + \dfrac{1}{3} + \cdots + \dfrac{1}{8}$
c. $2^1 + 2^2 + 2^3 + \cdots + 2^9$
d. $0^2 + 1^2 + 2^2 + 3^2 + 4^2$
e. $\dfrac{1}{1^2 + 1} + \dfrac{1}{2^2 + 1} + \dfrac{1}{3^2 + 1} + \cdots + \dfrac{1}{11^2 + 1}$
f. $X_1 + X_2 + X_3 + \cdots + X_{27}$

M–10. Write some questions designed to determine whether your students

    a. can use the commutative law of addition (see Exercises 1–1.2, Problem M–10) even if they cannot describe it.

    b. can describe the commutative law symbolically.

For further thought and study: Find regularities in three elementary textbook series and describe each of them verbally, iconically, and symbolically.

4.3 Further Observations

Descriptions are frequently intended to convey meaning to someone else—that is, they are used for communication. If possible, one should try to judge the listener's or hearer's ability to interpret and formulate his descriptions accordingly. Occasionally, however, it is so difficult to find a concise and accurate description of even very simple ideas that an alternative means of communication is tried (consider Example (6)). Thus, for example, teachers may introduce new ideas by constructing a display (set of instances), as described in Section 2, and letting the children discover the ideas for themselves. Children too often attempt to communicate ideas by simply pointing to, or constructing, instances but for an entirely different reason. They may merely be unable to describe the ideas.

Learning how to describe mathematical ideas increases not only the child's ability to communicate. It also may help him to solve problems that he could not otherwise solve. In solving a verbal problem, for example, it is frequently helpful to reformulate the problem in terms of mathematical equations. The first step in doing this is to interpret the statement of the problem. The second is to describe the intended meaning as a set of equations. Consider once again the verbal problem of Section 3.4. Suppose we interpret the statement to mean that, having obtained change, Jeanne is to subtract 60 cents of the dollar, divide the remaining amount in two equal parts, add one part to her 60 cents and give one part to Julie. When expressed in this form it may be hard to see exactly how to solve the problem. After the meaning has been formulated (described) as a set of equations, however, the solution procedure becomes quite obvious (for anyone who has had ninth-grade algebra):

> Let x represent the difference between what Jeanne received and what Julie received. Then,
>
> $$x = (100 - 60) \text{ cents}$$
> $$\text{Jeanne's share of the dollar} = (60 + x/2) \text{ cents}$$
> $$\text{Julie's share of the dollar} = x/2 \text{ cents}$$

Solution of the problem is now a simple matter of solving for x ($= 40$ cents), dividing it by 2, and substituting into the last two equations—all routine, mechanical steps. (Be sure to complete the problem yourself. How much does Julie get? Jeanne?)

Recent research also suggests that verbalizing a given rule may facilitate its retention by tending to "fix" it in the learner's mind. This is a positive result and suggests that people should make it a practice to describe rules or ideas which they have learned. On the other hand, premature verbalization or, rather, verbalization that occurs before an adequate rule or idea has been formulated (i.e., learned) may have the effect of *fixing the inadequate rule*. A study by Hendrix[12] provides general support for this interpretation by showing that premature verbalization may have a negative effect on transfer to new instances. Hence, extreme care should be taken to avoid this possibility by providing the learner with a sufficiently diverse set of test items before having him attempt to verbalize, in order to ensure that he really knows the rule in question.

The same sort of premature closure should be guarded against in situations where a child is able to learn a specific variant of a more general and desirable rule presently beyond his capability. Under these conditions, the short-term gains of verbalization (i.e., increasing the child's ability to remember the more specific rule) must be weighed against possible long-term losses (such as making it more difficult to learn the more general rule later on).[13]

4.4 Summary and Specific Techniques

In summary, a learner may know and be able to use mathematical ideas and rules but be unable to describe them. The ability to describe things mathematically may be important in communication, in solving problems, and even in helping the learner to remember by fixing ideas in his mind.

Mathematical descriptions may be formulated in words, symbols, and/or icons, forming part of what may be called the *language* of mathematics. How one learns this language and what specific techniques may facilitate its use are important questions that concern psychologists and linguists as well as mathematics teachers. So far, however, we have few hard answers and much research remains to be done. Nevertheless, we shall suggest a few specific components of the ability to formulate mathematical descriptions, and some related techniques.

Perhaps the major requirement in learning to use a language is to learn the vocabulary—the meaning of the words, symbols, and/or icons of which descriptions are to be composed. Without this one could not hope to describe anything in the language. Equally important, however, the user must also know how to put these signs together; that is, he must know the rules of the grammar. Learning to use parentheses correctly, for example, is or should be an important objective of elementary school mathematics.

To see how parentheses may be involved, suppose we have a rule in mind for writing triples of numbers, like (3, 7, 5), (8, 2, 5), (6, 12, 9), (7, 7, 7), and that we want to describe it mathematically (i.e., symbolically). (The rule we have in mind may be described verbally as the third number in each triple is the average

---

[12] G. Hendrix, "A New Clue to Transfer of Training," *The Elementary School Journal*, **48**, 1947–1948, pp. 197–208.

[13] W. G. Roughead and J. M. Scandura, " 'What Is Learned' in Mathematical Discovery," *Journal of Educational Psychology*, **59**, 1968, pp. 283–289.

of the first two.) Assuming that we have learned the technique of letting letters of the alphabet represent indeterminate numbers (called *parameters*), we might express the rule as $c = \frac{1}{2} \times a + b$, where a, b, and c represent the first, second, and third numbers of an arbitrary triple. But this equation is not precisely correct. It is ambiguous in that it does not tell whether the multiplication or the addition is to be performed first. (This is important since, for example, $\frac{1}{2} \times 3 + 3$ equals $4\frac{1}{2}$ if we multiply first and 3 if we add first.) We clearly intend for the addition to be done first and parentheses allow us to indicate this in our description: $c = \frac{1}{2} \times (a + b)$ (see section 3.4).

Iconic descriptions present a different kind of difficulty. There are a few commonly used icons whose meanings may be learned and then used as standard representations. Tally marks, arrows, geometric shapes, and the number line are examples. Often, however, iconic descriptions must be constructed "from scratch." In this case, a helpful procedure is first to decide exactly what features of the idea to be represented by the icon must be retained and which may safely be ignored. Then decide what *type* of icon might be used to represent these features. Third, determine which features of the icon correspond naturally to aspects of the idea to be represented and use this correspondence to construct an icon. Finally, any irrelevant features of the icon may be removed so that what remains represents only that which is critical.

Let us see how we might use this procedure to construct a graph which represents the *proportion* (step 1) of a university's annual income ($30,000,000) coming from student fees ($6,000,000). Suppose we decide on a bar graph (step 2). In this case, we might let the whole bar represent the university's annual income. Then we could divide the bar into 30 equal parts so that each part represents $1,000,000 and let the first six parts represent the amount coming from student fees (step 3). This gives us the icon represented in Figure 1–4.

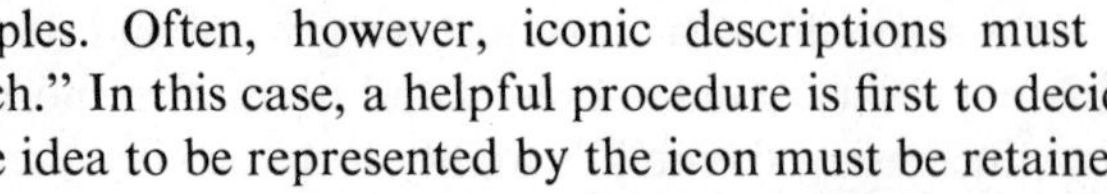

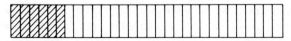

Figure 1–4

Because we are interested only in the *proportion* of fees to total income, and not in specific amounts, the subdivisions are irrelevant and so we discard them in the final graph (step 4), as shown in Figure 1–5.

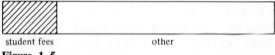

student fees   other

Figure 1–5

See if you can think up any other techniques for describing mathematical ideas that might be helpful to your students. Keep this in mind as you plan your instruction and, indeed, as you teach. Sometimes our very best ideas come when we least expect them.

EXERCISES 1–4.4

S–1.   Write an equation to represent each of the following verbal statements:

    a. x is 3 more than y.
    b.   The sum of the first n positive odd integers is n^2.
    c.   The sum of the squares of the legs of a right triangle equals the square of the hypotenuse.
    d.   □ and △ are the same numbers.
    e.   Area equals base times height.
    f.   The product of 4 and 6 is 24.
    g.   The difference of 10 and 4 is 6.
    h.   □ is the average of △ and ◇.

S–2.   Follow the four-step procedure on page 40 to construct iconic descriptions of each of the following:

    a.   a closed cardboard carton
    b.   an open cardboard carton
    c.   an open soup can and a closed tuna-fish can
    d.   three apples
    e.   a bed and a chair
    f.   a four-door sedan and a convertible

S–3.   Use graphs to represent each of the following situations:

    a. $\frac{2}{5}$ of the national budget goes to the Defense Department.
    b.   Student tuition pays just under half of the costs at State College.
    c.   The city of Fictionville has the following yearly expenses:

Police	$ 5 million
Fire	4 million
Nonuniformed employees	7 million
Schools	13 million
Administration	4 million
Miscellaneous	7 million

For further thought and study: (a) What are some ordinary classroom objects which children can represent iconically?

    (b) Read the article by Hendrix referenced in Section 4.3.

    (c) How do bar graphs, line graphs, and circle graphs differ in the way they represent data?

5.   THE ABILITY TO MAKE LOGICAL INFERENCES

The four kinds of processing skills considered so far are equally relevant in all areas of education. Detecting regularities, for example, is important in science as well as mathematics. Similarly, one could hardly be said to have a program in reading and the communicative arts which did not make provision for interpretative and describing skills. Our major concern in the preceding sections was to help clarify the nature of these kinds of processing skills as they pertain to mathematics.

We now turn our attention to the deductive process, that form of reasoning which, although not unique to mathematics, certainly attains its clearest and most profound expression there.

5.1 What Is Involved in Making Logical Inferences?

In symbolic logic, inference has a very precise meaning and is defined in terms of the way abstract symbols may be manipulated. In most logics,[14] for example, a rule exists which, given any pair of symbols of the form "$X \supset Y$" and "X," allows one to infer another symbol of the form "Y." "Y" is called a conclusion and "$X \supset Y$" and "X," premises. Thus, given "$(A \cup B) \supset C$" and "$(A \cup B)$" one can conclude "C" and given "$P \supset Q$" and "P," one can conclude "Q."

Logical inference may also be viewed as a higher-order skill that operates not on strings of symbols but on ideas, facts, and rules. In particular, by drawing inferences one may "deduce" (i.e., arrive at) new ideas (rules) on the basis of given ones. Such skills are used every day by most people to draw conclusions about their worlds from things they already know or believe to be true. Thus, if we know that the zoo is always closed on holidays, and that today is Thanksgiving, we would immediately conclude that visiting the zoo is not a possible after-dinner activity. The conclusion follows naturally and without effort. We do not need to go to the zoo to double-check (although, of course, there is always the chance in real life that someone might break the rules). No one who is able to reason deductively would seriously question the conclusion given the same facts. In short, the process of making logical inferences provides a third, and fundamentally different, way to acquire new knowledge.

EXERCISE 1–5.1

S–1.   Deduction is one way to acquire new knowledge. What are the other two ways?

5.2 Some Examples

The following examples illustrate additional situations in everyday life and in mathematics where the elementary school child might need to make logical inferences.

Consider first three conversations between adult and child in which conclusions, obvious to the reasoning adult, are left unstated under the apparent assumption that they are also obvious to a child (see Section 5.3 where the validity of this assumption is questioned).

---

[14] Logicians have invented many different logical systems.

(1) *Adult:* "You may go with us to the zoo if your mother has signed your permission form. So bring this slip back tomorrow after your mother has signed it."

Child (next day): "Here it is. My mother signed it."

What would you, and presumably the teacher, conclude? May he go to the zoo? Do you think that a kindergarten child would make the same inference? How might you tell without asking the child? What would a happy face probably indicate? What if he presents the slip matter-of-factly or waits expectantly for a reply? In the latter case, would you feel that the logical implications of the situation are clear to him?

(2) *Adult:* "This is where the panther lives. He has an indoor cage and a fenced-in area outside."

Child: "He isn't in his cage! Let's go outdoors."

Do you think that the child has drawn the natural conclusion? What is the natural conclusion?

(3) *Adult:* "Reptiles are cold-blooded animals."

Child: "Is an alligator a reptile?"

Adult: "Yes."

Child: "Is an alligator cold-blooded?"

What is the obvious conclusion? Do you think the child knows this conclusion and is confident of it? Why? Why not?

Make up some other hypothetical situations like these—where the child knows the natural conclusion(s)—where he does not. Why not try out some of these situations with some young children you know? Be sure to include situations of varying complexity and work with preschool (ages 2, 3, 4) as well as school-age children.

(4) There are many opportunities, when children are first exposed to arithmetic, to discover new ideas not only inductively but also by drawing inferences from things he already knows. Suppose that a child has learned (either by discovery or by exposition) that the order in which any two given numbers are added is immaterial:

(a) $\square + \triangle = \triangle + \square$ (e.g., $\boxed{5} + \triangle\!\!\!3 = \triangle\!\!\!3 + \boxed{5}$)

and that in adding any three given numbers, either the first pair or the last pair may be added first:

(b) $(\square + \triangle) + \bigcirc = \square + (\triangle + \bigcirc)$

e.g., $(\boxed{4} + \triangle\!\!\!6) + \textcircled{2} = \boxed{4} + (\triangle\!\!\!6 + \textcircled{2})$, where the parentheses indicate the order in which the additions are to be performed.[15]

On the basis of (a) and (b), the child could then conclude that the sum of any three given numbers may be obtained by adding *any* two of these numbers first. That is, in addition to (a) and (b) above, the following also holds:

(c) $(\square + \triangle) + \bigcirc = \square + (\bigcirc + \triangle)$

(Compare (c) with (b) and note the difference in order to the right of $=$.) Although (c) could be discovered or learned by exposition, just as (a) and (b), the important difference is that (c) may also be learned by making a logical inference (from (a) and (b)). We will examine this process in more detail in Example (6).

(5) As mathematicians use the term, deductive reasoning refers primarily to the process of constructing mathematical *proofs*. In your study of high school mathematics, particularly in geometry (and algebra if you had a modern course), you may recall that most proofs are essentially logical arguments by which certain statements, called *conclusions*, are shown to follow from other statements, called *premises*. A *theorem* is a statement that certain given premises imply some given conclusion. For example, (1) "If p is a factor of whole numbers n and m, then it is a factor of $n + m$," is a theorem in which "p is a factor of n" and "p is a factor of m" may be viewed as premises and "p is a factor of $n + m$" as the conclusion. Theorems like "There is no largest prime number," have unstated premises. In this case, the conclusion "There is no largest prime number" follows from the *unstated* premise "If we are in the system of natural numbers." (Though important, the precise meanings of *prime number* and *system of natural numbers* are not important for present purposes. These concepts are discussed thoroughly in Chapters 5 and 6.)

A proof itself is a list of statements, ending with the conclusion, in which the earlier statements are either premises (or assumptions) or follow directly from preceding statements by logical inference. For example, one proof of Theorem (1) above is as follows:

STATEMENT	REASON
1. p is a factor of n and p is a factor of m	Premises
2. Therefore, there are numbers q and q' such that $n = p \cdot q$ and $m = p \cdot q'$	Premises: by the definition of what a factor[16] is—that is, by the *meaning* of the premises
3. Hence, $n + m = (p \cdot q) + (p \cdot q')$	Logical inference: by substitution of equals—e.g., $p \cdot q$ for n

[15] Regularities (a) and (b) are called (a) the commutative law of addition and (b) the associative law of addition, respectively. See Chapter 4, Section 2, and Chapter 5, Section 7, for more detailed discussions. Also, see Section 4.4 for discussion of the use of parentheses.

[16] p is said to be a factor of another number, n, if there is a number q such that $n = p \cdot q$. For example, 3 is a factor of 6 since there is a number, 2, such that $6 = 3 \cdot 2$.

4. But $(p \cdot q) + (p \cdot q') = p \cdot (q + q')$ Implicit premise: by the distributive law of multiplication over addition in the system of natural numbers (see Chapter 5)

5. Therefore, $n + m = p \cdot (q + q')$ Logical inference

6. Thus, p is a factor of $n + m$ Premise: by definition of a factor

What is the logical inference rule used in step 5? Compare with step 3. Compare step 6 with step 2. Do they both follow from the definition of a factor in exactly the same sense?

(6) Returning to the subject of Example (4), we note that (c) above may be regarded as a conclusion that follows from premises (a) and (b). Putting these together, we can write the theorem: "If $\square + \triangle = \triangle + \square$ and $(\square + \triangle) + \bigcirc = \square + (\triangle + \bigcirc)$, then $(\square + \triangle) + \bigcirc = \square + (\bigcirc + \triangle)$." A proof of this theorem is easy to find.

STATEMENT	REASON
1. $(\square + \triangle) + \bigcirc = \square + (\triangle + \bigcirc)$	Premise (b)
2. $\triangle + \bigcirc = \bigcirc + \triangle$	Premise (a) (renaming variables)
3. $(\square + \triangle) + \bigcirc = \square + (\bigcirc + \triangle)$	Inference: by substitution of $\bigcirc + \triangle$ for $\triangle + \bigcirc$ in the right side of step (1)

It would be a rare child indeed who would or could reason this formally without explicit instruction. Nonetheless, one of the important things to come out of curriculum development in school mathematics in recent years is that many children in the fifth grade or thereabouts take quite naturally to this type of activity.[17] They enjoy proving simple theorems about numbers and, apparently, it doesn't take very long to get them to the point where they can do it. We would encourage this type of activity but would also emphasize that, in our opinion, elementary school teachers should place relatively more emphasis on teaching children how to make inferences in increasingly complex situations of the sort indicated in Examples (1)–(3) which they are likely to meet almost every day.

EXERCISES 1–5.2

S–1. State a conclusion that follows by logical inference from each of the following sets of premises:

    a. All students enrolled in Psychology 1 are sophomores. Alice is enrolled in Psychology 1.

[17] For some sample activities, see R. B. Davis, *Discovery in Mathematics*, Reading, Mass., Addison Wesley, 1964.

    b. If the teacher is sick, we always have a substitute. Whenever we have a substitute, the class is noisy.

    c. All queeks have green teeth. No one with green teeth has red hair.

    d. If Santa Claus came down the chimney, then he must be sooty. Santa Claus has a clean suit.

    e. If a is not larger than b, then $a - b$ is not a natural number. $a - b = 5$.

S–2. Assuming the associative law, $(\Box + \triangle) + \bigcirc = \Box + (\triangle + \bigcirc)$, and the commutative law, $\Box + \triangle = \triangle + \Box$, prove that $(\Box + \bigcirc) + \triangle = \bigcirc + (\triangle + \Box)$.

S–3. a. Assume we have learned that $7 + 3 = 10$, $3 + 5 = 8$, $10 + 5 = 15$ and the associative law for addition, i.e., $\triangle + (\Box + \bigcirc) = (\triangle + \Box) + \bigcirc$. Show that it must be true that $7 + 8 = 15$.

    b. Assume we have learned that $6 + 4 = 10$, $4 + 3 = 7$, $10 + 3 = 13$, and the associative law for addition. Show that it must be true that $6 + 7 = 13$.

S–4. a. Prove that the sum of any two even numbers must be even. *Hint:* Any even number must be a multiple of 2. Let $2m$ and $2n$ be the two arbitrary even numbers and show that their sum must be a multiple of 2.

    b. Prove that the sum of two odd numbers must be even. *Hint:* Any odd number must be of the form $2m + 1$.

    c. What can you say about the sum of an odd number and an even number? Prove your answer.

S–5. a. When two even numbers are multiplied, the product is an even number (e.g., $2 \times 4 = 8$, $4 \times 6 = 24$, $6 \times 2 = 12$). Prove that this must be true for any two even numbers. *Hint:* Let $2m$ and $2n$ be two arbitrary even numbers and show that their product must be a multiple of 2.

    b. What can we say about the product of an even and an odd number? Prove your answer.

    c. What can we say about the product of two odd numbers? Prove your answer.

5.3 Further Observations and Specific Techniques

It cannot automatically be assumed that the beginning elementary school child can make a particular inference, as obvious as it may seem to the teacher. In a recent research study with beginning kindergarten children,[18] for example, three out of four were *unable* to make very simple inferences. The children were told a fictitious story about the Gruundians, who were described as "*some* being tall, *all* having big feet, *some* wearing tin hats," and so on. Only one of the attributes was possessed by *all* of the Gruundians. The experimenter then said that he had seen a Gruundian on the way to school that day and asked each child what he

[18] J. Scandura and R. McGee, "An Exploratory Investigation of Basic Abilities of Kindergarten Children," *Arithmetic Teacher*, 1971 (in press).

could say for sure about what the Gruundian looked like. If the child said he was certain only that the Gruundian had big feet, then he had drawn the correct logical conclusion. (If *all* Gruundians have big feet, then the one seen must certainly have had big feet; but if only *some* Grundians are tall then the one seen could have been short.)

The work of Jean Piaget over the past four decades provides even stronger support for the contention that young children are not always able to make even the simplest inferences and the teacher must be aware of this, particularly when trying to explain something verbally. If the discussion calls for inferences like those of Examples (1)–(3), which the child cannot make, the explanation, no matter how clear it seems to the teacher, will fall on deaf ears.

The ability to infer develops gradually as the child becomes familiar with specific instances of various deductive patterns. Such patterns are called rules of inference, and they are the means by which valid conclusions may be drawn logically from given premises. Three of the most basic rules of inference are illustrated by Examples (1), (2), and (3) of Section 5.2. When represented formally, the deductive patterns involved in these examples are clearly exposed:

STATEMENTS		PATTERNS
1. *If* your mother has signed the permission form, *then* you may go to the zoo with us. My mother has signed the permission form.	premises	If p, then q p
You may go to the zoo with us.	conclusion	q
2. The panther must be in his cage *or* the panther must be outdoors. The panther is *not* in his cage.	premises	p or q not p
The panther is outdoors.	conclusion	q
3. All reptiles are cold-blooded animals. An alligator is a reptile.	premises	All p's are q's x is a p
An alligator is a cold-blooded animal.	conclusion	x is a q

Compare pattern 3 with the Gruundian illustration above. What rule of inference was involved there? Can you think of any other patterns of inference? What about substitution of equals? Can you represent the rule formally?

Basic patterns such as these form the building blocks of more complicated rules of inference. The proofs in Examples (5) and (6), for example, involve a number of such rules (put together in a certain way).

Can you identify all of the rules of inference used in each of these proofs?

More complete discussion of rules of inference is beyond the scope of this text. For a more rigorous and psychologically oriented discussion of inference rules and deductive reasoning, the interested reader is referred to the author's book, *Mathematics and Structural Learning*.[19]

As indicated in Section 1.1, there is an important difference between the

[19] Englewood Cliffs, N.J., Prentice-Hall, to be published. See particularly Chapter 13 by John Corcoran and Chapters 3–5 by the author.

ability to detect regularities and the *ability to use* a particular regularity. A similar distinction can be made with respect to logical inference.

In this section we have discussed the ability to *use* established rules of inference. This does not imply, however, that one cannot become aware of (detect) new patterns of inference (e.g., new combinations of known inference rules). On the contrary, research mathematicians, in their search for new proofs, sometimes develop patterns of inference never used before. Presumably this is accomplished by combining known inference rules in new ways, but no one knows for sure how this is accomplished. All that can be said with any certainty is that heuristics are involved in one form or another. (For further discussion of heuristics and related matters of proof, see Section 7 and *Mathematics and Structural Learning.* Also see Polya's well-known books on mathematical problem solving [references given in Section 7].)

It is also meaningful for logicians (mathematicians do not normally do this) to talk about the ability to *formalize* (i.e., *describe*) deductive processes (or methods of proof) in terms of known inference rules and the *ability to interpret given formalizations.* To be sure, although secondary school and college students may not be asked to devise new rules of inference, they may be required to formalize those rules of inference they have already learned how to use. For example, a student taking logic would almost certainly have to be able to formalize simple deductive patterns such as that inherent in:

> All four-sided figures are quadrilaterals.
>
> A rectangle has four sides.
>
> Therefore, a rectangle is a quadrilateral.

This deductive pattern is commonly known as *syllogistic reasoning* and may be represented by

> All P's are Q's
>
> a is a P
>
> Therefore, a is a Q

where a, P, and Q may denote whatever one wants them to. The same pattern may be represented even more succinctly by

$$\forall\, x,\, P(x) \supset Q(x)$$
$$P(a)$$
$$\therefore Q(a)$$

In the elementary school, we might be content to concentrate on simply improving the child's ability to *use* the more common rules of inference—and, perhaps, to formalize a few of these inference rules during the fifth and sixth years of study. We feel that this is well within the reach of most elementary school children. In Chapter 3, for example, it is shown that the so-called statement logic has exactly the same structure as does the system of sets and set operations. And, as any modern teacher knows, sets typically cause little difficulty for most elementary school children, even very young ones. Furthermore,

OBJECTIVES AND PROCESS ABILITIES IN MATHEMATICS **49**

some innovators have even been successful in teaching *formal* logic to gifted fourth- and fifth-graders.[20]

EXERCISES 1–5.3

E–1. We may symbolize certain statements as follows:

"If p, then q" may be symbolized $p \supset q$.
"p or q" may be symbolized $p \lor q$.
"not p" may be symbolized $\sim p$.
"All P's are Q's" may be symbolized $\forall x, (P(x) \supset Q(x))$.
"x is a P" may be symbolized $P(x)$.

Write each of the following statements symbolically:

a. If p or q, then r e. If p, then s or t
b. Not p, or not q f. All R's are T's.
c. If q, then not r g. y is a T.
d. If not p, then s

E–2. Use the symbols given in Problem E–1 to express symbolically each of the deductive patterns given on page 47.

6. THE ABILITY TO AXIOMATIZE

An idea that can be logically derived from others is said to be logically *dependent* (on them); otherwise, it is said to be logically *independent*. For example, consider the set of ideas expressed by equations (a), (b), and (c) of Examples (4) and (6) of Section 5.2. In Example (6), given equations (a) and (b) (i.e., on the commutative and associative laws), which served as premises, we were able to prove equation (c). Therefore, (c) is logically dependent on (a) and (b). (Although we shall not prove it, (a) is logically dependent on (b) and (c) as well; (b) is also dependent on (a) and (c).)

If we consider only equations (a) and (b), on the other hand, then each one is logically independent of the other. That is, neither equation can be deduced from the other by means of inference rules. We can demonstrate this independence conclusively by constructing what mathematicians call *counterexamples*, displays in which (a) holds but (b) does not, or in which (b) holds but (a) does not. Consider, for example, the display

∘	a	b	(c)	d
a	a	b	c	d
(b)	b	c	(b)	a
c	c	b	a	b
d	d	a	b	c

---

[20] P. Suppes, "Mathematical Logic for the Schools," *Arithmetic Teacher*, **9**, 1962, pp. 396–399. *See also* P. Suppes and C. Ihrke, *Accelerated Program in Elementary School Mathematics—The Fourth Year*, Technical Report No. 148, Stanford University, Institute for Mathematical Studies in the Social Sciences, 1969.

where ∘ indicates that each letter in the first column (to the left of the vertical line) can be combined with each letter in the top row (above the horizontal line) to produce (→) the letter in the corresponding row *and* column.[21] For example, b ∘ c → b. In this case, (b ∘ c) ∘ d → (b) ∘ d → a, whereas b ∘ (c ∘ d) → b ∘ (b) → c. Hence, (b ∘ c) ∘ d ≠ b ∘ (c ∘ d) and the display does not satisfy the associative property (i.e., equation (b) in Example 4). Check *each* pair of letters to make sure that the display has the commutative property (i.e., equation (a)). For example, b ∘ d → a and d ∘ b → a, so b ∘ d = d ∘ b. (There are five other pairs to check but perhaps you can find some short cuts.)

Do you think it is possible to prove the associative property from the commutative property? That is, does the associative property hold whenever the commutative property does? How do you know?

Does this display demonstrate *conclusively* that the *commutative* property cannot be derived from the *associative* property? Why not? Could there possibly be a display that satisfies the associative property but not the commutative property?

Can you construct a display that satisfies the associative but not the commutative property? If not, be sure to work Problems E–2 and E–3. (Also see Problem S–4 of Exercises 4–1.1 and Problem S–7 of Exercises 4–2.)

EXERCISES 1–6

S–1. For each of the following pairs of statements, determine which one(s) is (are) dependent on the other or whether the two statements are independent.

a. A. x is even.
    B. x is a multiple of 4.
b. A. Tim is a girl.
    B. Tony is a boy.
c. A. Sally is Tom's wife.
    B. Tom is Sally's husband.
d. A. Bill is 6 and Fred is 3.
    B. Bill is 3 years older than Fred.

e. A. Bob is 3″ taller than Trudy.
    B. Trudy is 5′4″.
f. A. x is an odd number.
    B. y is an even number.
g. A. $\square = 4$ and $\triangle = 5$.
    B. $\square + \triangle = 9$.
h. A. $\square \times \triangle = 12$.
    B. $\square = 4$ and $\triangle = 3$.

E–2. Suppose we define an operation ∗ on the whole numbers as $x ∗ y = y$; e.g., $2 ∗ 3 = 3, 4 ∗ 7 = 7, 3 ∗ 1 = 1$.

a. Evaluate each of the following:

   i. $2 ∗ 4$
   ii. $4 ∗ 3$
   iii. $0 ∗ 7$
   iv. $4 ∗ 0$
   v. $2 ∗ 1$

   vi. $(3 ∗ 1) ∗ 4$
   vii. $2 ∗ (3 ∗ 6)$
   viii. $3 ∗ (0 ∗ 4)$
   ix. $3 ∗ (1 ∗ 4)$
   x. $(2 ∗ 6) ∗ 3$

[21] See Section 2.1 of Chapter 4 for a concrete interpretation of this display. In this interpretation the letters correspond to moves of an elevator and ∘ to the operation of first making one move and then the other.

b. This operation is associative, i.e., for any numbers x, y, and z,
$$x * (y * z) = x * z = z \text{ and } (x * y) * z = y * z = z.$$

Show that this operation is *not* commutative, i.e., find two numbers x and y such that $x * y \neq y * x$.

E–3. Consider the display

·	a	b	c	d	e	f
a	a	b	c	d	e	f
b	b	a	e	f	c	d
c	c	f	a	e	d	b
d	d	e	f	a	b	c
e	e	d	b	c	f	a
f	f	c	d	b	a	e

a. Check several examples to convince yourself that the associative law holds (it can be proved that it holds for *all* elements in the display).
b. Find a counterexample to show that the commutative law does *not* hold for the operation represented by this display.

6.1 What Axiomatization Is and Some Examples

For many sets of mathematical ideas (properties), there is often a subset from which all of the other ideas (properties) may be deduced. (Indeed, there may be many such subsets for each set.) In this case, *any* display with all the properties in that subset will also have all the properties in the set. Such a subset is said to form an *axiom system*, and the ideas or properties belonging to this subset are called *axioms*. *Axiomatization* is the process of selecting just such a subset from a given set of mathematical properties.

To determine whether a person can axiomatize a set, of course, we need to expose him to a number of ideas and see how well he can identify those which are crucial—to see how well, so to speak, he can separate the chaff from the wheat. A better idea of what this involves and why this ability is important in everyday life can be obtained by going through each of the examples of Section 5 once again. This time, however, consider the premises together with the conclusions and see how many different (proper) subsets you can find from which the other statement(s) may be derived.

In considering Examples (1), (2), and (3), we refer freely to the discussion in Section 5.3.

(1) In Example (1), the two given premises clearly serve as an axiom set. The premise "p," together with the conclusion "q," constitute another axiom set. (Perhaps you can show this with the help of your instructor.) The premise "p," however, is not logically dependent on the conclusion and the other premise. (That is, "q" and "If p, then q," does not imply "p.")

(2) The given set of Example (2) can be represented, {*p* or *q*, not *p*, *q*}, where "*p* or *q*" and "not *p*" are the original premises. Is "*q*" logically dependent on these premises? Do the premises form an independent axiom set?

What about "not *p*" and "*q*"? Can you deduce "*p* or *q*" from these statements? That is, is "*p* or *q*" necessarily true if "not *p*" and "*q*" are? To satisfy yourself that it is, let "not *p*" mean "John did not get an A" and "*q*" mean "Bill got an A" and see if the corresponding conclusion "John or Bill got an A" necessarily follows.

What about "*p* or *q*" and "*q*"? Do they form an axiom set? Here you may note that the answer depends on whether or not "*p* or *q*" is taken to be true when both "*p*" and "*q*" are true. (In logic we usually assume that it is, unless otherwise indicated—that is, "*p* or *q*" is true if "*p*" is true, or if "*q*" is true, or if "*p* and *q*" is true.)

(3) Satisfy yourself that in Example (3) only the original premises form an axiom set.

(4) In Examples (4) and (6) of Section 5.2 we saw that (c) $(\square + \triangle) + \bigcirc = \square + (\bigcirc + \triangle)$ can be derived from (a) $\square + \triangle = \triangle + \square$ and (b) $(\square + \triangle) + \bigcirc = \square + (\triangle + \bigcirc)$. It is also the case that (b) can be proved from (a) and (c); this gives an alternative axiom set.

This can be shown as follows.

STATEMENT	REASON
1. $\triangle + \bigcirc = \bigcirc + \triangle$	Assumption (a)
2. $(\square + \triangle) + \bigcirc = \square + (\bigcirc + \triangle)$	Assumption (c)
3. $(\square + \triangle) + \bigcirc = \square + (\triangle + \bigcirc)$	Substituting equals in step 1 in the right side of step 2

(5) Because Example (5) is more difficult, we leave it as an optional exercise for the ambitious reader.

(6) As a final example, consider the following list of equations, which represent properties of whole numbers. All of these equations can be derived from just two. See if you can find them. (It is not as hard as it might first appear.)

$$(\square + \triangle) + \bigcirc = (\bigcirc + \square) + \triangle$$

$$(\square + \triangle) + \bigcirc = \square + (\bigcirc + \triangle)$$

$$5 + 7 = 7 + 5$$

$$3 + 4 = 4 + 3$$

$$(\square + \triangle) + \bigcirc = \square + (\triangle + \bigcirc)$$

$$(7 + 6) + 2 = 7 + (6 + 2)$$

$$2 + \triangle = \triangle + 2$$

$$\square + (\triangle + (\bigcirc + \heartsuit)) = (\square + \triangle) + (\bigcirc + \heartsuit)$$

$$\square + \triangle = \triangle + \square$$

$$(1 + (2 + 3)) + 4 = (1 + 2) + (3 + 4)$$

.

.

.

Identify an axiom set with just two axioms. Is there another set of two that would work just as well?

EXERCISES 1-6.1

S-1. Consider the information given by the following sentences:

(1) I have three coins.
(2) One coin is a nickel.
(3) One coin is a dime.
(4) I have 25 cents in coins.
(5) I have no pennies.

From what smaller set of these statements (axioms) can you obtain the same information by deductive reasoning?

S-2. A stranger in town learns the following from the passengers on his north-south bus:

(1) The streets of the town are numbered from 1–100, north to south.
(2) The first stop on the bus line is 7th Street.
(3) If the bus stops at a street, then it also stops at every tenth block thereafter.
(4) The bus stops at 91st Street.
(5) 33rd Street is the fourth stop.
(6) The bus does not stop at 49th, 50th, or 51st Streets.
(7) 97th Street is the last stop.
(8) There are 18 stops on the route.

These facts are enough to enable him to construct the entire route, but he could do the same even if some were eliminated. Write the minimum list of these facts he needs to remember.

For further thought and study: To see how children practice axiomatizing in the Madison Project, consult Davis, *Discovery in Mathematics,* Chapters 27 and 28.

6.2 Further Observations

Together with the discussion of Section 5, the examples of Section 6.1 make clear that deductive reasoning can affect the acquisition and structuring of knowledge in two ways. The learner may acquire new knowledge by making inferences of the sort described in Section 5. He may also reorganize his existing knowledge by going in the opposite direction. That is, having already acquired a number of ideas, he may be able to identify a smaller axiom set from which the others may be generated. In effect, deductive reasoning provides the "glue" that may hold together a number of different ideas.

Mathematicians usually require that a set of axioms be mutually independent, that is, that no axiom can be derived from any of the others. We consider various sets of mutually independent axioms which characterize given mathematical systems in Chapter 4, and further discussion is deferred until then.

In everyday affairs, however, we often make use of axiomatization more informally. We are axiomatizing, in effect, whenever we seek to identify the underlying principles on which various conclusions are based. Such a process is essential in any kind of rational debate. If someone claims, for example, that children of poor families are less intelligent than children from middle-class families, as indicated by the results of IQ tests, then his underlying assumption is that performance on IQ tests and intelligence are synonymous. This is clearly true if one defines intelligence to be whatever IQ tests measure, as is frequently done by measurement specialists. If, however, we have a different conception of intelligence in mind, we would probably want to question his conclusion.

Axiomatization may also help to reveal inconsistences in arguments. Thus, a politician may claim, for example, that he will lower taxes and increase governmental services. This claim is inconsistent with (i.e., does not follow from) the implicit "axioms" that (1) increased services require increased expenditures and (2) increased expenditures require increased taxes. Either the politician's assumptions differ from these—for example, he may reject (1), assuming instead that expenditures may be lowered by increasing efficiency—or his promises are false. In order to vote intelligently, one must be alert to possible inconsistencies of this sort. In effect, axiomatization, loosely defined, may help to clarify arguments and in the process provide a basis for further discussion and/or lead to wiser decisions.

Finally, axiomatization makes it possible to reorganize knowledge so that a large number of facts may be available with a minimum of memorization. Thus, remembering a set of basic ideas that form an axiom set may be a sufficient basis for an individual to recover other information through logical reasoning, even though the information may be temporarily forgotten. On a very simple level, a child who knows that the commutative law applies to all whole numbers does not need to remember, for example, that $1578 + 2867 = 2867 + 1578$. (Nor does he need to show that they are equal by adding.) He simply deduces it from his knowledge of the commutative law.

A less trivial example is available to anyone who has had a basic course in trigonometry. The student in a traditionally taught course typically has to learn a good many formulas. The student who "sees" the relationships among these formulas, however, may have to remember only five or six basic ideas, whereas the so-called rote learner may perhaps have to remember a hundred different formulas. (For more details, see the example in Chapter 3, pages 108–109.)

Unfortunately, there is relatively little general guidance one can give about how to axiomatize. About all one can say is that with experience people seem to get better at identifying "key" ideas. In view of its importance in both mathematics and in everyday life, it is clear that such experiences should be provided whenever possible.

7. HEURISTICS AND COMBINING PROCESS ABILITIES

The ability to detect regularities, to construct examples, to interpret, to describe, to make inferences, and to axiomatize is of great importance, both in doing mathematics and in dealing with everyday affairs; a good mathematics program should make explicit provision for teaching all of them. As we indicated before, it is unfortunate that so few specific techniques have been identified to date. But one thing is quite clear: The only way students will develop such abilities is if we provide them with ample opportunities to do so.

This does not imply, however, that children should only be presented with problems of the six types given in the preceding sections. These problems were selected carefully in order to clearly illustrate the kind of processing under consideration. Many problems, if not most, involve more than one kind of processing skill for their solution. Others involve the same kind of processing skill applied in two or more different ways.

Obviously, there are many ways in which such processing skills may be combined. Some of the combinations have been given special names and, generally, have gone under such labels as *heuristics*, searching behaviors, and learning to learn.

In this section we shall present some illustrative problems, identify heuristics (rules of thumb) which may be useful in solving them, and show how these heuristics can be viewed as combinations of the six basic types of processing skill identified above. No attempt, however, is made to demonstrate conclusively that *all* so-called heuristics are simple combinations of these six types. Do not hesitate to try out new heuristics which come to your attention to see how well they work, even if you cannot see how they relate to the six kinds of processing skill discussed.

7.1 Translation

In problem solving, one is frequently advised to first *translate* the problem into mathematical or symbolic form. This advice often takes one of two forms: *Heuristic (1)*—Given a verbal statement of a problem, represent the basic relationships as a set of equations; *Heuristic (2)*—Given a concrete display, construct a symbolic one which exhibits the same regularities.

Heuristic (1) involves two steps: first, *interpreting* the verbal statement, and, second, *describing* the underlying meaning as a set of equations.

This heuristic stems largely from the deep concern many mathematics educators have had for teaching students how to solve verbal problems. Thus, some have felt that it is best to have the child translate the verbal statement of a problem directly into algebraic equations and then solve the equations. For example, given the problem, "If Janie had $10 when she left home and had $4.75

left after buying a blouse, how much did she spend for the blouse?" some educators would have the child let x = the cost of the blouse, translate the problem (statement) directly, giving $10 − x = 4.75, and then solve the equation.

Others have felt that the child should go directly to the subtraction. There is evidence,[22] in fact, that the latter procedure may be more efficient in the middle elementary grades.

In education, however, a single experiment rarely settles an issue once and for all. Thus, for example, it is easy to conceive of problems which readily lend themselves to algebraic treatment but which become unwieldly when a direct arithmetic attack is used. Some of the time-work and mixture problems of elementary algebra provide ready examples. "Suppose Joey can do $1\frac{1}{2}$ times as much work per hour as Julie and that he starts work 4 hours after Julie. How many hours will they have to work together before they both do the same amount of work?"

Can you solve the problem without first writing an equation? If so, you probably used a good deal of "trial and error."

Now solve the problem by first expressing it as an equation. (*Hint:* Let x represent the number of hours they have to work together. This gives $(4 + x) = 1\frac{1}{2} x$, where one side represents the amount done by Julie and the other, by Joey. Which is which?)

What kinds of processing skill do you feel were involved in using this latter procedure? Could you tell what each side of the equation (in the *hint*) stood for without first knowing what the problem statement meant? (You may not have been able to write the equation without some help but you can identify the kinds of processing skill required. What are they?)

Heuristic (2) is somewhat more difficult to analyze than the first because it involves more different kinds of processing skill. In this case, the problem is one of translating a given concrete display into symbolic form while preserving the original regularities.

In order to accomplish this, first decide what aspects of the display are to be preserved. This already involves a form of abstraction or *detecting regularities*. For example, consider a situation in which a group of Boy Scouts is learning how to carry out the instructions: "Attention," "Right face," "(Turn) about face," and "Left face." Suppose further that the scoutmaster decides to introduce a new twist by giving his commands in *pairs* with the instruction that the scouts are to make only that move which results in the same position as first carrying out one command and then the other.

This concrete display is too complex to consider very effectively but, nonetheless, a few features (regularities) do stand out: the four commands, pairs of commands, and commands equivalent to pairs of commands. Anyone who has has experience in working with such situations recognizes that these features may be represented (i.e., *described*) as a table of the form

[22] J. Wilson, "The Role of Structure in Verbal Problem Solving," *Arithmetic Teacher*, **14**, 1967, pp. 486–497.

	A	R	T	L
A	A	R	T	L
R	R	T	L	A
T	T	L	A	R
L	L	A	R	T

where A stands for "Attention," R for "Right face," T for "(Turn) about face," and L for "Left face." The move equivalent to any two commands can be found by looking in the row corresponding to one command and the column corresponding to the other.

Translating a concrete display into a symbolic one is not, of course, an end in itself. The whole point of heuristic (2) is that it is frequently easier to *detect* (*new*) *regularities* in a symbolic display than in one which is concrete and thereby difficult to think about as an entirety. In the table (symbolic display) above, for example, it is comparatively easy to see that the order in which two commands are carried out is immaterial. The equivalent single move is the same. For example, "R followed by L" is equivalent to A and so is "L followed by R." We can represent this regularity more generally by $X \circ Y = Y \circ X$, where "$\circ$" means "followed by." Other regularities that can readily be detected are:

1. For every move, X, there is another move, X', such that $X \circ X' = A$ (e.g., If X = R, then X' = L and $R \circ L = A$; note also that $T \circ T = A$ and $A \circ A = A$).
2. Any combination of A's and T's will result in an A or a T. (It is impossible to get an R or L.) For example, $T \circ T \circ A \circ T = T$.

Can you detect any other regularities? What happens when you combine A with any other command?

In effect, then, detecting regularities enters into this type of task in two ways. The heuristic itself involves, first, detecting relatively simple regularities in a concrete display and then representing these regularities in a symbolic display. Finally, new regularities in the symbolic display, which were obscured in the concrete situation, may be detected.

It would be convenient if some simple rules existed for translating displays from one form into another and particularly for knowing when to represent a concrete situation in an iconic form and when in symbols. We know of no answers to these difficult questions. Nonetheless, icons seem better able to represent geometric or spatial situations—situations in which the various elements may vary *continuously*. Thus, situations involving lines, areas, volumes, or time lend themselves readily to iconic representation. Where the major elements in a concrete situation are discrete, symbols tend to provide the better alternative.

Translating displays does not always facilitate detecting new regularities, however. Suppose, for example, that we are working with sets and want to talk about combining sets, taking one set away from another, and so on. In this case, we might be tempted, say, to represent the sets by letters, such as *A*, *B*, *C*, and "combining" sets, for example, by "+." If so, we would be led to such poten-tially confusing equations, as, $A + A = A$. (Clearly, combining a set (A) with itself does not change anything, whereas *adding* a number to itself does.)

EXERCISES 1–7.1

S–1. Translate each of the following word problems into an equation. (Be sure to state what the variables in each equation stand for.)

    a.   Joey bought a model airplane for $1.37 and has $2.14 left. How much money did Joey start with?

    b.   A house is twice as old as the furnace in it. If the total age of house and furnace is 48, how old is the house?

    c.   Bobby gave 11 marbles to Terry and 13 marbles to Jerry. If Bobby still has half of his marbles left, how many marbles did he start with?

    d.   Janie has twice as many dolls as Julie. Janie and Julie together have four times as many dolls as Jeanne. If Jeanne has three dolls, how many dolls does Janie have? How many dolls does Julie have?

S–2. Complete the following display for addition on a clock face, using only the hours 3, 6, 9, and 12.

+	3	6	9	12
3	6		12	3
6			3	
9	12			
12			9	

S–3. Construct a display describing "addition" on a clock face using only the hours 2, 4, 6, 8, 10, 12.

S–4. Construct a display describing "multiplication" on a clock face using only the hours 3, 6, 9, 12.

S–5. Consider the instructions: attention (0), turn right 60° (A), turn right 120° (R), turn left 60° (B), turn left 120° (L), and about face (T). Assume that for each pair of commands given by a drill sergeant, a recruit is to execute the single command equivalent to executing the two commands in order. Construct a table (display) similar to the one on page 57 showing the result of each possible pair of commands.

7.2 Other Heuristics and Problem Solving

We now consider the not-too-easy task of determining the number of regions into which five arbitrarily arranged planes divide three-dimensional space.[23] This problem is difficult to visualize directly; it calls for some alternative procedure.

---

[23] Our discussion is based on the film, "Let Us Teach Guessing," featuring George Polya, which was produced by the Mathematical Association of America.

One of Polya's[24] pet tricks (a heuristic) is to consider a simpler problem of the same sort to see if any insight can be obtained into the problem at hand. (This involves keying on particular aspects of the given problem (e.g., number of planes and number of regions) and *particularizing* on the basis of these aspects.) Thus, one plane divides space into two regions, two planes into four regions, and three planes into eight regions. The last fact can be seen by taking one plane as the plane of this sheet of paper and having the other two planes cut this "paper" plane as indicated.

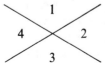

There are clearly four regions on top of this sheet of paper and four below (for a total of eight). Noting that *no* (0) plane divides space into just one region (itself), we represent this set of facts symbolically (recall heuristic (2) of the previous section) as:

NUMBER OF CUTTING PLANES	NUMBER OF REGIONS (3-SPACE)
0	1
1	2
2	4
3	8
4	(16)?

Extending the most evident *regularity* in this display we suppose that four planes would cut 3-space into 16 regions. (Going one step further, of course, we might guess that the answer to our problem involving five planes would be 32.)

Let us look more closely at the four-plane problem. Although not exactly trivial, a solution can be visualized (Figure 1–6) by noting that four arbitrary cutting planes enclose a tetrahedron (i.e., a solid bounded by four planes). (No *closed* solid can be formed by fewer than four planes.)

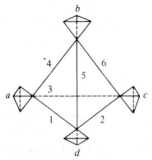

Figure 1–6

Notice that constructing this solid involved another instance of *particularization*.

[24] *Ibid.*

In addition to the *interior* region formed by the cutting planes, there is one *exterior* region which corresponds to each of the four vertices of the tetrahedron (i.e., the regions labeled *a, b, c, d*), one exterior region to each of the six lines (labeled 1, 2, 3, 4, 5, 6), and one exterior region to each of the four faces of the tetrahedron (unlabeled for simplicity). Thus, four planes cut 3-space into $1 + 4 + 6 + 4 = 15$ distinct regions. This rather upsets our guess of 16 and bids us to try something else.

Luckily, this particular problem can be simplified (particularized) in still another way, this time by considering analogous problems in 2-space (i.e., the plane) and 1-space (i.e., the line). Applying the same heuristic, we first notice that one point divides a line into two regions (segments), two points divide it into three regions, and so on. Notice also that three lines divide 2-space into seven regions.

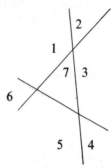

These facts may be summarized by:

NUMBER OF CUTTING PLANES (3-SPACE) LINES (2-SPACE) POINTS (1-SPACE)	NUMBER OF REGIONS		
	3-SPACE	2-SPACE	1-SPACE
0	1 +	1 +	1
1	2 +	2 +	2
2	4 +	4 +	3
3	8 +	7 +	4
4	15	(11)	5
5	(26)		6

Can you see any *regularity* or pattern in this display? Note that the number at the tail of each arrow and the number in the middle sum to the number next to the arrow head. Extending this pattern (dotted lines) we might guess that four lines would cut 2-space into 11 regions, and if so, that five planes might cut 3-space into 26 regions. (What process does extending the pattern involve?)

We will not prove logically (i.e., deduce) that this latter guess is indeed true. But let us pause long enough to show that this is certainly a plausible guess by showing that four lines do indeed divide 2-space into 11 regions. To see this, observe in Figure 1–7 that no matter where a fourth line cuts 2-space it can add at most four new regions to the seven already formed by three lines. The additional regions (labeled 1, 2, 3, 4 in each figure) are formed by dividing four of the original seven regions into halves. (What process is involved here?)

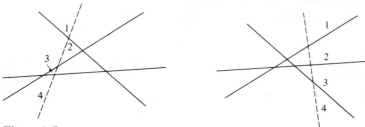

Figure 1–7

When analyzed in detail as we have done here, this rather simple-sounding heuristic, "Consider a simpler problem," involves a number of different kinds of processing skill. Particularization, detecting regularities, and representing (describing) were among the more obvious ones. Furthermore, each of these kinds of skill entered into the solution at more than one point.

Although we will not go into detail here, any number of other heuristics can be analyzed in the same general way. Rearranging data in an orderly form, for example, is another favorite heuristic of Polya and this one seems to correspond directly to detecting regularities.[25]

8. EPILOGUE

As suggested, there are relatively few specific techniques that have been identified for improving one's processing ability. Some were identified, however, and we can expect to find more and more as we begin to search in earnest. You are in a particularly good situation to do this in your everyday work with children. Keep on the lookout for new ideas which seem to work.

Do not expect miracles, however. It is doubtful that we shall ever come up with prescriptions for solving all of our problems. The difficulty is that, even after useful techniques are identified, it is difficult to tell ahead of time the situations in which they will apply and those in which they will not. In fact, the logician Alonzo Church has proved that there are some classes of problems for which no mechanical solution procedures exist. This does not mean that particular

[25] The interested reader is encouraged to analyze the other heuristics identified in G. Polya, *Mathematical Discovery* (Volume I), New York, Wiley, 1962. The so-called "pattern of two loci," for example, is closely related to making logical inferences, the "Cartesian pattern" to translating problem statements into a system of equations (see Section 7.1), "recursion" to detecting regularities, and "superposition" to making inferences and detecting regularities. You should not be surprised to find any number of different processes involved in each particular problem.

problems cannot be solved but that no one procedure will solve all of them. Looking for procedures in such cases would be a waste of time.

EXERCISES 1–8

M–1. In this chapter, six basic processes are described. List them and give illustrations of each taken from elementary school textbooks. State the source used.

S–2. The following exercises are taken from various elementary school textbooks. For each of these, state which of the six processes is being tested and briefly justify your choices. (More than one process may apply to some of these exercises.)

  a. (From the Greater Cleveland Mathematics Program (GCMP): Kindergarten, page 17.) Repeat the pattern:

  b. (From Modern Mathematics Through Discovery: Third Grade, page 12.) On a number line, start at 0 and move to the right 4 spaces. Then move to the right 5 more spaces. What numeral is at the end of the second move? Is the following number sentence true: $4 + 5 = 9$?

  c. (GCMP: Kindergarten, page 11.) Mark the triangle:

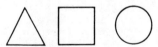

  d. Given:
$$1 = 1$$
$$1 + 3 = 2^2$$
$$1 + 3 + 5 = 3^2$$
$$1 + 3 + 5 + 7 = 4^2$$
construct a similar series whose sum is 6^2.

  e. (Maneuvers on Lattices—David Page, University of Illinois Committee on School Mathematics.) The following refers to the lattice as shown where $\uparrow$, $\downarrow$, $\leftarrow$, and $\rightarrow$ are operations or shifts in the prescribed directions, e.g., $12 \rightarrow = 13$, $12 \downarrow = 2$, etc.

30	31	32	33	34	35	36	37	38	39
20	21	22	23	24	25	26	27	28	29
10	11	12	13	14	15	16	17	18	19
0	1	2	3	4	5	6	7	8	9

$$5 \uparrow \rightarrow = 16$$

$$16 \uparrow \rightarrow \downarrow = 17. \qquad\qquad 24 \uparrow \downarrow \downarrow \downarrow \uparrow \uparrow \downarrow = \underline{\quad}$$

f. Refer to the above lattice. Wherever two frames of the same shape occur in one equation, whatever number is chosen to fill one of the frames must also be used to fill the other frames of that same shape.

$$\square \uparrow \rightarrow + \square \rightarrow \uparrow \uparrow = 38$$

g. (Suppes, *Sets and Numbers.*)

$$^-3 \times {}^+3 = {}^-9$$
$$^-3 \times {}^+2 = {}^-6$$
$$^-3 \times {}^+1 = {}^-3$$
$$^-3 \times 0 = 0$$
$$^-3 \times {}^-1 = ?$$
$$^-3 \times {}^-2 = ?$$

h. (From *Discovery in Mathematics: A Text for Teachers*, page 167.) Can you shorten this list?
    (a) $\square + \triangle = \triangle + \square$
    (b) $\square \times \triangle = \triangle \times \square$
    (c) $A + (B \times C) = (C \times B) + A$

S–3. a. Which of the six basic process abilities is being tested when you have pupils "discover" some rule, property, or relationship?
      b. Which are being tested in the following situations:
          (1) Mary drew 10 different triangles and constructed the three angle bisectors in each triangle. She noted that these angle bisectors met in a point and therefore concluded that the angle bisectors of a triangle are concurrent.
          (2) The class is presented with this information:
               A family has a weekly income of $110.
               The money is budgeted as shown.

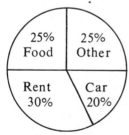

How much a week is spent on the car?

S–4. Briefly sketch how children might describe the regularities found in determining the perimeters of the following:

a. a rectangle   c. a square
b. an equilateral triangle d. an isosceles triangle

S–5. Sketch how a child might describe the procedure for changing percent to a common fraction.

S–6. Describe the regularity illustrated and list some of the prerequisites needed for each.

a.

$$\frac{1}{2} + \frac{1}{2} = \frac{4}{4}$$

$$\frac{1}{2} + \frac{1}{3} = \frac{5}{6}$$

$$\frac{1}{2} + \frac{1}{4} = \frac{6}{8}$$

b.

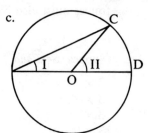

AB is a diameter.

c.

$\overset{\frown}{CD}$ is 50°.

0 is the center.

Angle I = 25°.

Angle II = 50°.

Two
Basic Ideas in Mathematics and Logic

Sets, Relations, and Operations

Because mathematics is so much a study of regularities, it is hardly surprising that mathematics is applicable in so many different situations. How different our world would be if there were no mathematics. There would be few technological devices of any kind, not to mention the lack of money (or taxes).

In this chapter, we shall identify those mathematical ideas *currently* deemed to be most basic, in the sense that they are most widely applicable, both within mathematics itself and in the real world. We have emphasized the word "currently" because mathematicians are, in fact, continually searching for new and better ways of organizing mathematics. Research in the foundations of mathematics is concerned largely with the search for increasingly primitive ideas from which all of the others may be logically deduced. Because of this ongoing activity, what is considered fundamental at one stage in the development of mathematics may not be considered equally fundamental at a later date. There was a time, for example, when *number* (arithmetic) and *space* (Euclidean geometry) were considered to be most fundamental. During most of the present century, *sets* have tended to occupy the central role. Today an increasing number of research mathematicians have tended to adopt the notions of function and relation (i.e., correspondence) as equally, if not more, basic.

The mathematical ideas that were being taught in even the best elementary schools until very recently were far different from those prevalent at the frontiers of mathematical knowledge. This, of course, was not all bad because other considerations besides mathematical elegance must be taken into account in planning a mathematics curriculum for the elementary school. The present and future needs of society and the principles by which learning takes place are perhaps more crucial.

Nevertheless, we are not at all convinced that arithmetic should constitute the bulk of the mathematics taught in the elementary school. Although arithmetic certainly has a large number of practical applications, more general objectives (e.g., those discussed in Chapter 1) are equally, if not more, important. Furthermore, there is no reason why a curriculum in which such general objectives are met cannot also do a satisfactory job of presenting arithmetic.

Accordingly, the following sections are devoted to those mathematical ideas which are felt (1) to be most fundamental, (2) to be psychologically suited for presentation during the elementary school years, and (3) to provide a sound foundation for those mathematical concepts that traditionally have been thought to have the greatest social value. No attempt is made to go into more detail than necessary. The emphasis is placed on how the various mathematical ideas fit together. Our aim is to help the elementary teacher see her daily mathematics lessons or weekly units not as isolated bits of mathematical knowledge but as parts of an integrated whole. The mathematics is presented in the sort of informal manner that might well be effective in the elementary school. Formal definitions, theorems, and proofs are to be found only where they serve a distinct pedagogical purpose.

We have taken this approach because we feel that the elementary teacher has too much to do to expect him or her to translate a formal presentation of mathematics into a form suitable for the elementary school pupil. Professional curriculum writers have found this to be an extremely challenging and difficult task, and to ask the elementary school teacher to assume this burden would be unfair.

1. THE NATURE OF THE REAL WORLD: THINGS, RELATIONS, AND OPERATIONS

The world about us is composed of *things*, *relations*, and *operations*. By *things* are meant physical entities such as animals, buildings, atoms, marks on paper, and so on which are meaningful in themselves and can be considered without reference to anything else. *Relations*, on the other hand, draw their meaning from two or more things. The relative size of two (or more) people, and the relationship of one line being perpendicular to another provide examples of relations. The statements, "*b* is longer than *a*," and "*b* is below *a*," describe two relational properties of the display

$$\overline{\hspace{2cm} a \hspace{2cm}}$$

$$\overline{\hspace{3cm} b \hspace{3cm}}$$

Operations refer to "actions" in the real world. Running, rearranging a collection of things, and the movement of the second hand on a watch are all operations. In some sense, operations may be most important in education since we are normally concerned with what children can *do*. Doing things involves action.

It is important to emphasize that the things, relations, and operations listed above have meaning only with respect to entities that can be directly perceived. In mathematics, things, relations, and operations are viewed more generally to relate to abstract entities as well.

EXERCISES 2–1

S–1. Classify each of the following as a thing, relation, or operation:

    a.   adding
    b.   is the father of
    c.   your grade in this class
    d.   your rank in this class
    e.   getting dressed

M–2. Give three relations and three operations that are important to children in the grade you are teaching (or will be teaching). What prerequisite knowledge must the child have before he or she can understand each of these?

[M–3.] What are some relations elementary school children might find which involve the numbers 11 and 17? How could you guide the children in detecting these relations?

2. SETS AND SUBSETS

The word "set" is usually taken as undefined by mathematicians. Nonetheless, each of us has an intuitive concept of what we mean by a set. We indicate our understanding by such statements as "a bunch of grapes," "the people in this room," "these five objects," and so on. As suggested above, however, mathematical concepts are not necessarily tied to physical reality. The basic "objects" of mathematical study may be made as abstract as we care to make them. Of course, these objects must be reasonably well defined. In this sense, the philosophies of Jean Paul Sartre would probably not constitute as unambiguous a basis for mathematical study as would the movements of an elevator[1] or the points and lines of some geometric display. Much more could be said about this topic and it is a fascinating area of study but the interested reader must be referred elsewhere for further discussion.[2]

2.1 Sets and Their Properties

In this section, we shall show how properties of sets (acting as elements) define higher-order sets at arbitrary levels of abstraction. We first consider an arbitrary collection of objects in the real world. These will serve as our basic elements of study. This collection may include chairs, balls, rotations, marks on paper, and

---

[1] See J. M. Scandura, "Concrete Examples of Commutative Non-Associative Systems," *The Mathematics Teacher*, **59**, 1966, pp. 735–736.
[2] Cf. J. M. Scandura, *Mathematics and Structural Learning*, Englewood Cliffs, N.J., Prentice-Hall, to be published.

so on. Given any property—for example, the property of being a chair, of being red, of being a mark on paper—we can distinguish between objects that have this property and those that do not. In effect, a property may be used to identify a class of elements. The property "red," for example, may be equated with the class of all things which have the property of redness. As we shall see, the property defines a set of elements. Not all sets, however, contain elements. The set of all 12-foot giants, for example, is empty (as far as we know).

Having introduced sets, it is natural to talk about properties of sets. Perhaps the most familiar *set property* is that of number. Any set (no matter what property of *elements* was used to define it) has a property of numerosity—the number of elements in it. For example, the property, "is a Senator of the United States," defines a set of men and women, and this set normally has numerosity 100.

Just as properties of elements may be used to define sets, set properties, in turn, may be used to define higher-order sets (i.e., sets of sets). Thus, the number 5 is often defined as the collection of all sets containing five elements, the number 37, as the set of all sets containing 37 elements, etc. In turn, these higher-order sets (i.e., numbers) also have properties (e.g., they may be odd or even) and these properties may be used to define still higher-order collections (e.g., set of all odd numbers). Beyond this there is a fourth level and a fifth. In general, properties of sets at any level define new sets at the next higher level. The process has no end.

When mathematicians refer to sets of elements, the only thing that remains invariant is the relationship between the corresponding sets and their elements. That is, the only thing that *all* sets have in common is the set-element relation of *membership*. A set contains its elements or the elements belong to the set. We denote this by $a \in S$, where a denotes an element, S, a set, and $\in$, the relation of membership.

A learner may be said to have acquired the "idea" of a set if he or she is able to construct the set in some way. Hence, one way to find out whether he has acquired the idea is to ask him or her to sort the various elements in a given *universe* of elements into two "piles," one containing the elements that belong in the set and the other containing those that do not. For example, a child might be asked to identify the set of square objects contained in some universe (e.g., pile) of objects. Of course, it is not always necessary, or even possible, to physically manipulate the elements of a set. In this case, the child may be asked to represent the set in some way. If the group of objects is represented pictorially, for example, the child might be asked to circle the required set as shown in Figure 2–1.

Figure 2–1

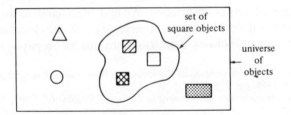

Another way to represent the set above would be to simply list the elements, {▨, □, ▦}. In this representation, the elements are icons but these could easily be replaced by symbols, for example, by assigning a, b, and c to ▨, □, and ▦, respectively.

In each of the cases listed above, each element of the set has been represented individually. Of course, when a set is very large, or if it contains an infinite number of elements, this will not be possible. Sets of this type must be represented in some other way. One way is to give a verbal description: for example, "the set of natural numbers" or "the set of square objects." Another way is to use the ellipsis "..." after a few members of the set have been listed. For example, {1, 2, 3, ...} can be used to represent the set of natural numbers. In more advanced mathematics, we frequently use what is called the *set-builder notation:* $\{x \in I | x$ is a square$\}$, where I is the set of all icons. This notation is read "the set of all icons, x (denoted $x \in I$), such that (denoted |) x is a square."[3]

Fortunately, most sets may be defined in terms of common properties.[4] For example, the set of triangles may be defined by the property of three-sidedness, and the set of even numbers, by the property of divisibility by two. These sets may be described in *set-builder notation* by $\{x \in P | x$ has three sides$\}$, where P is the set of all polygons, and $\{n \in N | n$ is divisible by 2$\}$, where N is the set of natural numbers.

How would you use set-builder notation to represent the set of all pentagons? The set of all odd numbers? The set of all red icons?

In order to help children learn to abstract, they should be given opportunities to consider sets as elements of higher order collections (i.e., higher-order sets). Families, for example, may be viewed as elements themselves as well as sets composed of people. Residence, telephone number, church affiliation, and so on are frequently considered to be characteristics of families as well as of the individuals in them. Thus, children may be asked to identify which families live on particular blocks in the school district or which families belong to particular churches. Each of the groups described in this way is a higher-order set (i.e., a set of sets). Note in particular that if there is only one family on a particular block, then the set of families on the block (i.e., the higher-order set) has only one element—the family, and *not* the members of the family.

Consider another example: The property of having *three* digits is a property of the numeral, "346," taken as a whole and has nothing whatever to do with the individual digits comprising it (including the "3" in the numeral "346"). Notice that "346" may be thought of as a set containing the three digits, "3," "4," and "6." In this case, children might be asked to classify given numerals according to the number of digits they have. Other examples might involve classifying (1) sets of numbers according to whether the "number of numbers"

---

[3] Although we do not necessarily recommend it, the set-builder notation has on occasion been successfully taught to some elementary school children.

[4] Not all properties, however, define sets. For example, the property, "Is a set which does not contain itself," does not define a set as the term is used by mathematicians. In this case, the set (of all sets which do not contain themselves) is neither a member of itself nor *not* a member of itself—a contradiction.

in each set is odd or even, or (2) sets of pairs of numbers according to whether or not the difference between the numbers in each pair is a constant (see Chapter 8).

2.2 Subsets

It is often necessary to talk of a set that is a part of another set. For example, the set of all squares is part of the set of all polygons, and the set of even numbers is part of the set of all whole numbers. Whenever one set is part of a second, we say the first set is a *subset* of the second.[5] That is, the set of squares is a subset of the set of all polygons, and the set of even numbers is a subset of the set of all whole numbers. Can you think of any other subsets of the set of polygons? Of the set of icons?

If A and B are sets, and set A is a subset of set B, then we write $A \subset B$. For example, we write $\{a, b\} \subset \{a, b, c\}$, and $\{x \in P | x$ is a square$\} \subset \{x \in P | x$ is a quadrilateral$\}$.

EXERCISES 2–2

[S–1.] Indicate (by listing) the elements in the following sets:
    a. the set of the first three months of the year
    b. the set of all words in this phrase which have more than four letters
    c. the set of all even numbers between 5 and 13

[S–2.] Indicate (by listing) the elements in the following sets:
    a. the set of members in your immediate family
    b. the set of common vowels in the alphabet
    c. the set of all one-letter words
    d. the set of all odd numbers from 2 to 10
    e. the set of all prime numbers that are even

S–3. What does the word "universe" mean, as used on page 70? What could you select for the universe for each of the sets in Problem [S–1]?

S–4. The elements of the set {January, February} have the property: "is one of the first two months of the year." In the following examples, state a property of the elements that defines the given set adequately and unambiguously:
    a. $\{w, x, y, z\}$
    b. $\{\triangle, \square, \heartsuit\}$
    c. $\{1, 2, 3, 4, 5, 6\}$

S–5. State a property of the elements which defines each of the following sets adequately and unambiguously (see Problem S–4).
    a. {Tuesday, Thursday}   d. $\{2, 4, 6, 8, 10\}$
    b. $\{a, b, c\}$ e. {red, yellow, blue}
    c. {red, yellow, green}

[5] See Chapter 3, Section 6.1, for a further discussion of subsets.

S–6.   State a property of elements that defines the empty set. What is the numerosity of this set?

[S–7.]   Represent each of the sets in Problem S–4 in set-builder notation.

[S–8.]   Give an example of a set (other than the ones in these exercises) containing

    a.   three elements
    b.   one element
    c.   no elements
    d.   an infinite number of elements

E–9.   The property "is a positive real number" defines a well-defined set, because it is clear to everyone with sufficient background what is meant by a positive number. On the other hand, the property "is a small fraction" does not define a well-defined set, because it is ambiguous, even to a person with sufficient background. What is "small" to one person may not be small to another. Which of the following properties defines a well-defined set?

    a.   is an even number
    b.   has passed his third birthday
    c.   is healthy
    d.   is a beautiful girl
    e.   is a student at the University of Pennsylvania
    f.   is a large number
    g.   is a month with 27 days

[S–10.]   In learning to use set-builder notation, what process abilities are also being acquired?

E–11.   Once you establish a universe and define a set with reference to this universe, another set is automatically defined. For example, if S is the set {Monday, Tuesday, Wednesday, Thursday} and the universe consists of the days of the week, then it is easily seen that the elements of the universe that are *not* in S form the set {Friday, Saturday, Sunday}. This set is called the *complement* of S. For each of the following, find the complement of S, relative to the given universe.

    a. $S = $ {Monday, Tuesday, Friday}
        universe $= $ {days of the week}
    b. $S = $ {even numbers}
        universe $= $ {all natural numbers}
    c. $S = $ {numbers divisible by 5}
        universe $= $ {all natural numbers}
    d. $S = $ {1, 2, 4, 5}
        universe $= $ {1, 2, 3, 4, 5, 6, 7, 8}

E–12.   Find the complement of each set S, relative to the given universe (see Problem E–11).

    a. $S = $ {January, February, March}
        universe $= $ {months of the year}

b. $S = \{1, 3, 5, 7\}$
   universe = {whole numbers between 1 and 10 inclusive}
c. $S = \{1, 3\}$
   universe = {1, 2, 3}
d. $S = \{a, c, e\}$
   universe = {a, b, c, d, e}

S–13. a. For the set {4, 5, 6, 2, 3, 7, 8}, give the subset of even numbers.
   b. Give the subset of odd numbers.
   c. Give two other subsets.

M–14. Football teams can be considered as sets (the members of the team are the elements) or as elements of sets (several teams make up a league). What are some sets that elementary school children might consider as elements of higher order sets? How could you help children identify these sets as elements of higher order sets?

For further thought and study: What methods of defining sets would you use with first-graders? Specify several methods you could use to determine whether a child can identify a set.

3. CORRESPONDENCE

Sets are indeed fundamental to mathematics. It is hard to conceive of any branch of mathematics that does not concern itself with sets. Equally fundamental, however, is the idea of a *correspondence* or mapping between sets of elements. Any situation in which a child is required to match elements from two different sets involves forming a correspondence. Setting a table, organizing teams, and assigning books to individuals provide ready examples.

In fact, the very act of sorting elements in some universe into a set and its *complement* (those elements of the universe not in the set) is a many-to-one correspondence between the elements in the universal set and that *set of sets* that includes only the given set and its complement (which is also a set). We illustrate this by referring to Figure 2–2, in which a set of icons (the universe) is

Figure 2–2

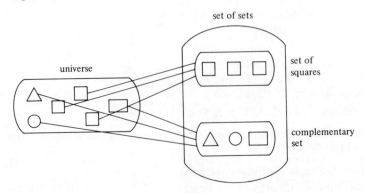

mapped into a set of sets. In particular, each square in the universe is matched via the correspondence with the set of squares. Each of the other icons is matched with the complementary set.

Notice that the universe contains all of the basic objects. The set of sets has exactly two elements, the set of squares and the complementary set of non-squares.

We have included this particular example for two reasons. First, it shows how closely connected the idea of a set and the idea of a correspondence really are. *While correspondences are defined between sets (so that sets are needed to define correspondences), sets themselves may be defined in terms of correspondences.*[6]

Second, this example illustrates that the elements of a set can be sets and that correspondences need not be limited solely to relations between sets which contain the same "sort" of elements. In the above illustration, for example, we consider both objects and *sets* of objects.

There are, of course, many examples of correspondence that do not involve this kind of complication. In fact, the idea of a correspondence between two sets is one of the very first mathematical ideas introduced to children in schools today. Thus, we teach children that the number of elements in any two sets may be compared by pairing the elements on a one-to-one basis. *If there are no elements in either set left over as a result of such pairing, then both sets are said to have the same number of elements.* Otherwise, the set with the extra elements is said to have *more* elements. The following pairs of sets, for example, may be paired in one-to-one fashion: cars and their rear license plates, children and their cartons of morning milk (or their names), U.S. workers and their Social Security numbers.

Defining "same number" in terms of one-to-one correspondences has the advantage of being equally as applicable to sets containing an infinite number of elements as to finite sets. For example, you might like to consider the question of which is larger, the set of natural numbers (i.e., $\{1, 2, 3, \ldots\}$) or the set of even natural numbers (i.e., $\{2, 4, 6, 8, \ldots\}$). Can the elements be paired in one-to-one fashion so that each number in the first set "goes with" a particular number in the second? What about the reverse? A little thought should convince you that the answer is yes.

There are many other situations where the correspondence between two sets is not one-to-one. People, for example, may be paired on a *many-to-one* basis with their family names. Although each person has only one family (last) name, a family name usually corresponds to more than one person. Conversely, the set of natural numbers can be paired on a one-to-many basis with (the set of) their factors. For example, 12 would correspond to each of 1, 2, 3, 4, 6, and 12. The term *relation* (i.e., mathematical relation) is also used to refer to all kinds of correspondences—one-to-one, many-to-one, and one-to-many. Relations which

[6] See, for example, M. H. Stone, "Learning and Using the Mathematical Concept of a Function." In S. N. Morrisett and J. Vinsonhaler, (eds.), *Mathematical Learning*, Chicago, Society for Research in Child Development, 1965. A growing number of mathematicians, in fact, seem to feel that the idea of a correspondence may, in some sense, be more basic than that of a set. Some of my theoretical work in psychology suggests that the basic unit of behavior may most profitably be viewed as a certain kind of correspondence called a function.

are one-to-one (e.g., children to their seats) or many-to-one (e.g., children to their teacher) have been of special interest to mathematicians and have therefore been given a special name. They are called *functions*. Thus, we see that all functions are relations but not all relations are functions.

In the elementary school, children are often asked to match corresponding elements *in two sets by drawing* lines between them (see Figure 2–3). Corre-

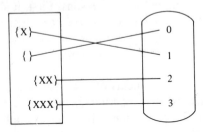

Figure 2–3

spondences may also be represented symbolically by sets of ordered pairs whereby elements of the first set are coupled with elements of the second set. Thus, the correspondence given above could be represented symbolically as:

$$\{(\{\ \}, 0), (\{X\}, 1), (\{XX\}, 2), (\{XXX\}, 3)\}$$

There is still a third basic way to represent a correspondence: State a rule that tells how the elements are to be paired. One such rule that works in the present case may be stated, "Pair each element (e.g., $\{XX\}$) in the first set with the numeral (e.g., "2"), which represents the number of X's." Other rules for forming correspondences might be stated: "Name the child (i.e., pair the child with his name)," "Double the number," "Add ten," and so on. Defining relations in terms of rules is particularly useful in dealing with correspondences involving *infinite* sets.

EXERCISES 2–3

S–1. In each of the following examples, a universal set U is given. The elements of the universe U are sorted into the sets S and T. Map each element of U into an element of the set of sets $\{S, T\}$, as shown in the text.

    a. $U = \{a, b, c, 1, 2\}$, $S = \{a, b, c\}$, $T = \{1, 2\}$
    b. $U = \{\infty, !, ꟼ, N\}$, $S = \{\infty, !, ꟼ\}$, $T = \{N\}$
    c. $U = \{1, 2, 3, 4, 5, 6\}$, $S = \{1, 3, 5\}$, $T = \{2, 4, 6\}$
    d. $U = \{\bigcirc, \triangle, \square, \heartsuit\}$, $S = \{\bigcirc\}$, $T = \{\triangle, \square, \heartsuit\}$

S–2. Pair the following sets in a one-to-one fashion. (Draw lines to indicate the correspondence.)

    a. the set of even integers from 2 to 20 inclusive and the set consisting of the first ten letters of the alphabet
    b. $A = \{\triangle, \square, \bigcirc\}$ and $B = \{ ⚥ , \cap, \wedge\}$
    c. the set of natural numbers and the set of positive multiples of 7
    d. $R = \{w, x, y, z\}$ and $S = \{1, 2, 3, 4\}$

S–3. a. Construct the "natural" correspondence between the sets $A = \{1, 2, 3, 4, 5, \ldots\}$ and $B = \{2, 4, 6, 8, 10, \ldots\}$

    b. Which element in set B is paired with 5 in set A?
    c. Which element in set A is paired with 4 in set B?
    d. Which element in set B is paired with 240 in set A?
    e. Which element in set A is paired with 240 in set B?
    f. Is this correspondence one-to-one, many-to-one, or one-to-many?

[S–4.] For each of the examples in Problem S–2 write a set of ordered pairs to represent the correspondence.

S–5. Give examples of the following:

    a. two sets that can be put in one-to-one correspondence
    b. two sets that cannot be put in one-to-one correspondence

S–6. Classify each of the following correspondences as one-to-one, many-to-one, or one-to-many:

    a. the set of living people and the set of their fingers, each person corresponding to his own fingers
    b. the tires of an automobile to the steering wheel
    c. the set of chairs and the set of chair legs, each chair corresponding to its own legs
    d. a set of jars and the set of their lids, each jar corresponding to its own lid
    e. the strings on a guitar to the guitarist
    f. a set of burning candles and their flames, each candle corresponding to its own flame

S–7. Which of the following correspondences are functions (i.e., one-to-one or many-to-one relations)?

    a. a set of books and the set of their pages, each book corresponding to its own pages
    b. a set of cups and the set of their saucers, each cup corresponding to its own saucer
    c. the earth to the moon
    d. a set of automobiles to their makes, each automobile corresponding to its own make
    e. the leaves of a four-leaf clover to its stem
    f. a set of birds to the set of their feathers, each bird corresponding to its own feathers

[S–8.] a. Represent the following correspondences by stating a verbal rule which tells how the elements are to be paired:

    1.

2.

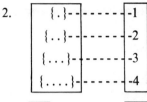

3.

4.

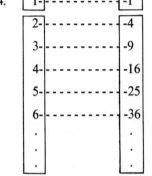

    b. Give an iconic or symbolic representation of each of the rules above.

S–9. The set {is older than, sits next to, is a descendant of} is a set whose elements are relations. Construct another set whose elements are relations such that the two sets can be put in one-to-one correspondence.

S–10. The elements of the set {travel 3 miles east, open the door} are operations. Construct another set whose elements are operations such that the two sets *cannot* be put in one-to-one correspondence.

∗[S–11.] a. Give a general rule for the one-to-one correspondence between the sets: {2, 4, 8, 16, ...} and { $\triangle$, $\square$, $\heartsuit$, $\bigcirc$, ...}
      b. Was your answer to part *a*, a verbal, iconic, or symbolic representation of the rule?
      c. Give another type of representation of the rule in part *a*.

∗E–12. A fraction is a pair of natural numbers of the form a/b. Can the set of all fractions be put in one-to-one correspondence with the set of natural numbers? *Hint:* Write the fractions in the following fashion:

1/1	1/2	1/3	1/4	1/5 ...
2/1	2/2	2/3	2/4	2/5 ...
3/1	3/2	3/3	3/4	3/5 ...
4/1	4/2	4/3	4/4	4/5 ...

Can you trace some sort of diagonal path from fraction to fraction so that each fraction in the path can correspond to some natural number?

4. RELATIONS AS STATES AND FUNCTIONS AS OPERATORS

Virtually all primary school mathematics texts give a good deal of attention to correspondences per se. For this reason, we have bowed to convention in our treatment of the topic in the previous section. The notion of correspondence, however, actually has its roots in two very different kinds of situation in the real world: (1) relations between things, and (2) operations or functions which act on things and relations. This fact has a number of important implications, and the teacher should be aware of some of them.

We first note that relations do not necessarily have to involve pairs of elements. There is no reason why relations cannot exist among three, four, or, for that matter, any number of elements. The proverbial "triangular love affair," for example, defines a *ternary* (three-element) relation, which is a set of triples and can be denoted, $\{(a, b, c) | a$ is a man, b is a woman, and c is the other man$\}$. "Same difference as" (i.e., $a - b = c - d$) is a *four-ary* relation because it defines a set of *quadruples* of elements. The relations (correspondences) discussed in Section 3 involved sets of pairs (of elements) and are called *binary* relations. Even sets of single elements can be viewed as *unary* (one-ary) relations. When looked at in this way, it is hardly surprising that there is such a close relationship between sets and relations. The second thing we note about relations stems directly from this close relationship. It is simply that sets (unary relations) and n-ary relations (where n may be any number) are typically used to refer to static situations and, thus, may be called *states*.

The terms *function* and *operation*, on the other hand, are usually used to denote some action on a class of states. For example, "take a quarter turn to the right," "flip the switch," and "add the numbers" all refer to an action on a class of initial states (positions, switches, pairs of numbers, respectively). Because application of such procedures (actions) invariably results in a *unique* output, the term *function*, as defined in Section 3, is quite appropriate. Consider, for example, what happens when you add a pair of numbers. Unless you make a mistake, in which case you are not really adding, you will always get a unique sum (output). In later sections, we shall use the term "operator" when we wish to look at the action implied by some mathematical idea.

Obviously, there is a very close relationship between relations and operations (functions). Logically, in fact, there is no distinction at all. *Relations can be defined in terms of functions and functions can be defined in terms of relations. The only difference is in the way one looks at them, as states or as operators.* You should be aware of this fact, as it may help you to understand how children look at the world and, thereby, to improve your teaching.

What other relations can you think of in mathematics? Functions (Operations)? What about the real world? Can you classify the relations as to numbers of elements? Can you think of a parallel way of classifying functions (see Section 7 on binary operations)? Ask yourself these questions when you are preparing to teach these topics to your students. Young children are not only able to respond to such questions but, if properly prepared, they usually do so with vigor and enthusiasm.

EXERCISES 2–4

S–1. In the statement *Sally is taller than Jane*, "is taller than" is an example of a relation which involves the elements "Sally" and "Jane." On the other hand, in the statement *Sally is a member of the Jones family*, "is a member of" is an example of a relation which involves the element "Sally" and the set "the Jones family."

    a. Give two examples of relations that involve pairs of elements.

    b. Give two examples of relations that involve an element and a set.

[S–2.] a. Addition, a familiar arithmetic operation, is a *function* that maps each pair of numbers into a unique third number called the sum. If the input is 3 and 5, for example, the *unique* output is 8. Name three other familiar arithmetic operations.

    b. What is a common symbol used to represent each?

S–3. "Close the window" is an example of an operation in the real world. Give three other familiar operations in the real world.

E–4. Give an example of an operation which generates output elements that are pairs of objects.

EQUIVALENCE RELATIONS, ORDER RELATIONS, AND BINARY OPERATIONS

Although most of the other ideas basic to mathematics can be defined in terms of relations and operations (or sets and correspondences), several other notions are often introduced as "basic" in their own right. From a psychological point of view, it is not entirely clear whether it is preferable to define these other notions in terms of relations and operations, or to treat them as "primitives."

In the next three sections, we shall briefly consider both alternatives. The advantage of defining everything in terms of relations and operations is, of course, that the student might see the interrelationships involved and thus have less to learn. On the other hand, the notions of binary operations, order, and equivalence are so evident in the world around us that to treat them as derivatives may do them a disservice. It may be more efficient (and natural) to view the ideas themselves as fundamental. We shall try to give you some of the facts and let you make up your own mind.

5. EQUIVALENCE

What does it mean to say that two entities are equivalent? In certain special cases two "things," "actions," or "relations" may actually be identical. Thus, Jeanne Marie Scandura and the oldest daughter of the author are one and the same person. The labels "Jeanne Marie Scandura" and "the oldest daughter of the author," however, are *not identical*. The labels are said to be *equivalent* in the sense that both labels have the same referent. In the same way, 1 foot is *equivalent* to 12 inches with respect to over-all length, but they are not identical. For example, one could use a 1-foot ruler divided into inches to do much more than a 1-foot ruler with no subdivisions. Furthermore, when we say that 1 foot is equivalent to 12 inches we are implying that the 12 1-inch units must be arranged end-to-end along a straight line. Twelve 1-inch units might conceivably be used in other ways—say, to measure simultaneously the length of 12 discrete objects (by using one measuring unit per object). Consider one more example: A dollar bill can be used to purchase the same item(s) in a dime store as four quarters or two 50-cent pieces, but it would *not* serve equally well in a coin-vending machine.[7] *All this can be summed up by saying that the question of whether two entities are equivalent or not depends on the basis for comparison.* Two or more entities might be equivalent in one sense but not in another. Any particular entity, of course, is always equivalent, and in fact identical, to itself.

EXERCISES 2–5

S–1.   If by equivalent words we mean "are pronounced the same," then the words "blue" and "blew" are equivalent, but if by equivalent words we mean "mean the same," then "blue" and "blew" are not equivalent. Give three more examples of pairs of entities that are equivalent in one sense but not in another.

E–2.   a.   In modular 3 arithmetic, two numbers are equivalent, or congruent, if they leave the same remainder when divided by 3. Thus, 5 and 17 are congruent (mod 3) because they each leave a remainder of 2 when divided by 3. Give at least three numbers which are congruent to 4 (mod 3).
   b.   In modular 4 arithmetic, how would you define the equivalence or congruence of two numbers?
   c.   Give at least three numbers that are congruent to 4 (mod 4).
   d.   Give at least three numbers that are congruent to 4 (mod 5).

5.1 Equivalence Relations and Equivalence Classes

A relation that divides a universal set into exhaustive, mutually disjoint subsets by putting all related elements into the same subset is called an equivalence relation. "Exhaustive" means that each element belongs to some subset; "mutually disjoint" means that no subset has any element in common with any other subset. So we can say that an equivalence relation divides a universal set into a certain number of subsets such that each element of the set is in one and only one subset. Any two elements in the same subset are related; any two elements in

---

[7] See Z. P. Dienes, *Mathematics in the Primary School*, Adelaide, Australia, Macmillan, 1964, for constructions of similar examples based on the British and "decimal" coinage systems.

different subsets are not related. Thus, in the diagram below, the set $U = \{a, b, c, d, e, f, g\}$ is divided into four exhaustive and mutually disjoint subsets: $\{a, b\}$, $\{c\}$, $\{d, e, f\}$, $\{g\}$, where a is related to a and to b, but a is not related to c, d, e, f, or g. Each subset is called an *equivalence class*.

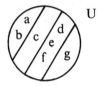

We are all familiar with equivalence relations; buy the same as, same class as, same color as, is equal to ($=$), is congruent to ($\cong$) are all examples. Thus, the equivalence relation, "buy the same as," divides the universal set of all coin combinations into equivalence classes of the same purchasing power. For example, two quarters and a half-dollar would be assigned to the same equivalence class since two quarters "buy the same as" a half-dollar. *Each equivalence class would include all and only those combinations with the same purchasing power.* Since each combination of coins can belong to only one equivalence class, the equivalence classes are *disjoint* (i.e., they do not have any coin combinations in common). Thus, for example, a combination consisting of a quarter, two dimes, and a nickel belongs in the 50-cent class and in no other. This state of affairs can be represented by Figure 2–4, where $1.23, $.50, $.65, $.01, and so on represent equivalence classes formed by applying the relation *buy the same as* to the set of all possible coin combinations.

To illustrate further, consider the equivalence classes formed in using the relation "is the same *shape* as" to partition the (universal) set of polygons given

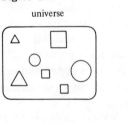

Figure 2–4

Figure 2–5

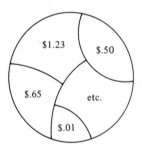
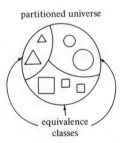

in Figure 2–5. A still simpler (i.e., more familiar) example is provided by the equivalence relation "in the same class as," which partitions the children in a typical elementary school into self-contained classrooms. The self-contained classrooms form the equivalence classes.

If we look at any one equivalence class (e.g., Mrs. Smith's third grade, in the universal set of all children in a given elementary school), we see that each element is related to every other element in the class by the equivalence relation and that all elements which are related in this way are in the equivalence class.

More precisely, we observe that the relation "in the same class as" has the following properties:

1. Each child is in the same class as himself.
2. If one child is in the same class as another, then the second child is in the same class as the first.
3. If one child is in the same class as a second, and the second is in the same class as a third, then the first child is in the same class as the third.

These three points illustrate what mathematicians call the (1) *reflexive*, (2) *symmetric*, and (3) *transitive* properties of equivalence relations.

It was probably some such motivation that led mathematicians to define an *equivalence relation* as a relation between the elements of any universal set with the following properties:

1. The *reflexive property*—every element A of the set is related to itself (denoted: $A \equiv A$).
2. The *symmetric property*—if element A is related to element B, then B is related to A (denoted: If $A \equiv B$, then $B \equiv A$).
3. The *transitive property*—if element A is related to element B and B is related to element C, then A is related to C (denoted: If $A \equiv B$ and $B \equiv C$, then $A \equiv C$). (*Note:* A and C may be the same element.)

You should check to see that the relation "is the same shape as" satisfies the reflexive, symmetric, and transitive properties.[8]

How about the relation "is the same age as" (referring to people)? The relation "is perpendicular to" (referring to lines)? In general, any relation that satisfies these properties divides or partitions the universal set into *disjoint* subsets called equivalence classes.

To prove this, we first define what is meant by an equivalence class defined by an equivalence relation and a given element, a. The equivalence class defined by a is that subset of the universal set which contains those and only those elements related to a by the equivalence relation. We complete our argument by showing that any two equivalence classes, with at least one element in the first class not equivalent to at least one element in the second class, are necessarily

---

[8] Equivalence relations may be thought of as collections of sets of ordered pairs, where each set in a collection is an equivalence class. For example, $\{(x, y) \in C | x$ buys the same as $y\}$, where C is the set of all pairs of coin combinations, represents the equivalence relation, "buys the same as," because it defines a collection of sets of ordered pairs. Each particular "amount" chosen defines a different equivalence class (set). One such equivalence class, for example, is $\{(x, y) \in C | x$ and y buy \$1 worth$\}$.

disjoint. Let *A* and *B* be equivalence classes defined by *a* and *b*, respectively, where *a* is not equivalent to *b* (i.e., $a \not\equiv b$). We want to show that for all *x* in *A* and all *y* in *B* that $x \not\equiv y$. To show this, we argue as follows:

(1) Suppose some element *a'* in *A* was equivalent to some element *b'* in *B* (i.e., $a' \equiv b'$).

(2) Recalling that $a \equiv a'$ (by definition of *A*), we see that $a \equiv a'$ and $a' \equiv b'$ (by assumption) imply that $a \equiv b'$ by property 3 above.

(3) Now, since $b' \equiv b$ (by definition of *B*), $a \equiv b'$ (just proved) and $b' \equiv b$, imply that $a \equiv b$ (again by property 3). But, this (i.e., $a \equiv b$) cannot be, because we said $a \not\equiv b$.

(4) Our sound reasoning can only mean that our assumption (i.e., that $a' \equiv b'$) was in error. This means that equivalence classes *A* and *B* do not share any elements in common and, hence, are disjoint, which is what we wished to prove.

We included this reasoning, not because the proof itself is so important, but because it provides a good example of indirect reasoning: Make an assumption; reason from it and reach a contradiction; finally, infer that the assumption is (necessarily) false. You should be alert to this sort of reasoning and, as appropriate examples come to your attention, you may wish to suggest them to your pupils. See Chapter 3 for a more complete discussion of deductive reasoning.

Of course, not all relations are equivalence relations. Consider, for example, the set of students in *high school* together with the relation "has the same teacher as." As illustrated in Figure 2–6, one high school student (e.g., Sally) can have two different teachers (e.g., Mr. Math and Miss Homec), because high school students typically attend more than one class. This means that the relation "has the same teacher as" is *not* transitive. For example, George has the same teacher as Sally and Sally has the same teacher as Susan, but George and Susan do *not* have the same teacher. Thus, the relation "has the same teacher as" is *not* an equivalence relation and, therefore, does not divide the set of students (Fred, George, . . ., Susan) into equivalence classes.

In everyday life, we continually organize our world into equivalence classes in order to better describe and understand it. We may sort people according to sex, age, or family; objects according to size, shape, or color; and so on. In school, children should also be given opportunities to organize (i.e., partition)

Figure 2–6

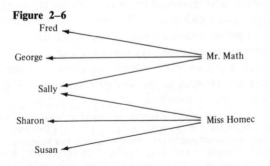

sets of *abstract* elements, particularly (but not exclusively) during mathematics lessons. In Chapter 7, we will study several situations where partitioning of this sort is crucial. We shall see, for example, that the rational number $\frac{1}{2}$ may be viewed as an equivalence class of fractions $\{\frac{1}{2}, \frac{2}{4}, \frac{3}{6}, \frac{4}{8}, \ldots\}$.

A person may be said to have acquired a given equivalence class if he or she uses "equivalent" entities in situations where they are equivalent, but not otherwise. Thus, a child who "knows" that a nickel buys the same amount as five pennies might not hesitate to exchange his or her five pennies for a nickel (unless he or she had some other reason for wanting pennies) but might react violently if told that we wanted to exchange three pennies for her nickel. A child who had not acquired this equivalence class would be easier to dupe.

Although it is more difficult to pin down precisely what it means for a child to know an equivalence relation, it certainly implies the ability to: (1) sort a collection of objects (in some universe) into equivalence classes according to the equivalence relation, and (2) identify the equivalence relation used to sort a given collection of objects (which has *already* been sorted into equivalence classes). Consider, for example, the equivalence relation, "has the same shape as." One way to determine whether a child has acquired this equivalence relation would be to present the child with a collection of objects (e.g., cut-outs of various shapes and colors) to see whether or not he or she can sort them correctly. Another type of test would involve presenting the child with a collection after it has already been partitioned into equivalence classes. In this case, the test might require the child to describe the basis for sorting or require him to sort correctly some additional items of the same type.

As a final note, we mention that equivalence classes may be partitioned into other, lower-order, equivalence classes. Consider a "tasty" example involving the following "edibles": desserts, vegetables, meats, fish, and so on. Each edible defines an equivalence class. Thus, all desserts are equivalent in a sense that only a growing child can fully appreciate. The class of all desserts, of course, can be further partitioned into equivalence classes described by "cakes, cookies, pies, ice cream, and so on." Each of these equivalence classes, in turn, can also be partitioned. Ice cream, for example, can be naturally classified according to flavor. There is no reason to stop there, of course. The process might be continued indefinitely. Brands of ice cream might do for the next level.

EXERCISES 2–5.1

S–1. a. Consider the people in your class. The property of "wearing glasses" divides this group into equivalence classes (those with glasses and those without glasses). Find two other ways of dividing the class into equivalence classes.

b. Consider any other set. Divide the set into equivalence classes by two different methods.

S–2. Find an example, such as the one above, where one equivalence class is subdivided into further equivalence classes.

S–3.   Consider set $S = \{a, b, c, x, y, z\}$. S may be divided into the equivalence classes $E_1 = \{a, b, c\}$ and $E_2 = \{x, y, z\}$. These equivalence classes may have been formed on the basis of the relation "in the same (first or second) half of the alphabet." Divide the following sets into equivalence classes and give the basis for your division.

    a. $\{1, 2, 3, 4, 5, 6, 7, 8, 9, 10\}$
    b. $\{red, large, white, blue, small\}$
    c. $\{\triangle, 2, 10, \square, 3, 9, \heartsuit\}$
    d. $\{1, 2, 3, \ldots\}$
    e. $\{January, March, May, June, July\}$

[S–4.]   The set $S = \{2, 3, 4, 5, 6, 7, 8, 9\}$ may be divided into the equivalence classes $E_1 = \{2, 4, 6, 8\}$ and $E_2 = \{3, 5, 7, 9\}$ according to whether a number is even or odd, respectively. Set S may likewise be divided into the equivalence classes $E_1' = \{2, 3, 5, 7\}$ and $E_2' = \{4, 6, 8, 9\}$ according to whether a number is prime or not. These are only two possible ways of dividing the set into equivalence classes. There are other ways. Divide the following sets into equivalence classes in two different ways and state the basis for each division.

    a. $\{10, 11, 12, 13, 14, 15\}$
    b. $\{a, A, b, B, 1, 2\}$
    c. $\{March, April, May, August\}$
    d. $\{Monday, Friday, Sunday, Saturday\}$
    e. $\{cat, chair, mouse, hat\}$

S–5.   State the reflexive, symmetric, and transitive properties (known as the R, S, T properties) of any equivalence relation.

S–6.   a.   Show the relation "is the mother of" is not an equivalence relation by stating which of the R, S, T properties it fails to satisfy.
    b.   Is $\leq$ ("less than or equal to") an equivalence relation? Check the three properties. Which ones are satisfied?
    c.   Is the relation "is perpendicular to" an equivalence relation on the set of all lines in the plane? On the set of all lines in three-dimensional space? Which properties are not satisfied?

S–7.   The relation "leaves the same remainder when divided by 5" is an equivalence relation on the set of all natural numbers. There are five equivalence classes. One of them is $\{1, 6, 11, 16, \ldots\}$. List at least three elements in each of the other equivalence classes.

S–8.   Give two other examples of relations which are equivalence relations.

S–9.   State whether the following relations are reflexive, symmetric, and transitive by putting "yes" or "no" in the appropriate column:

RELATION	R	S	T
a. is parallel to (lines in plane)			
b. is the brother of (people in the United States)			
c. is not equal to (whole numbers)			
d. is a descendant of (people)			
e. is a subset of (sets of numbers)			
f. is taller than (people in the United States)			
g. is greater than or equal to (whole numbers)			
h. is a multiple of (whole numbers)			
i. is congruent to (triangles in plane)			

S–10. Find a relation that is:

    a. reflexive but not symmetric (besides "is less than or equal to")
    b. symmetric but not transitive
    c. symmetric but not reflexive (besides "is not equal to")
    d. transitive but not reflexive
  *e. symmetric and transitive but not reflexive

[M–11.] How would you have your children demonstrate that they have acquired the ability to detect whether or not an arbitrary relation is an equivalence relation?

*E–12. Professor Bufflegab presented the following argument to his class to show that the reflexive property is unnecessary when considering equivalence relations. He said, "If you have the symmetric and transitive properties, the reflexive property can be deduced from them, as follows:

    1. $a \sim b$ implies $b \sim a$, by the symmetric property
    2. But if $a \sim b$ and $b \sim a$, then $a \sim a$, by the transitive property
    3. Hence, $a \sim a$ (reflexive property) can be deduced from the other two."

Where did Professor Bufflegab go wrong?

*E–13. a. Consider the relation of implication: "A implies B." Does this relation possess the reflexive property, i.e., is "A implies A" true for all A? The symmetric property? The transitive property?
    b. Consider the relation of conjunction: "A and B." Does this relation possess the reflexive property? The symmetric property? The transitive property?

6. ORDER RELATIONS

Order is another major kind of relation. A child encounters the idea of an order relation long before he enters school. His world is filled with large toys and small toys, with going to bed before his older brother, with taking turns, with getting a smaller piece of cake than his sister, and so on. Furthermore, early in his school experience the child learns to line up according to height, anticipate his name being called in alphabetical order, and arrange sticks according to their lengths. In this section, we will look more specifically at what is involved in these and other order relations. We will look at the properties *any* order relation must have and then consider the properties of particular kinds of order relations.

For instance, consider the order relation "is taller than." In this case, if *a* is taller than *b*, then *b* is *not* taller than *a*. This is called the *antisymmetric* property. Likewise, if *a* is taller than *b* and *b* is taller than *c*, then *a* is taller than *c*. This is called the transitive property. Similarly, in the alphabetical order relation (1) If *a* precedes *b* alphabetically, then *b* does not precede *a*, and (2) if *a* precedes *b* and *b* precedes *c*, then *a* precedes *c*. That is, the alphabetical order relation has the antisymmetric and transitive properties.

By definition, any relation that has the antisymmetric and transitive properties is called an order relation. This means, of course, that if a relation does not have both of these properties, it cannot properly be called an order relation. (Does the relation "is longer than" have the antisymmetric and transitive properties? Is it an order relation?)

There are four particular kinds of order relations: *strict linear order, strict partial order, nonstrict linear order,* and *nonstrict partial order.* Each of these four relations has the two properties given above and each has some additional distinguishing property or properties. We will look at some examples of each kind of order relation and discuss their properties.

Perhaps the most familiar example of an order relation involves number and is called *greater than.* Although we have not considered the idea of number in any depth, your present intuition will not be misleading. Notice, first, that if *a* and *b* are numbers and *a* is greater than *b*, then *b* is not greater than *a*. We denote this: if $a > b$, then $b \not> a$. For example, $12 > 7$ but $7 \not> 12$. Second, if one number is greater than a second and the second number is greater than a third, then the first is greater than the third. This is denoted: if $a > b$ and $b > c$, then $a > c$. Third, a number cannot be greater than itself. We denote this: $a \not> a$. Finally, given any two distinct numbers *a* and *b*, either *a* is greater than *b* or *b* is greater than *a* (but, by our first property, not both). This is denoted $a > b$ or $b > a$.

Because it satisfies these four properties, the order relation "greater than" is called a *strict linear order relation.* In general, a relation is a *strict linear order relation* if it has each of the following properties (compare these properties with those given for "greater than"):

1. *The antisymmetric property:* If two elements of a set are distinct and the first element precedes the second element, then the second element does *not* precede the first element.
2. *The transitive property:* If one element precedes a second and the second precedes a third, then the first also precedes the third.
3. *The irreflexive property:* An element cannot precede itself.
4. *The linear property:* Given *any* two distinct elements of a set (exactly) one precedes the other.

The passage of *time* (before–after) is another example of a strict linear order relation. Before going on, convince yourself that this order relation satisfies each of the above four properties.

A second type of order relation is illustrated by the relation "is an ancestor of," which defines an order on the set of all people. This relation does not satisfy the linear property. Thus, the maternal grandmother of a child is not related by "ancestor of" to the father of the child.[9] In general, then, given any two persons in the set of all people, neither one is *necessarily* the ancestor of the other. Direct parallels do hold for the other three properties:

1. *The antisymmetric property:* If one person is an ancestor of a second, then the second person is *not* an ancestor of the first.
2. *The transitive property:* If one person is an ancestor of a second, and the second is an ancestor of a third, then the first is also an ancestor of the third.
3. *The irreflexive property:* No one is his own ancestor.

The "ancestor of" order is a *strict partial ordering.* In effect, a *strict partial ordering* is a relation which has the *antisymmetric, transitive,* and *irreflexive* properties, but where the elements need not be linerally ordered.

We should note that a strict linear order relation is (i.e., satisfies the properties of) a strict partial order, but that a strict partial order need not be a strict linear order relation. The relationship between the two is the same as between squares and rectangles: every square is a rectangle but not all rectangles are squares.

Another order relation that does not satisfy all the requirements of a strict linear ordering is the "greater than or equal" relation on the set of numbers. We denote this relation by the symbol $\geq$. The "greater than or equal" relation does not satisfy property (3), *the irreflexive property.* Instead, it has what is called the *reflexive property:*[10] *Every* number is "greater than or equal" to itself (denoted $a \geq a$). Properties (1), (2), and (4) still hold, however. Thus:

1. *The antisymmetric property:* If a and b are distinct numbers and $a \geq b$, then $b \not\geq a$ ($b \not\geq a$ is read "b is not greater than or equal to a," which is equivalent to saying that b is less than a).
2. *The transitive property:* If $a \geq b$ and $b \geq c$, then $a \geq c$.
4. *The linear property:* Given any two distinct numbers a and b, either $a \geq b$ or $b \geq a$.

"Greater than or equal," then, is an order relation, which has properties (1), (2), and (4) together with:

3'. *The reflexive property:*[10] An element can precede itself (i.e., $a \geq a$).

Any order relation which satisfies properties (1), (2), and (4) is called a *nonstrict linear ordering.* Property (3') is optional.

---

[9] That is, the maternal grandmother is not an ancestor of the child's father.
[10] This is the same reflexive property discussed in the section on equivalence relations.

We complete our discussion by considering one final type of order relation, called a *nonstrict partial order*. A nonstrict partial order need not satisfy the linear property (4). Inclusion of sets is an example of a partial order which is not linear. (*Note:* Set *A* is said to be included in set *B*, if *A* is a subset of *B*; we write this $A \subset B$.) Thus, the partial order relation of set inclusion has the following properties:

1. *The antisymmetric property:* If *A* and *B* are distinct sets[11] and *A* is included in *B*, then *B* is not included in *A*. (This is denoted: If *A* and *B* are distinct and $A \subset B$, then $B \not\subset A$.)
2. *The transitive property:* If set *A* is included in set *B* and *B* is included in set *C*, then *A* is included in *C*. (This is denoted: If $A \subset B$ and $B \subset C$, then $A \subset C$.)

Set inclusion also satisfies the *reflexive property* (3′): A set is included in itself (denoted $A \subset A$). But, this is not true of all nonstrict partial orders.

Consider some new order relations, such as height, weight, supervision of (*note:* a supervisor may be supervised), distance, is a subcategory of (e.g., spinach is a subcategory of vegetable), and see which properties they satisfy.

The properties of the four distinct types of order relations are summarized in Table 2–1.

Table 2–1 Properties of Order Relations

	Strict	Nonstrict

Linear	1. antisymmetric
	2. transitive
	3. irreflexive
	4. linear
	e.g., "is less than"
	    "is greater than"
	    "comes before" (in time)
		1. antisymmetric
		2. transitive
		
		4. linear
		e.g., "is greater than or equal to"

Partial	1. antisymmetric 2. transitive 3. irreflexive e.g., "is the ancestor of"	1. antisymmetric 2. transitive e.g., "is a subset of"

Order relations lend themselves naturally to graphical representation. Such representations seem to capture and retain most of their important features. Graphical representations of the order relations "ancestor of" and "greater than" are shown on the next page.

---

[11] Two sets are said to be *distinct* if at least one element of one set is not an element of the other. This does not necessarily imply that the sets are disjoint. For example, {1, 2, 3} and {1, 2, 3, 4} are distinct sets which are *not* disjoint. *Disjoint* sets have *no* elements in common whatsoever.

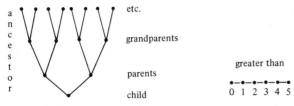

etc.

grandparents

greater than

parents

child

Figure 2–7

Order relations may also be represented by simply listing the elements of the set in order. Thus

$$\left\{\text{child}, \begin{Bmatrix}\text{father}\\\text{mother}\end{Bmatrix}, \{\text{grandparents}\}, \{\text{great grandparents}\}, \text{etc.}\right\}$$

and

$$\{0, 1, 2, 3, 4, 5, \ldots\}$$

are alternative symbolic representations of "ancestor of" and "greater than." In each case "greater than" is simpler to represent since it is strictly linear—that is, there are no "branches."[12]

The simplest way to tell whether a student has acquired a particular order relation is to see if he can generate the "successors" of any given element in the ordered set. A child who has truly learned the order "greater than" (as applied to the set of natural numbers up to, say, 100) would be able to state any and all of the numbers (up to 100) which are greater than any given number (up to 100). Thus, something more is involved than simply being able to repeat the chant, "1, 2, 3, 4, 5, ..., 100." Testing for the inverse order relation, "less than," of course, would involve the ability to go in the reverse direction. Although these abilities are closely related, the teacher should not automatically assume that having learned one of the order relations necessarily guarantees that the other has also been learned.

EXERCISES 2–6

S–1. Let a, b, c be any elements of a set S. Let R denote a relation on set S. If a is related to b, we write aRb. Represent the following properties using this notation. For example, the irreflexive property would be represented: For all a, $a\cancel{R}a$, where $a\cancel{R}a$ means that a is not related to a.

    a. reflexive
    b. symmetric
    c. antisymmetric
    d. transitive
    e. linear

---

[12] As with all (binary) relations, an *order relation* may be represented as a set of ordered pairs. Thus, $\{(x, y) \in P \mid x$ is the ancestor of $y\}$, where P is the set of all pairs of people, and $\{(x, y) \in N^2 \mid x$ is greater than $y\}$, where N^2 is the set of all pairs of natural numbers, denote the order relations, "ancestor of" and "greater than," respectively.

S–2. Give an example of a relation that has each of the properties listed in Problem S–1 (one relation per property).

S–3. What is the difference between a strict linear order relation and a strict partial order relation? Give an example (one not in the text) of each.

S–4. Show that the relation "is a divisor of" for natural numbers is a partial order relation. Is it a strict or a nonstrict partial order relation?

S–5. What two properties are common to all four types of order relations? (The term *order relation*, in its most general form, refers to any relation with these properties.)

S–6. Let the letters R, S, T, I, A, L stand for the reflexive, symmetric, transitive, irreflexive, antisymmetric, and linear properties, respectively. Match each relation in column 2 with the appropriate properties in column 1.

PROPERTIES	RELATION
1. A, T	( ) a. strict partial order
2. R, S, T	( ) b. nonstrict partial order
3. A, T, I	( ) c. order
4. A, T, L	( ) d. strict linear order
5. A, T, I, L	( ) e. equivalence
	( ) f. nonstrict linear order

S–7. State whether the following relations are reflexive, irreflexive, symmetric, antisymmetric, transitive, and linear on the indicated sets by putting a "yes" or "no" in the appropriate column.

RELATION	R	I	S	A	T	L
a. is a descendant of (people who existed)						
b. is a subset of (sets of whole numbers)						
c. comes after (letters of alphabet)						
d. is twice as large as (whole numbers)						
e. is shorter than (people in the United States)						
f. is less than or equal to (whole numbers)						
g. is similar to (triangles in plane)						
h. is the same color as (crayons)						
i. has the same teacher as (boys at U. of P.)						

S–8. Which of the relations in Problem S–7 are:

    a. strict partial order relations
    b. strict linear order relations
    c. nonstrict partial order relations
    d. nonstrict linear order relations
    e. not an order relation at all.
    f. equivalence relations

[S–9.]   Relations may be represented symbolically. Match the following relations with their symbols:

SYMBOLS		RELATIONS	
a.	$>$	( ) 1.	is a subset of
b.	$\leq$	( ) 2.	is less than or equal to
c.	$\perp$	( ) 3.	is perpendicular to
d.	$\approx$	( ) 4.	is greater than
e.	$\subset$	( ) 5.	is similar to

[S–10.]   Using the ordinary meaning of $\geq$ and $\leq$, put a correct symbol between each of the following pairs:

a.	2	3	d.	38	28
b.	6	6	e.	38	38
c.	7	4	f.	28	38

[S–11.]   Detect a regularity in the following displays:

   a. $2 < 3$ and $2 \cdot 2 < 2 \cdot 3$
   $2 < 3$ and $3 \cdot 2 < 3 \cdot 3$
   $2 < 3$ and $4 \cdot 2 < 4 \cdot 3$
   $2 < 3$ and $5 \cdot 2 < 5 \cdot 3$
   b. $2 < 3$ and $^-2 \cdot 2 > {}^-2 \cdot 3$
   $2 < 3$ and $^-3 \cdot 2 > {}^-3 \cdot 3$
   $2 < 3$ and $^-4 \cdot 2 > {}^-4 \cdot 3$
   $2 < 3$ and $^-5 \cdot 2 > {}^-5 \cdot 3$

[M–12.]   How would you have your pupils demonstrate that they understand order relations?

For further thought and study: Examine several textbook series to see how inequality of numbers is introduced.

7.   BINARY OPERATIONS

In Section 4, we gave examples of binary, ternary (three-ary), four-ary, and even unary (one-ary) relations and indicated that particular relations may involve any number of elements. We also suggested that functions (operations) may be similarly classified. There are, in effect, unary operations, binary operations, ternary operations, and so on. Respectively, they map sets of (single) elements into sets (of elements), sets of pairs of elements into sets, sets of triples of elements into sets, and so on.

Recall that operations refer to an action, a "doing something." Suppose you were asked to map a whole number into its negative. (This involves doing something.) We can represent this action by $n \rightarrow {}^-n$ where n is any whole number. Look at the symbol (usually called a variable) to the left of the arrow. Because there is one such symbol, the arrow represents a unary operation. Suppose you were asked to find the perimeter of a rectangle. To perform this action you would have to know at least two facts about the rectangle, namely, its length and its width. If we represent its length by *l* and its width by *w*, we can represent this action by the mapping $(l, w) \rightarrow 2l + 2w$. So for example, a rectangle having a length of 4 and a width of 3 is mapped into $2 \cdot 4 + 2 \cdot 3$ or 14, its perimeter. Because there is an ordered *pair* to the left of the arrow, this arrow represents a *binary* operation. Consider another example: If you wanted to know the volume of a certain carton, you would have to know its length, width, and height. Representing the last three entities by *l*, *w*, and *h*, respectively, we can represent this action by $(l, w, h) \rightarrow l \times w \times h$. This is an example of a ternary operation because there is an ordered *triple* to the *left* of the arrow.

The operations most frequently associated with arithmetic (and elementary school mathematics in general), however, are *binary* operations and in this section we shall concentrate on them. We are all familiar with the binary operations of addition, subtraction, multiplication, and division. In each case, given any two numbers, we can uniquely specify a third number called a sum, difference, product, and quotient, respectively. Addition, for example, is a binary operation because pairs of numbers *a* and *b* are mapped into their sums. We can represent this by $(a, b) \rightarrow a + b$. Division is likewise a binary operation since $(a, b) \rightarrow a \div b$. Thus, for example, 2 and 3, when added, yield the sum 5, and 6 and 2, when divided, yield the quotient 3.

Of course, binary operations are not limited to $+, -, \times$, and $\div$, nor do they necessarily involve numbers. Consider, for example, the set of rotations of a circle (e.g., a rotation about the center of a circle of, say, 74°). Here, the binary operation, "followed by," associates a unique composite rotation to any pair of given rotations, namely that (composite) rotation which is equivalent to the first given rotation "followed by" the second. Thus, a rotation of 50° followed by one of 30° is equivalent to one of 80°. By analogy, we can define a whole class of similar binary operations: one movement of an elevator followed by a second (see Figure 2–8), one symmetry transformation (any movement of an object that leaves its shape and size unchanged) of a square followed by another, one flip of a light switch followed by another, and so on. Similarly, Figure 2–8 shows how one single move (the dotted line) of an elevator may end up at the same floor as two moves taken successively.[13] (The examples of embodiments that follow in Chapter 4 provide additional examples.) Before reading on, make

Figure 2–8

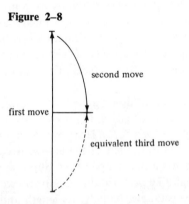

second move

first move

equivalent third move

[13] For more details see Chapter 4.

sure that carrying out the instructions in each case yields exactly one element of the same type.[14] Also, devise some examples of your own.

Clearly, the order in which certain pairs of actions are carried out does not make any difference. Thus, for example, a rotation of 30° followed by one of 50° is the same as a rotation of 50° followed by one of 30°. Similarly, ingesting meat and potatoes, in either order, has the same (unfortunate) effect (on one's waistline). The order in which we put on our shoes and socks, however, could determine whether we are laughed at or admired. More painful consequences may result from attempting to look outside before opening the window. The reverse order of actions might save us from an unwanted headache. This question of order will come up frequently in Chapters 4 and 5. It relates to the property of commutativity about which the teacher has heard much during the past few years.

Like all operations, and sets and relations as well, binary operations may be viewed as basic entities which are so intuitively simple that no further definition is needed. Binary operations may also be defined in terms of sets and relations in a very natural way. The "set" definition is as follows: *A binary operation is a set of ordered triples of elements such that there are no two triples such that the first two elements are the same and the third one different. In effect, the first two elements of any triple specify a unique third element.* For example, the binary operation of addition may be defined in terms of the set $\{(a, b, c) \mid a, b,$ are numbers and c is their sum$\}$. The binary operation of finding the perimeter of a rectangle may be defined in terms of the set $\{(l, w, p) \mid l$ and w are the length and width and $p = 2l + 2w$ is the perimeter$\}$.

Sets of ordered triples, of course, are nothing but ternary relations. Hence, *binary operations* may be defined as (certain) *ternary relations.* In most advanced research on the foundations of mathematics, it is general practice to represent all operations in terms of relations: unary operations as binary relations, binary operations as ternary relations, ternary operations as four-ary relations, and so on. For example, the binary operations of addition and (one rotation) "followed by" (another) may be represented by (1) $\{(x, y, z) \mid x, y, z$ are numbers and $x + y = z\}$; and (2) $\{(r_1, r_2, r_3) \mid r_1, r_2, r_3$ are rotations and r_1 "followed by" $r_2 = r_3\}$, respectively.[15]

A student may be said to have mastered a given binary operation if, given any two elements associated with the operation, he or she can invariably determine the appropriate third element. Of course, the number of possible test instances of this sort may be indefinitely large (e.g., in addition) so that it will be impossible to test exhaustively. In such cases, we should introduce a variety of key test items (e.g., addition problems) to determine which kinds the child can solve and which kinds he cannot. There is much more that we could say on

---

[14] The elements on which binary operations act do not necessarily have to be of the same type. Witness the example given above of a ternary operation where the elements in the triples (lengths, widths, and heights) differ from one another as well as from the outputs (volumes). We do not consider further examples of this type, but the teacher should be aware of such examples so that children do not acquire too narrow a view of operations.

[15] As suggested previously, it is also possible to represent relations in terms of operations. In fact, many contemporary mathematicians feel that this way of viewing mathematics has important conceptual advantages.

this question of how to determine what a child knows, as it is an area in which the author has done a good deal of research. Unfortunately, discussion here would take us much too far afield.[16]

EXERCISES 2-7

S-1. Suppose we define an "averaging" operation by $a \circ b = \dfrac{a+b}{2}$. For example, $7 \circ 5 = \dfrac{7+5}{2} = 6$, $13 \circ 3 = \dfrac{13+3}{2} = 8$.

    a. Is this operation unary, binary, ternary, etc.?
    b. Compute $5 \circ 7$, $8 \circ 10$, and $12 \circ 14$.
    c. Compute $7 \circ 5$, $10 \circ 8$, and $14 \circ 12$.
    d. Compare your answers to part b with your answers to part c. What would you *guess* about $a \circ b$ and $b \circ a$?
    e. Compute $(5 \circ 13) \circ 5$. Remember to do the operation within parentheses first.
    f. Compute $5 \circ (13 \circ 5)$.
    g. What can you say about $(a \circ b) \circ c$ and $a \circ (b \circ c)$?
    h. Compute $(5 \circ 7) \circ 12$ and $5 \circ (7 \circ 12)$. Now answer part g.

S-2. As described in the text, any formula, such as $A = \frac{1}{2}bh$ where b is the base, h is the height, and A is the area of a triangle, may be represented by a correspondence. Thus $(b, h) \rightarrow A = \frac{1}{2}bh$ can be used to classify this formula as a binary operation (consider (b, h)) and as a ternary relation (consider (b, h, A)). For each of the following, a formula is given as a correspondence. State what kind of operation and relation (unary, binary, ternary, etc.) each represents.

    a. area of a rectangle $(l, w) \rightarrow A = lw$
    b. area of a circle $(r) \rightarrow A = \pi r^2$
    c. volume of a cylinder $(r, h) \rightarrow V = \pi r^2 h$
    d. volume of a pyramid with a rectangular base $(l, w, h) \rightarrow V = \frac{1}{3}lwh$
    e. volume of a cube $(e) \rightarrow V = e^3$

S-3. Define the average of 3 quantities by $(a, b, c) \rightarrow Av = \dfrac{a+b+c}{3}$. For example, the average of 7, 12, and 3 $= \dfrac{7+12+3}{3} = \dfrac{22}{3} = 7\frac{1}{3}$.

    a. Is this operation unary? Binary? Ternary?
    b. Compute the average of 17, 4, 9.
    c. Compute the average of 30, 2, 14.

[S-4.] Considering Problems S-1 and S-3, how would you define the average of four quantities?

S-5. Classify each of the following operations as unary, binary, ternary, ..., n-ary.

    a. a man and woman being married by a minister

---

[16] This question is discussed at length in *Mathematics and Structural Learning*. Englewood Cliffs: Prentice-Hall, 1972.

b. finding the complement of a set A in a fixed universe U (see Exercises 2–2, Problem E–11)

c. making a pie crust, if the only ingredients are flour, salt, water, and shortening

[S–6.] Represent each operation in Problem S–5 as a correspondence.

S–7. If admission to a theater is $1.50 per adult and $.75 per child, write an equation to represent the operation of calculating the total admission for a group of n adults and m children. What kind of operation does this equation represent?

S–8. Any binary operation can be written as a ternary relation, as described in Problem S–2. Thus the equation $12 - 10 = 2$ can be represented as $(12, 10, 2)$, an instance of a ternary relation. Write each of the following as instances of ternary relations.

a. $7 + 5 = 12$
b. $90 = 15 \times 6$
c. left face "followed by" about face = right face
d. $12 = 60 \div 5$
e. $20 - 5 = 15$
f. rotating a circle $30°$ "followed by" rotating the circle $45°$ = rotating the circle $75°$

E–9. Suppose we define a "$*$" operation by $A * B = 2A + B$, where A and B are any numbers. For example $2 * 3 = 2 \cdot 2 + 3 = 4 + 3 = 7$.

a. What kind of operation is $*$?
b. Compute $7 * 5$ and $5 * 7$.
c. Is "$*$" a commutative operation? (i.e., does $A * B = B * A$ for all numbers A and B?)
d. Find three numbers A, B, and C such that $(A * B) * C \neq A * (B * C)$.

For further thought and study: Probably the most familiar binary operation in arithmetic is addition. For insight into five possible algorithms for adding small numbers, read P. Suppes and G. Groen, "Some Counting Models for First Grade Performance Data on Simple Addition Facts." In J. M. Scandura (ed.), *Research in Mathematics Education*, N.C.T.M., 1967.

8. SUMMARY

Let us summarize what we have said so far. Sets, relations, and operations (correspondences) appear to be the most basic mathematical ideas. Equivalence relations, binary operations, and order relations can be viewed as special cases. These are just a few of the basic ideas which pervade all of mathematics, from its very beginnings in the elementary school through the university. More important, all of these ideas find realizations everywhere in the world around us. Even a young child can hardly help but be familiar with them on an intuitive level, even before he enters the elementary school. *The teacher's primary job is to make these understandings more explicit.*

3

Logic
and Set Operations

Deductive reasoning provides a basic means for acquiring new knowledge, not only in mathematics, but in all walks of life. Accordingly, instruction in this area should begin at the earliest possible time. In our view, this means that even elementary school students should be provided with a wide variety of such experiences. Providing such experiences, however, does *not* necessarily imply that young children should be introduced to the study of formal logic. What is important is that they have an opportunity to develop their ability to reason deductively.

In this chapter we will first consider the general nature of deductive reasoning. Then we will examine some ideas about sets and, particularly, the close relationship between operations on sets and a restricted form of logic, called the statement logic. Finally, we will show how the statement logic may be enriched by introducing the quantifiers, "some" and "all."

We do not propose that this material should be presented to children in the form given. But we believe that the teacher should understand the basic ideas involved so that she can modify and present them to her classes in an appropriate form.

1. LOGICAL POSSIBILITIES

In the words of one author,[1] "One of the most important contributions that mathematics can make to the solution of a scientific problem is to provide an

---

[1] J. G. Kemeny, J. L. Snell, and G. L. Thompson, *Introduction to Finite Mathematics*, Englewood Cliffs, N.J.: Prentice-Hall, 1957.

exhaustive analysis of the logical possibilities for the problem." Suppose, for example, that we are interested in the current state of the weather. A crude breakdown of the possibilities might be: There is precipitation; there is *no* precipitation. Each of these possibilities, of course, could be broken down still further: It is raining; it is snowing; it is hailing; it is clear; and so on. Depending on the detail desired, the process might be continued by accounting for such variables as temperature, humidity, percent of cloud cover, and pollution index. In other examples, the level or levels of analysis are more clearly defined. For example, if one is concerned with the names of the months, then the only logical possibilities are January, February, March, April, May, June, July, August, September, October, November, and December. Similarly, if one is concerned with the sum on two dice, the set of logical possibilities is {2, 3, 4, 5, 6, 7, 8, 9, 10, 11, 12}. (For other purposes, it might be important to distinguish between the ways in which the particular sums are obtained. Thus, for example, it is "harder" to roll an 8 with a 4 on each die than to roll an 8 with a 5 on one die and a 3 on the other.)

A *statement* is any sentence that can be classified as either true or false. "George Washington was the King of England," "If it is over 90°F., then it is sunny," "John went to the store," all qualify as statements. The following do not: "Goodbye everyone," "Hurry up," "Nuts." Thus, any *assertion* (statement) about the state of the weather would be true or false depending on the *actual* state of the weather. In general, the assertion would be true for any one of a number of (actual) weather conditions and false for others. Thus, "There is precipitation" would be true whether it was raining, snowing, hailing, or whatever, so long as some form of H_2O was descending from the sky, but would be false if the weather was clear and dry. Similarly, "This month has 30 days" would be true or false depending on which month we are referring to, and "Sam rolled a (sum of) 7" would be true or false depending on what sum Sam actually rolled.

Other assertions may be either true or false for all possibilities. For example, consider the following assertions: (1) "The name of the month begins with J or it does not begin with J"; (2) "The sum in rolling two dice is either an even number or is not an even number"; (3) "If it is raining, then there is precipitation"; (4) "The name of the month starts with J and has six letters"; (5) "The sum on the dice is 6 and the sum on the dice is not 6"; (6) "It is raining and there is no precipitation."

The first three assertions are true. There are no logical possibilities which make any of them false. (In interpreting these and the other illustrative sentences, we use their commonly accepted meanings to determine whether they are true or not.)[2] For the third assertion to be false, for example, it would have to be raining with no precipitation, which is absurd.

The last three assertions are false. For example, there is no English name for a month which begins with J and has six letters. Similarly, the sixth assertion

---

[2] The first two statements are also what logicians call *logically true*. That is, they would be true regardless of the meaning attached to the nonlogical words. For example, the second statement is true no matter what meaning might be attached to the word "even."

is false because it is impossible to have rain and at the same time no precipitation.[3]

EXERCISES 3–1

S–1.   Determine whether each of the following statements is always true, always false, or may be either true or false.

    a.   He is a policeman.
    b.   The mystery person is either a male or a female.
    c.   □ is a number greater than 4.
    d.   It is snowing, and it is sunny and over 50° outside.
    e.   This animal is a cat with six legs.
    f.   This triangle has four sides.
    g.   3 feet is longer than 2 feet.
    h.   Three of these weigh more than two of those.

S–2.   a.   Write three statements which are always true.
       b.   Write three statements which are always false.
       c.   Write three statements which may be either true or false.

E–3.   A statement is *logically true* if it is true regardless of the meaning attached to the nonlogical words ("all," "some," "and," "or," "if-then," "not" *are* logical words; John, son, dog, *are not* logical words). For example, "He is tall or he is not tall" is logically true because it is true regardless of the meaning attached to the words "tall" or "he." But the statement "15 inches is more than 1 foot" is not logically true because its truth depends on the meaning of the words "1," "15," "inches," and "foot" (see footnote 2). Determine which of the following statements are logically true and which are true only with the usual interpretation of the nonlogical words.

    a.   A yard is longer than a foot.
    b.   If it is snowing, then it is snowing.
    c.   If it is sunny and warm, then it is sunny.
    d.   The number 17 is odd.
    e.   The number I am thinking of is even or it is not even.

E–4.   A statement is called *logically false* if it is false regardless of the meaning attached to the nonlogical words. For example, "He is strong and he is not strong" is logically false because it must be false regardless of the meaning attached to the word "strong" or "he." But the statement "2 feet equals 1 yard" is not logically false because its falsity depends on the meaning of the words "feet" and "yard" (see footnote 3). Determine which of the following statements

---

[3] The fifth assertion is also *logically false*. To say that the sum of two dice is 6 runs counter to the possibility that it is not 6—this compound statement cannot be true regardless of what the words "sum," "dice," or "6" mean.

are logically false and which are false only with the usual interpretation of the nonlogical words.

    a.   There are 11 eggs in a dozen.
    b.   Mr. Smith is tall and not tall.
    c.   All dogs have three legs.
    d.   All men are animals, and no men are animals.

2. OBSERVATION VERSUS DEDUCTION

There are two fundamentally different ways in which we may determine the truth or falsity of an assertion (statement)—by observation or by deduction (proof).

Observation: Where possible, perhaps the simplest way is to make a direct observation. Thus, the truth or falsity of the statement, "It is over 90°F outside now," can be easily checked by reading a reliable thermometer. Obviously, there are any number of possibilities for this statement to be true. It could be raining and the temperature over 90°F, the humidity might be 10 percent and the temperature over 90°F, etc. There are many other situations where the statement would be false.

Deduction: Of more direct concern here is the situation where the truth or falsity of an assertion may be determined on strictly logical grounds, independent of any observation.[4]

The truth or falsity of many assertions may be determined logically (i.e., deductively) from a knowledge of the truth or falsity of other assertions. When this can be accomplished, we say that the assertion has been deduced from the other assertions. For example, the familiar assertion, "Socrates is mortal," may be deduced from the assertions, "All men are mortal" and "Socrates is a man," by use of an inference rule called the syllogism. Thus, if "All men are mortal" and "Socrates is a man" are true, then "Socrates is mortal" must be true. The same inference rule is involved in deducing "Sam has a tail" from "All dogs have tails" and "Sam is a dog." Using other inference rules, if "$X = 5$ or 6" and "$X \neq 6$" are known to be true, then the assertion "$X = 5$" must also be true, and if "It is raining" is false, then "It is *not* raining" must be true.

We must emphasize that if we assume certain statements to be true and argue deductively to a conclusion, and then discover that one or more of our assumptions is actually false, the conclusion need not be true. Consider the second and third examples. If Sam is actually the neighbor's 3-year-old child, so that "Sam is a dog" is false, the conclusion "Sam has a tail" is certainly false. If "$X = 5$ or 6" is false (e.g., if $X \geq 7$), then knowing that "$X \neq 6$" does not allow us to conclude that "$X = 5$."

---

[4] We have already considered two cases where this is possible. Thus, the truth of a logically true statement follows because it must be true under *all* possible interpretations. The falsity of a logically false statement follows for the reverse reason—there are no possible circumstances (interpretations) under which it might be true.

EXERCISES 3–2

S–1.   Which of the following statements are true by observation and which may be deduced logically?

    a. The temperature is 75° in this room.
    b. The square of an even number is even.
    c. The milk is frozen.
    d. My raincoat is larger than yours.
    e. Judge Smith is a lawyer. (*Hint:* Assume all judges are lawyers.)
    f. Any collection of piles of marbles, each pile containing four marbles, can be combined to give an even number of marbles.
    g. If it is raining, then it must be precipitating.

S–2.   a.   Write two statements (appropriate to the grade level you will teach) whose truth or falsity can be determined by observation.
    b. Write two statements (appropriate to the grade level you will teach) whose truth or falsity can be determined by deduction.

M–3.   All of the boys in a small high school are on the football team, and all of the students know this fact. You ask a girl, "Is Tom Smith (one of the boys) on the football team?" She answers "I don't know." How could you convince her that Tom Smith is on the team without having her look at the roster?

M–4.   Two ways are given in this section for determining the truth or falsity of a statement: observation and deduction. Suppose I wished to determine whether Lew's Volkswagen had an engine in the back. I am given two facts. (1) Lew's Volkswagen is next to me; (2) all Volkswagens have an engine in the back. Describe two different methods of determining where the engine is.

M–5.   If the children in your class know that all rabbits like carrots, and if the class has a pet rabbit, what are two different ways you could convince the children that the pet rabbit likes carrots?

3.   FALLACIES

In addition to having some experience in making deductions, elementary school children should have experience in detecting fallacious arguments. The importance of such training cannot be underestimated when one considers the effects it could have on the course of world (as well as everyday) affairs. The effects of often misleading Madison-Avenue-type advertising might well be diminished, and future Hitlers might find it more difficult to promote the idea of a master race. Perhaps most important, the average consumer of information would be less likely to tolerate a one-sided argument without raising the important point, "But what if . . .?" In short, although deductive reasoning clearly has application in mathematics, its importance certainly cannot be limited to it. I might even make the blasphemous (for a mathematics educator) statement that, for most people, deductive reasoning is far more important outside of mathematics than it is within. Training in deductive thought should be an important part of *general education*.

Let us consider a couple of examples of fallacious reasoning: "Diamonds have traditionally been used as mere baubles and playthings, as shiny toys for the idle rich to play with. It is obvious, therefore, that diamonds can be of no use in any serious undertaking"; and "All children have equal rights before the law. Therefore, schools should teach the same things to all children." In both cases, the speaker has drawn an invalid conclusion. In neither case can his final statement be deduced logically from those preceding it. In order to make the desired deductions, additional premises would need to be made explicit. *In the process of making them explicit, of course, the conclusion may no longer appear valid (because the hidden premises may not be tenable)*. The first conclusion is based on the implicit premise that, "Anything that the rich have played with can be of no use in any serious undertaking." This is hardly a tenable premise, for there are too many obvious counterexamples. The second argument involves more circuitous reasoning. Few would question the major premise, "All children have equal rights before the law"—it is basic to our conception of democracy. The conclusion is also somewhat tempting. Let us look at the hidden premises. (There may be more than one set but only one is considered here.) First, "Schools should teach the same things to all children with the same capabilities." This premise is questionable, but perhaps not too bad. The second hidden premise, however, "All children with equal rights before the law have equal capabilities," is clearly in error.

Another type of fallacious reasoning is called "reasoning from the converse." We will demonstrate, using two examples: "All numbers which are divisible by 4 are even. Because I am thinking of an even number, I must be thinking of a number divisible by 4"; and "All of the intelligent people in town are supporting the bond issue. Mr. Jones is supporting the bond issue, so he must be a smart man." It is quite easy to see the fallacy in the first example. Although it is true that any number divisible by 4 must be even, the *converse*, "All even numbers are divisible by 4," may not be. In fact, it is clearly false, because 6, 18, 46, and so on are all even numbers not divisible by 4. Hence, the reasoning is fallacious and the conclusion is invalid. The second example is similar. The speaker has also reasoned from the converse, "All of the people who support the bond issue are intelligent." But the original statement does not imply the converse, and it may not be true. According to what has been said, it is quite possible that unintelligent, as well as intelligent, people are supporting the bond issue. Therefore, the conclusion is invalid.

If we desire a world in which politicians will have to lay their cards on the table or be thrown out of office (because their public reasoning is so ridiculously bad), then teachers can help by exposing their students to fallacious arguments. Better still, students should be given experience in detecting fallacies in a wide variety of situations. Teachers may find it fun, too.

EXERCISES 3-3

S-1.   Write the hidden premise(s) in each of the following arguments.

   a.   One make of automobile is the country's best seller. Therefore, it is the one you should buy.

b. The latest model of our car is longer, wider, and heavier than ever before. Therefore, it must be the best we have ever made.

c. Tommy received more toys for Christmas than Pete did, so Tommy must have been a better boy than Pete.

d. The individual suburbs have better schools than the large cities. Therefore, we should decentralize the large city school systems and make them more like the suburban schools.

e. This model appliance costs more than that one, so this one must be better.

S–2. Each of the following arguments results in a false conclusion. Find the fallacy in each.

a. All good mathematicians are hippies. Sir Isaac Newton was a good mathematician. Therefore Newton was a hippy.

b. All sets containing more than three elements are non-empty. The set $\{\square, \bigcirc\}$ is non-empty. Therefore the set $\{\square, \bigcirc\}$ contains more than three elements.

c. Automobiles cause the deaths of thousands of people each year. Therefore automobiles should be illegal.

d. All college professors are smart. The president of General Motors is smart. Therefore the president of General Motors is a college professor.

For further thought and study: Find an article or advertisement in a newspaper or magazine which contains a fallacious argument. Explain the fallacy involved.

4. AXIOMATIZING

The ability to reason deductively may make it possible to reduce the number of assertions one must remember. Thus, a person may have access to a large number of assertions by virtue of the fact that he can derive (deduce) each of them, using a much smaller set of assertions which he may have memorized. Since there is a limit to how many discrete bits of information we can store in our finite minds, reducing the number of things we have to remember can be a great asset. Ideally, the number of assertions may be reduced to some logically determined minimum.

The problem, then, is to determine some subset of a larger collection of "facts" from which *all* of the facts may be derived. This is what mathematicians do every time they axiomatize a theory. Theoretical scientists have the same problem, although in this case their "axioms" (i.e., basic assertions) must be in agreement with the empirical world. Most important for present purposes, Davis[5] and a number of other investigators have demonstrated that elementary school students (fifth-graders) not only can solve such tasks, but that they truly enjoy them. Davis was concerned with open sentences of the forms $X + a = b$ and/or $X + Y = Y + X$. The latter statement should be interpreted to mean that if any number is substituted for X and any number for Y, then $X + Y = Y + X$.

[5] R. B. Davis, *Discovery in Mathematics*, Reading, Mass., Addison Wesley, 1964.

The general procedure Davis used was to have children construct a list of open sentences that are true no matter what numbers[6] are substituted for the unknowns (i.e., X and Y).

Consider the following selection from an actual list that has been compiled and axiomatized by children:

1. $X + Y = Y + X$

2. $X \cdot Y = Y \cdot X$

3. $X \cdot (Y + Z) = X \cdot Y + X \cdot Z$

4. $X \cdot 1 = 1 \cdot X = X$

5. $X \cdot 2 = 2 \cdot X = X + X$
   $X \cdot 3 = 3 \cdot X = 2 \cdot X + X = X + 2 \cdot X$
   $X \cdot 4 = 4 \cdot X = 3 \cdot X + X = X + 3 \cdot X$

Clearly, this list of assertions might be continued indefinitely. It turns out, however, that the first four of these assertions about numbers are basic to all of the others. To see why this is so, first notice that $X + Y = Y + X$ and $X \cdot Y = Y \cdot X$ mean that the order of adding or multiplying *any* two numbers is immaterial. Thus, we could substitute X and $2 \cdot X$, X and $3 \cdot X$, X and $4 \cdot X$, and so on, for X and Y in the "commutative" law for addition, $X + Y = Y + X$. This would show that

$$X + 2 \cdot X = 2 \cdot X + X$$

$$X + 3 \cdot X = 3 \cdot X + X$$

$$X + 4 \cdot X = 4 \cdot X + X, \text{ and so on.}$$

Similarly, substituting in the commutative law for multiplication, we would get

$$2 \cdot X = X \cdot 2$$

$$3 \cdot X = X \cdot 3, \text{ and so on.}$$

All that remains to be shown is that

$$2 \cdot X = X + X$$

$$3 \cdot X = 2 \cdot X + X,$$

and so on, can also be derived from the first four axioms. Here again, the first thing we do is substitute into one of the axioms, this time the "distributive" law, $X \cdot (Y + Z) = X \cdot Y + X \cdot Z$. Letting $Y = 1$, $Z = 1$, we get

$$X \cdot (1 + 1) = X \cdot 1 + X \cdot 1$$

Next, we note that the left side of the equation, $X \cdot (1 + 1)$, is the same as $X \cdot 2$ because "$1 + 1$" is simply another name for the number denoted by "2." Then,

---

[6] As we shall see in later chapters, the resulting list will depend on just what numbers are considered—the natural numbers (i.e., 1, 2, 3, . . .), the whole numbers (i.e., 0, 1, 2, 3, . . .), the integers (i.e., . . ., $^-3$, $^-2$, $^-1$, 0, $^+1$, $^+2$, $^+3$, . . .), and so on. For present purposes, consider whatever set of numbers is most convenient or familiar.

using assertion four, $X \cdot 1 = X$, we find that the right side, $X \cdot 1 + X \cdot 1$, is the same as $X + X$. Putting these facts together, we get

$$X \cdot 2 = 2 \cdot X = X + X$$

(Three facts are used. Can you find them?) We can do the same thing all over again, letting $Y = 2$ and $Z = 1$, to get the statement

$$X \cdot 3 = 3 \cdot X = 2 \cdot X + X$$

and similarly for the remaining assertions in the list. Strictly speaking, a proof by induction is called for, but we shall not go into this here, because it would only cloud the main point—it is often possible to find a distinguished (proper) subset of assertions, called *axioms*, from which all of the others may be deduced. (The word "proper" simply means that there are more assertions than axioms.)

As you may have observed, each of the *theorems* (i.e., assertions derivable from the axioms) in the preceding list can be derived in the same general manner. This is not surprising, of course, since many of the assertions are of the same general form. (This is the sort of list that children will tend to compile until they gain some sophistication in "theory construction." This should not be discouraged. In fact, teachers will often gain valuable insights into how children think by giving them free rein—and, it can be fun.)

A somewhat more "sophisticated" type of list is the following. Here, the logical interrelationships are a bit more varied.

1. $1 \cdot X = X \cdot 1 = X$
2. $X \cdot (Y + Z) = X \cdot Y + X \cdot Z$
3. $1 \cdot (X + Y) = X + Y$
4. If $X + Y = X + Z$, then $Y = Z$
5. $0 + X = X + 0 = X$
6. $X \cdot 0 = 0 \cdot X = 0$
7. If $Y = Z$, then $X \cdot Y = X \cdot Z$

We first note that assertion 3 follows from assertions 1 and 2.

$1 \cdot (X + Y) = 1 \cdot X + 1 \cdot Y$	by assertion 2
$1 \cdot X = X$	by assertion 1
$1 \cdot Y = Y$	by assertion 1—giving X the label Y does not change anything
Therefore, $1 \cdot (X + Y) = X + Y$	by substitution of equals—what are the equals being substituted?

The last equation also follows directly from assertion 1, substituting $X + Y$ for X. We can also show that assertion 6 follows from assertions 2, 4, 5, and 7.

$0 + 0 = 0$	by assertion 5—substituting 0 for X
So, $X \cdot (0 + 0) = X \cdot 0$	by assertion 7—substituting $0 + 0$ for Y and 0 for Z
But, $X \cdot (0 + 0) = X \cdot 0 + X \cdot 0$	by substitution of 0 for Y and Z in assertion 2

Hence, $X \cdot 0 + X \cdot 0 = X \cdot 0$	substitution of equals
Now, $X \cdot 0 + 0 = X \cdot 0$	by assertion 5—substituting $X \cdot 0$ for X
Hence, $X \cdot 0 + X \cdot 0 = X \cdot 0 + 0$	substitution of equals
Therefore, $X \cdot 0 = 0$	by "cancelling" $X \cdot 0$ from both sides of the previous step as indicated in assertion 4

(Check these proofs carefully to make sure that you understand them.)

For those who have had a little trigonometry, let us consider one more example. Suppose we know that

1. $\sin \theta = \dfrac{a}{c}$ 4. $\tan \theta = \dfrac{\sin \theta}{\cos \theta}$

2. $\cos \theta = \dfrac{b}{c}$ 5. $c^2 = a^2 + b^2$

3. $\tan \theta = \dfrac{a}{b}$ 6. $\sin^2 \theta + \cos^2 \theta = 1$

where a, b, c, and θ are as indicated below.

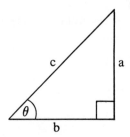

(The list could be made longer, if we introduced cotangents, secants, and cosecants, but this would leave the basic form of the argument unchanged.) In this case, the definitions of $\sin \theta$, $\cos \theta$, and $\tan \theta$ (assertions 1, 2, and 3) and the Pythagorean relation (assertion 5) may be considered basic. The other assertions (4 and 6) can be derived from them by performing simple algebraic manipulations. Thus, assertion 4 can be derived as follows.

$a = c \cdot \sin \theta$ (multiplying both sides of assertion 1 by c)

$b = c \cdot \cos \theta$ (multiplying both sides of assertion 2 by c)

Then, $\tan \theta = \dfrac{a}{b} = \dfrac{c \cdot \sin \theta}{c \cdot \cos \theta} = \dfrac{\sin \theta}{\cos \theta}$

To prove statement 6, we divide statement 5 by c^2 and substitute for $\dfrac{a}{c}$ and $\dfrac{b}{c}$.

$$\frac{c^2}{c^2} = \frac{a^2}{c^2} + \frac{b^2}{c^2}$$

$$1 = \left(\frac{a}{c}\right)^2 + \left(\frac{b}{c}\right)^2 = \sin^2 \theta + \cos^2 \theta$$

This example provides an excellent opportunity to make another point: Although certain axiomatizations may appear more natural than others, *a given set of assertions can generally be axiomatized in more than one way*. In this example, a second axiomatization is obtained by simply replacing the Pythagorean relation (assertion 5) with assertion 6. We shall leave it as an exercise for the reader to prove that assertions 4 and 5 can be deduced from the other four assertions.

Axiomatization is not restricted to mathematics. For example, consider the following verbal statements:

1. Everyone in my family is well.
2. I am well.
3. My father is well.
4. My mother is well.
5. My brother is well.

Clearly, the first statement is basic; all of the others follow from it. (Conversely, if there are only four people in the family, then the last four statements would together imply the first.)

The following list is slightly more complicated.

1. Patricia is Bobby's younger sister.
2. Nancy is older than Patricia.
3. Nancy is Patricia's sister.
4. Bobby and Nancy are twins.

In this case, statements 2 and 3 may be derived from statements 1 and 4. To see this, observe that because Bobby and Nancy are twins (4) and Patricia is younger than Bobby (1), Patricia must also be younger than Nancy. That is, Nancy must be older than Patricia (2). Furthermore, because Patricia and Nancy are both Bobby's sisters (1 and 4), Nancy must be Patricia's sister (3).

A similar sort of "axiomatic" reasoning is often advantageous in summarizing the important points of a discussion. The essential idea involved is that the main points are singled out for emphasis on the assumption that subordinate ideas may be derived as needed.

It would appear, then, that a teacher who gives his or her students the opportunity or, better, encourages them to *summarize* is in effect providing them with some of the kinds of experiences necessary to develop facility in axiomatization. This is not intended to imply that axiomatization in mathematics may be ignored. Mathematics undoubtedly provides the greatest variety and clearest experiences of this sort, and the teacher should make as much use of this medium as his or her background will allow.

EXERCISES 3–4

S–1. Shorten each of the following lists so that only the basic statements (axioms) remain, and all of the other statements can be derived from these basic statements. In each case there may be a single basic statement or several basic statements. For additional examples see R. B. Davis, *Discovery in Mathematics*, Reading, Mass., Addison Wesley, 1964.

 a. All pro football players are good athletes.
   My cousin is a pro football player.
   My cousin is a good athlete.

 b. Tommy and I are thinking of the same number.
   I am thinking of the number 20.
   Tommy is thinking of the number 20.

 c. One-third of the people in my class are boys.
   There is at least one boy in my class.
   There are more girls in my class than boys.
   There are at least two girls in my class.

 d. All good students get 3.5 averages or better.
   I got a 3.5 average or better.
   I am a good student.

 e. Sam is older than Tom.
   Tom and Elizabeth are twins.
   Sam is older than Elizabeth.
   Tom and Elizabeth are the same age.
   Sam is younger than Susan.
   Susan is older than Tom.
   Elizabeth is younger than Susan.

 f. $A + B = B + A$
   $A \cdot B = B \cdot A$
   $A + (B \cdot C) = (C \cdot B) + A$
   $A \cdot (B + C) = (C + B) \cdot A$
   $A \cdot (B + C) = (A \cdot B) + (A \cdot C)$

 g. In the National League, only the Phillies, Dodgers, Giants, Pirates, Cardinals, and Mets played yesterday.
   The Dodgers played the Phillies.
   The Pirates did not play the Cardinals.
   One sports writer picked the Dodgers, Pirates, and Cardinals to win their games.
   One sports writer picked the Dodgers, Mets, and Giants to win their games.

 h. $A + B = B + A$
   $A \cdot (B + C) = (A \cdot B) + (A \cdot C)$
   If $A + B = A + C$, then $B = C$
   $A + 0 = A$
   $0 + A = A$
   If $A + B = A$, then $B = 0$
   $A \cdot 0 = 0$

S–2. Make up a list of five statements such that two of them are basic (axioms) and the other three can be derived from the two basic ones.

M–3.   Give an example of a situation in elementary school mathematics where you might want your students to reduce a list (axiomatize).

5.   DEDUCTIVE PATTERNS

The ability to draw a correct inference—that is, to make a deduction—involves making a new assertion that follows logically from a set of given assertions. (This is a slight oversimplification and the reader is referred, for more detail, to the author's *Mathematics and Structural Learning*.)

To see what is involved let us consider a few examples. Suppose a young child is informed that, "It is raining or it is snowing," and that, "It is *not* raining." We shall assume that this has been done in a convincing manner. (Perhaps the first statement is made by one person and the second by another.) Suppose that the child is then asked what it is like outside. The typical youngster will want to run to the window and see. But let us assume that we had the foresight to draw the blinds. Under these conditions, we might reasonably expect a child to give the correct response, "It is snowing," (without guessing) *only if* he had previously acquired the following deductive pattern.

1. One or the other (or both) of two possible events (e.g., rain or snow) are known to have occurred.
2. One of the events has *not* occurred.
3. Therefore, the other event must have occurred.

When stated in this form, it is apparent that the same deductive pattern would be applicable in any number of different situations.

One such situation is provided by an episode witnessed by the author. The chairman of a mathematics department remembers approximately how old one of his employees (a logician) is but is not sure whether he is 25 or 26. So he asks, "Are you 25 or 26?" The logician answers "Yes." The chairman could have gotten angry, of course, or he could have asked the logician a direct question such as, "Are you 25?" Whatever the logician's answer, application of the deductive pattern above (perhaps in slightly modified form—if the answer were "yes") would settle the issue directly. Being of equally nimble mind, however, the chairman retorted, "Are you 24 or 25?" The logician answered "No." At this point the questioning ended, for the chairman had the information he desired. By his first admission the logician was either 25 or 26. Since his second reply indicated that he was neither 24 nor 25 (in view of his previous response, of course, 24 was not even a logical possibility), it was apparent to the chairman that the logician was 26.

The familiar syllogism provides another type of deductive pattern. The syllogistic argument can be summarized as follows.

Two statements are assumed to be true.

All P's are Q's.
x is a P.

We then conclude

x is a Q.

The following are three examples of this deductive pattern.

All men are mortal.
Socrates is a man.
Therefore, Socrates is mortal.

All intelligent people support better schools.
Joe Smith is intelligent.
Therefore, Joe Smith supports better schools.

All politicians are dishonest.
The President of the United States is a politician.
Therefore, ...

Although the third conclusion (implied) is open to serious question, the (deductive) argument is valid.

Contrast these arguments with the superficially similar, but *invalid*, argument:

All intelligent people support better schools.
Joe Smith supports better schools.
Therefore, Joe Smith is intelligent.

This argument does not fit the syllogistic pattern and may be called reasoning from the converse. It says, in effect, that

All P's are Q's.
x is a Q.
Therefore, x is a P.

We noted earlier that this sort of reasoning is fallacious.
Consider one more example, this involving simple algebra.

$X + Y = Y + X$
Therefore, $2 + 3 = 3 + 2$

In this case, both the premise and the conclusion appear valid, but the line of reasoning (i.e., the deductive pattern) appears to be quite different from those given above. In fact, there is a missing, or at best implicit, premise. A more complete statement of the deductive argument, however, bares its similarity to the syllogism.

All whole numbers (X and Y) satisfy the equation $X + Y = Y + X$.
2 and 3 are whole numbers.
Therefore, 2 and 3 satisfy $X + Y = Y + X$ (or $2 + 3 = 3 + 2$).[7]

---

[7] Many apparently different types of deductive argument are highly interdependent. In fact, each of the arguments illustrated above can be shown to be a consequence of a few logically true assertions and deductive patterns, called "modus ponens," "generalization," and "particularization" by logicians. The basic idea involved in modus ponens is that if "If assertion P implies assertion Q" and "assertion P" are true, then "assertion Q" must also be true. Generalization and particularization are in some sense opposites. The latter simply says that if an

Many other examples can easily be fabricated, and the reader should satisfy himself that he can do just this. *The main point to remember is that a deductive pattern may be applied in any number of situations.*

Pedagogical Comment: In spite of how simple these examples may appear, the teacher must not make the mistake of assuming that young children are capable of such reasoning. To the contrary, in some of our research we found that about three-fourths of the children in a kindergarten class (age, $5\frac{1}{2}$ years) were unable to make even the simplest deductions.[8] To obtain this information, we chose some material that we felt would gain the student's interest but, at the same time, would not conflict with any specific knowledge that the student might have had. Stories about imaginary people served this purpose well. One of our favorites was a story about the "Gruunda" tribe from a strange far-off land. The children were told, for example, that *all* of the Gruunda had big feet, that some had pink hair, that others always wore velvet vests, and so on. Next, they were told that the experimenter-teacher had just seen a "Gruunda" on his way to school that morning. The children were then asked if they could tell the teacher anything about this Gruunda. Only about one-fourth of the children were able to consistently say that the Gruunda had big feet (or whatever other properties

---

assertion, like $X + Y = Y + X$, is true for all values of its variables, then it is necessarily true for any particular set of values (e.g., $3 + 4 = 4 + 3$).

It has been shown, for example (see J. B. Rosser, *Logic for Mathematicians*, New York: McGraw-Hill, 1953, p. 56), that modus ponens, together with the following tautologies (logically true assertions), provides a sufficient basis for all possible logical interrelationships among statements (i.e., verbal assertions):

1. assertion p implies assertion p and p;
2. assertion p and q implies assertion p;
3. assertion p implies q, in turn, implies assertion not (q and r) implies not (r and p).

Let us emphasize, however, that whereas the logic resulting in this case would be sufficient for analyzing logical relationships between statements, it is nowhere nearly sufficient for mathematics in general. Do not feel that you should memorize these logically true assertions. We list them only to show how few really fundamental logically true assertions one really needs to assume. All of the others may be derived from these with the help of the deductive pattern known as modus ponens.

Nonetheless, it may be worthwhile noticing just how obvious some of these tautologies are. The first one, for example, amounts to nothing more than collectively asserting such things as, "It is raining," implies that "It is raining and it is raining"; "The moon is blue," implies that "The moon is blue and the moon is blue," and so on. How could it be otherwise? And that is exactly the point. Logical axioms *must* be so obvious that no one could possible question them.

To the logician, making a deduction (i.e., proving an assertion) in the statement logic amounts to finding a chain of assertions that are either (1) logically true (i.e., are logical axioms or logical theorems derivable therefrom), (2) assertions that may be assumed to be true (they constitute the conditions under which the conclusion is to be true), or (3) derivable from assertions satisfying (1) or (2) by the use of modus ponens. In actual practice, both in everyday life and in devising mathematical proofs, many steps are typically omitted. Not to do so would be to make even the simplest of proofs too long and cumbersome to be of any real value. The important thing is that the reasoning used can be shown to be sound. But, this is a task for the logician and not the elementary school teacher, or her pupils.

For a more thorough discussion of logical reasoning, refer again to *Mathematics and Structural Learning.*

[8] Although this research is described earlier, the main ideas are repeated here for the convenience of those who skipped Part I.

were associated with the word *all*). (We might note parenthetically that the other properties of the Gruunda were included as distractors to make sure that the children were not glibly repeating something that they had heard.)

EXERCISES 3–5

S–1. Assuming that each of the following pairs of assumptions is true, write a true conclusion for each pair.

    a. Mr. Smith is rich or a Rotarian.
       Mr. Smith is not rich.
    b. All little girls have blonde hair.
       Susie is a little girl.
    c. Bob or Dave is missing one mitten.
       Bob has both of his mittens.
    d. All redheads are very placid.
       Trudy is a redhead.
    e. All golguks have long hair.
       My pet has short hair.
       (Be careful on this one.)

S–2. a. Give two more examples using the syllogistic deductive pattern.
      b. Give two more examples using the deductive pattern at the beginning of Section 5.

S–3. Each of the following arguments uses the deductive pattern of S–2b with one assumption missing. Write the missing assumption.

    a. Tommy is not a girl.
       Therefore, Tommy is a boy.
    b. John is not 6 years old.
       Therefore, John is 7 years old.
    c. My automobile is red or blue.
       Therefore, my automobile is blue.

S–4. Each of the following uses the syllogistic argument with one assumption missing. Write the missing assumption.

    a. Mike is a boy.
       Therefore, Mike likes sports.
    b. All logic problems are interesting.
       Therefore, this problem is interesting.
    c. All boys like girls.
       Therefore, Bryan likes girls.

E–5. Each of the following arguments uses a deductive pattern similar to one discussed in this section, yet each one arrives at a false conclusion. Find the fallacy in each (see Section 3).

    a. All multiples of 3 are odd.
       $6 = 2 \times 3$ is a multiple of 3.
       Therefore, 6 is odd.

b. All multiples of 6 are even.
26 is even.
Therefore, 26 is a multiple of 6.
c. All U.S. Senators are Republicans or Democrats.
Senator Kennedy is not a Democrat.
Therefore, Senator Kennedy is a Republican.

E–6. The deductive patterns discussed in this section are by no means exhaustive. Here are three other deductive patterns with three examples of each. The conclusion(s) is given for the first example in each case; you should find the conclusion(s) for the other examples.

a. The first pattern is so simple (or intuitive) that we are not really aware we are reasoning when we use it. If A and B represent statements and we know that "A and B" is a true statement, we know, therefore, that the individual statement "A" and the individual statement "B" are both true.

Examples: i. It is sunny and warm.
Therefore, it is sunny.
Therefore, it is warm.
ii. It is cloudy and cool.
Therefore, _____.
Therefore, _____.
iii. This figure is a red square.
Therefore, _____.
Therefore, _____.

b. The second pattern is similar to the syllogistic deductive pattern. If we know

All P's are Q's.
All Q's are R's.

then we may conclude that

All P's are R's.

Examples: i. All squares are rectangles.
All rectangles are polygons.
Therefore, all squares are polygons.
ii. All mathematicians are scientists.
All scientists are eccentric.
Therefore, _____.
iii. All multiples of 8 are multiples of 4.
All multiples of 4 are multiples of 2.
Therefore, _____.

c. The third pattern is also similar to the syllogistic deductive pattern. If we know

All P's are Q's.
x is *not* a Q.

then we may conclude

x is not a P.

(You should convince yourself that this is a valid deductive pattern.)

Examples: i. All dogs have four legs.
            Mike does not have four legs.
            Therefore, Mike is not a dog.

      ii. All large primes are odd numbers.
            276 is not an odd number.
            Therefore, _____.

     iii. All pro football linemen are big men.
            Johnny Unitas is not a big man.
            Therefore, _____.

6. SET OPERATIONS AND VENN DIAGRAMS

Proving or disproving conjectures (statements) has been felt by many to be one of the most difficult things to do in mathematics. Many a geometry student must have found himself wishing that someone would come up with a systematic way to do this. All he would then have to do is learn the procedure and his troubles would be over. *Unfortunately, there is no systematic procedure that is sufficiently general to determine whether an arbitrary assertion about mathematics is or is not true.* More can be said than this. Not only has no such procedure been found, but the logician Alonzo Church[9] has proved that no such procedure will ever be found. Because computers must be given explicit instructions, Church's theorem suggests that the mathematician will never be replaced by a machine.

If we restrict our attention to a logic (called the Statement Logic) which is sufficient for analyzing logical relationships between statements, however, there *is* a systematic procedure for proving or disproving arbitrary assertions. This procedure may be thought of in terms of operations on sets. Before we will be in a position to show how this can be done, however, we first need to say more about sets and set operations than was feasible in Chapter 2.

6.1 Sets and Subsets

A set may be thought of as a collection of objects. Thus, we speak of: the set of all Americans; the set of all red blocks; the set consisting of the numbers 2, 4, 5, and 9; the set containing the symbols ·, /, and = ; the set of all sets containing two elements; and so on. A person may be said to have acquired the "idea" of a set if he can sort some larger "universe" of elements into two piles, one containing the elements of the set and the other containing those elements not in the set. In many cases, all of the objects in a set will have some easily recognizable property in common—e.g., the set of all *Americans*, the set of *green* triangles, the

---

[9] See J. B. Rosser, *Logic for Mathematicians*, New York, McGraw-Hill, 1953, pp. 161–162.

set of *English* vowels, and so on. But the elements of a set need not have any common property except, of course, the property of being a member of the set (e.g., the set consisting of 3, *h*, and +, or the set consisting of Sam, Fido, △, and *Gone with the Wind*).

With this idea of sets in mind, it is natural to talk of sets which are included in some parent set. Within the set of red blocks, for example, there might be both large (red) blocks and small (red) blocks. The set of large red blocks, then, would be a *subset* of the set of red blocks, namely that subset consisting of all of the large blocks within the set defined by the property red block. In general, a subset of a set consists entirely of elements in the original set.[10] If a subset does not contain all of the elements in its parent set, it is said to be a *proper subset*. If set *A* is a proper subset of set *B*, we write $A \subset B$.[11]

One of the most common notations for sets is the "brace" notation, in which the members of a set are written between braces. For example, {1, 3, 5}, {*a, e, i, o, u*}, { △, ☐, ○ } denote three sets containing the indicated elements. Sets may also be represented iconically (i.e., pictorially) in the form of *Venn diagrams*. In Venn diagrams, the universe is frequently represented by the set of points inside a rectangle and subsets of this universe by the set of points inside a closed curve in the rectangle. In the examples below, *A*, *B*, and *C* refer to sets in some universe (i.e., they are subsets of the universe).

 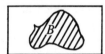 

In a Venn diagram, we indicate that set *A* is a subset of set *B* by drawing the closed curve for set *A* inside the closed curve for set *B*. For example

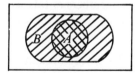 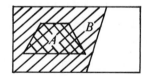

The idea of a subset is equivalent to Piaget's notion of inclusion. Not only is it necessary for the learner to be able to sort elements into sets and subsets, but he must also be able to recognize that a subset is necessarily included in its parent set. To find out if children have learned to *conserve* inclusion, Piaget has used such tasks as the following. A box of *wooden* beads is set before the child, most of which are painted *white* and the remaining few, *blue*. After familiarizing the child with these facts, the experimenter-teacher asks whether there are more *white* beads or more *wooden* beads. The nonconserver will almost invariably answer "white" while the conserver, of course, will typically say something like, "Well, they are all wooden." As surprising as the nonconserver's answer may seem, he will react this way consistently. If he is a "true" nonconserver (that is,

---

[10] Notice that this definition implies that every set is a subset of itself.

[11] The closed side of the symbol, $\subset$, may be thought of as pointing to the smaller set, in this case, *A*.

if he is not in a transitional stage), the child will react in the same way to even simpler tasks where *all* defining attributes of the subset are stated explicitly. Suppose, for example, that a set of large blocks is before a child, most of which are red, and the child is asked whether there are more large blocks or more large red blocks. Many nonconservers will still say there are more large red blocks. Those who do not are presumably on their way to understanding the general notion of subset inclusion (i.e., only certain subsets like those in the beads problem will cause them difficulty). Preliminary evidence suggests that we have been reasonably successful in helping kindergartners to conserve inclusion by introducing them to the idea with such questions as, "Are there more people in the room or more boys?" One child (a nonconserver according to the bead test but on his way to becoming a conserver) said, "Why did you ask me that? We are all people."

EXERCISES 3–6.1

S–1. Let the universe consist of the set of all natural numbers 1, 2, 3, 4, ... List all of the members of each of the following sets, using the "brace" notation:

    a. the set of all numbers less than 27 which are divisible (with no remainder) by 5

    b. the set of all even numbers which are divisible (with no remainder) only by themselves and 1

    c. the set of all numbers which when multiplied by themselves give an answer between 10 and 110

    d. the set of all numerals between 900 and 9900 that end with three zeros

    [e.] the set of all numbers divisible (with no remainder) by 5
        (*Hint:* You will not be able to list all of the members of this set. Why? Can you describe the numbers in this set, even though you cannot list all of them?)

S–2. List the elements of the following sets using the brace notation:

    a. the set of Great Lakes
    b. the set of states bordering on the Pacific Ocean
    c. the set of living former Presidents of the United States

S–3. a. List all of the subsets of the set {1, 2}.
    b. Give two subsets of the set of geometric figures.
    c. Give some subsets of the set of all even numbers.

6.2 Complements, Unions, Intersections, and Differences

The *complement* of a set is simply that subset of elements of the universal set which are *not* in the given set. Thus, the complement of the set of all even numbers (in the universe of all natural numbers, 1, 2, 3, ...) is the set of all *odd* numbers; the complement of the set of all red objects is the set of objects (in the universe) which are not red; and so on. Obviously, the ability to form the complement of a set and the ability to form the set itself are essentially the same. Both involve the ability to distinguish between the elements in the set and those not in the set. The only difference is in the way the question is posed, "Which of these elements is *not* in the set of ...?" The complement of the set *A* is denoted $\bar{A}$. The complements of sets *A* and *B* are represented by the shaded areas in the Venn diagrams below.

 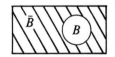

The *union* of two or more sets, denoted "$A \cup B$," is simply that set which includes all of the elements (and no others) that are in at least one of the original sets. For example, if one set consists of all of the blocks (in some universe) which are red, and another of all of the blocks which are large, then their union would consist of all of the blocks that are either red or large (or both). Notice, in particular, that any large red block would belong to both original sets, but this block would be counted as only one element in the union. Thus, if two sets overlap (i.e., if they have elements in common), the number of elements in the union will be smaller than the sum of the numbers of elements in the original sets. The number of elements in the union of two *non*-overlapping or disjoint sets, however, equals the sum of the numbers associated with each set.[12] For example, there are $3 + 2 = 5$ elements in $\{a, b, c\} \cup \{d, e\}$, but only 4 elements in $\{a, b, c\} \cup \{c, d\}$.

When working with sets of concrete objects, the fact that the number of elements in the union may be less than the sum of the number of elements in each set causes most young children little difficulty. To form the union all they need to do is "put the two (or more) piles together." In using "brace" notation, however, children should be introduced gradually to the *convention* that any two identical pictures really represent the same object so that only one such picture should appear in a brace representation of the union. Thus, the union of $\{\Box, \triangle, \bigcirc\}$ and $\{\Box, \triangle\}$ (denoted $\{\Box, \triangle, \bigcirc\} \cup \{\Box, \triangle\}$) is the set $\{\Box, \triangle, \bigcirc, \Box\}$ and *not* the set $\{\Box, \triangle, \bigcirc, \Box, \triangle\}$. In his elementary text book series, Suppes has succeeded in introducing young children to this convention. His basic procedure is one of getting the child accustomed to the idea that a picture of a bear, say, represents the same bear whether the bear is running, playing, or sleeping.

Depending on the particular sets *A* and *B* in question, the union $A \cup B$ may be represented by shaded areas in one of the following Venn diagrams.

inclusion   overlapping   disjoint

  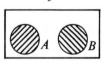

By the *intersection* of two or more sets, denoted "$A \cap B$" for sets *A* and *B*, is meant the set of all elements *common* to these sets. Here again, after they are able to form simple sets, most young children find it fairly natural to identify

---

[12] As we shall see, this fact can be used to define the operation of addition.

elements common to two or more sets. For example, once a child can reliably form the set of all red objects and the set of all large objects, it usually is not too difficult to get him to form the set of all large, red objects. Of course, in order to demonstrate his awareness of the interrelationship implied, the child must show that he is able to form the intersection set determined by two (or more) given sets. Specifically, he needs to realize that some of the elements belong at once to more than one set. As our discussion of inclusion suggests, most children acquire this ability *after* they enter school.

In working with concrete objects, Dienes has found it convenient to use hula-hoops to delineate sets in much the same way that Venn diagrams serve to represent sets of points. Again depending on the particular sets A and B, the intersection $A \cap B$ may be represented by one of the following Venn diagrams:

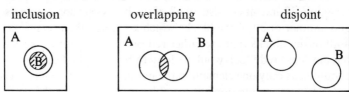

inclusion   overlapping   disjoint

Notice that when two sets are disjoint, their intersection (denoted $\emptyset$ or $\{\ \}$) is empty. The intersection of the set of all children and the set of all cookie jars, for example, is empty (in more than one sense). Most children find the idea of an empty set quite natural, and experience suggests that empty sets can and should be introduced from the very beginning.

Although it is possible to define all possible relationships between sets in terms of complements, unions, and intersections (in fact, intersections may also be defined in terms of complements and unions),[13] it is convenient to also introduce the idea of the *difference* between sets. As you might expect, the difference, $A - B$, between the sets A and B is simply the set of all elements in A that are *not* also in B. For example, let A be the set of all red objects and B be the set of all large objects. Then, $A - B$ is the set of all red objects that are not large (i.e., the set of all small red objects). That is, $A - B$ is the same as $A \cap \bar{B}$. In each of the Venn diagrams below, the shaded area represents $A - B$.

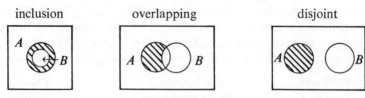

inclusion   overlapping   disjoint

EXERCISES 3–6.2

S–1. Let universe = $\{1, 2, 3, 4, 5, 6, 7, 8, 9, 10\}$

$A = \{1, 2, 3, 5, 7\}$
$B = \{2, 4, 6, 8\}$
$C = \{1, 3, 5, 7, 9\}$

---

[13] $A \cap B$ is the same as the complement of the union of the complements of A and B, denoted $\bar{A} \cup \bar{B}$.

Find each of the following:

a. $A \cap B$ h. $B - A$
b. $A \cup C$ i. $B - C$
c. $B \cap C$ j. $\bar{A} \cup \bar{B}$
d. $B \cup C$ k. $\overline{A \cap B}$
e. $\bar{A}$ l. $\overline{A \cup C}$
f. $\bar{B}$ m. $\bar{A} \cap \bar{C}$
g. $A - B$

S–2. Let universe = $\{1, 2, 3, 4, 5, 6, 7\}$

$A = \{1, 5\}$
$B = \{2, 3, 4, 5\}$

Find each of the following:

a. $\bar{A} \cup \bar{B}$
b. $\overline{A \cap B}$
c. $\overline{A \cup B}$
d. $\bar{A} \cap \bar{B}$
e. $\overline{\bar{A} \cup \bar{B}}$
f. $B - A$

S–3. In each of the Venn diagrams below shade the indicated set.

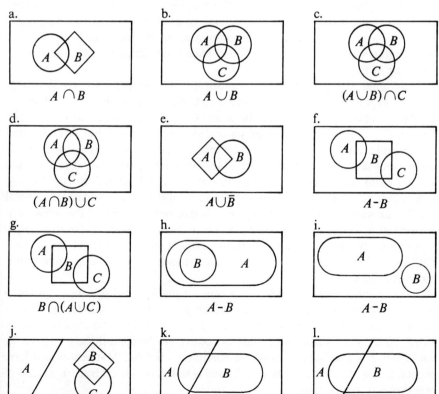

a.

$A \cap B$

b.

$A \cup B$

c.

$(A \cup B) \cap C$

d.

$(A \cap B) \cup C$

e.

$A \cup \bar{B}$

f.

$A - B$

g.

$B \cap (A \cup C)$

h.

$A - B$

i.

$A - B$

j.

$A \cup (B \cap C)$

k.

$\overline{A \cup B}$

l.

$\bar{A} \cap \bar{B}$

[S-4.] Can you detect a regularity from parts *j*, *k*, *l*, and *m* of Problem S-1, parts *a*, *b*, *c*, and *d* of Problem S-2, and parts *k* and *l* of Problem S-3?

7. THE STATEMENT LOGIC

As indicated above, the statement logic is concerned with the truth or falsity of statements taken as a whole without reference to their internal structure (e.g., the subject-predicate relationship). Recall that a statement is any assertion that can be classified as either true or false. Thus, the statement, "George Washington was President of the United States," is true while the statement, "General De Gaulle was the King of England," is clearly false.

The *negation* of a statement necessarily takes on that truth value which is opposite to the truth value of the original statement. Thus, the statement, "General De Gaulle was *not* the King of England" is necessarily *true* since "General De Gaulle was the King of England" is false. The negation of a statement, "*p*," may be symbolized "$\sim p$" or "not *p*."

The statement logic is also concerned with the truth or falsity of statements formed by combining two or more simpler statements. Compound statements of this sort may be formed by use of the words "*and*" and "*or*." Using the simple statements given above we can form the compound statements, (1) "George Washington was President of the United States *and* General De Gaulle was the King of England," and (2) "George Washington was President of the United States *or* General De Gaulle was the King of England." By introducing negations, the variety of compound statements can be further increased (e.g., "George Washington was President of the United States *and* General De Gaulle was *not* the King of England.").

In order to be true, both of the constituent statements in "George Washington was President of the United States *and* General De Gaulle was the King of England" would have to be true. Because "General De Gaulle was the King of England" is false, the compound statement must also be false. A compound statement involving the use of "or," however, is true if any one (or more) of the constituent statements is true. Thus, "George Washington was President of the United States *or* General De Gaulle was the King of England" is true. Finally, notice that "George Washington was President of the United States *and* General De Gaulle was *not* the King of England" is true because both of the constituent statements are true. Letting *p* and *q*, respectively, correspond to the two simple statements, these compound statements may be represented, (1) "*p* and *q*," (2) "*p* or *q*," and (3) "*p* and $\sim q$." For convenience, we shall use *p* and *q* in our subsequent discussions.

A *conditional statement* is a statement of the form "If *p*, then *q*," where *p* and *q* are statements, and is usually written $p \to q$. The conditional may be interpreted to mean that either statement *p* is false or statement *q* must be true. In other words, when statements *p* and *q* are both true, the conditional $p \to q$ is true, and when *p* is true but *q* is false, the conditional is false. Mathematicians have agreed that when *p* is false, the conditional is to be considered true, regardless of whether *q* is true or false.

To see why this definition of the truth of $p \rightarrow q$ is reasonable, suppose a friend tells you "If you beat me in a game of chess today, then I will give you \$5." If you beat him and he actually gives you \$5, you would agree that he made a true statement. If, however, you beat him in chess and he does not give you \$5, he made a false statement. But, suppose he beat you. Then, whether or not he gives you \$5, you must agree that he has not lied to you. Although you may be disappointed, you cannot claim that his statement was false and so must agree that it is true.

As another example consider the conditional, "If A is subset of B, then A is a set," which is known to be true. Then the logical impossibility would be for "A to be a subset of B" and "A *not* be a set." Notice, however, that if "A is *not* a subset of B" (or if "A is a subset of B" is false), then "A is a set" might be *either* true or false without contradicting the conditional. Thus, A could be an arbitrary set not contained in B (in which case "A is a set" would be true) or A might be something other than a set, say an automobile, in which case "A is a set" would be false.

We see that in order to show that a conditional statement "If p, then q" is true, one need only show that q must be true when p is true (because the conditional is automatically true when p is false). Consider the following two examples:

1. If today is the 31st of the month, then tomorrow is the first.
2. If tomorrow is the first of the month, then today is the 31st.

If it is true that "today is the 31st," then "tomorrow is the first" necessarily follows. Hence, the first conditional must be true. The second conditional, however, is false when tomorrow falls on March 1, May 1, July 1, October 1, or December 1. (It is true when tomorrow falls on the first of any other month.) Hence, the second conditional is not necessarily true. Similarly, one can show that "If $x = 2$, then x is even" is necessarily true, whereas "If x is even, then $x = 2$" is false for all even numbers except 2.

It can be shown that the conditional "$p \rightarrow q$" is logically equivalent to "$\sim p$ or q." That is, "$p \rightarrow q$" and "$\sim p$ or q" are both false when p is true and q is false, and are both true in all other cases. Thus, the two examples given above are equivalent to

1'. Today is *not* the 31st of the month or tomorrow is the first.
2'. Tomorrow is *not* the first of the month or today is the 31st.

The first statement is always true, while the second statement is false when tomorrow falls on March 1, May 1, July 1, October 1, or December 1.

So far we have been concerned with determining the truth or falsity of compound statements given the truth or falsity of simpler statements. In certain cases, it may also be possible to work in the opposite direction, that is, to determine the truth or falsity of simpler statements given the truth or falsity of compound statements.

As a quick example, note that if "p and q" is true, then "p" and "q" must also be true. If "p and q" is false, of course, we know that either "p" or "q" (or both) is false but we have no way of knowing which. Modus ponens, or what has been called the "law of detachment," also fits this pattern. Thus, if "$p \rightarrow q$" and "p" are known to be true, then "q" must also be true. If either premise is false, however, the situation regarding "q" requires a different analysis. The interested reader may wish to work out the solution to this problem for himself by considering, in turn, the various possibilities. (As a start, let $p \rightarrow q$ be false and p true.)

EXERCISES 3–7

S–1.   Assume that statements p and q are both true and that statements r and s are both false. Determine whether each of the following statements is true or false.

a. p and q	b. p and r	c. r and s
d. s and r	e. q or p	f. q or r
g. r or q	h. r or s	i. If p, then q
j. $p \rightarrow r$	k. If r, then p	l. $s \rightarrow r$

S–2.   a.   If "p and q" is known to be true, what do we know about the truth or falsity of statement p and of statement q?
   b.   If "p or q" is known to be true and so is statement p, what do we know about the truth or falsity of statement p and of statement q?
   c.   If "$p \rightarrow q$" is known to be true and so is statement p, what do we know about the truth or falsity of statement q?
   d.   If "p or q" is known to be true and statement p is known to be false, what do we know about the truth or falsity of statement q? (Does this deductive pattern look familiar?)
   e.   If "$p \rightarrow q$" is known to be true and so is statement q, what do we know about the truth or falsity of statement p?

S–3.   If each of the following pairs of statements is assumed to be true, what conclusion can be drawn from each pair?

   a.   If the sunset was red, then tomorrow will be clear.
       The sunset was red.
   b.   If two coins add to 55¢, then one must be a nickel.
       These two coins add to 55¢.
   c.   If two coins add to 35¢, then one must be a dime.
       One of these two coins is a dime.
       (Be careful.)
   d.   If today is July 3, then tomorrow is a holiday.
       Today is not July 3.
       (Be careful.)
   e.   If today is December 31, then tomorrow is a holiday.
       Tomorrow is not a holiday.
       (Be careful.)

    f.   If Mr. Jones is rich, then he owns a sports car.
        Mr. Jones is rich.
    g.   If it is raining, the ground is wet.
        It is not raining.

S–4.   Give two examples showing why a conditional $p \rightarrow q$ should be considered true whenever p is false, regardless of whether q is true or false.

[M–5.] How would you bring a class to see that a conditional $p \rightarrow q$ should be accepted as true when p is false, regardless of whether q is true or false.

8. THE RELATIONSHIP BETWEEN THE STATEMENT LOGIC AND OPERATIONS ON SETS

In a previous section, we promised to show how the truth or falsity of any statement could be *systematically* determined from a knowledge of the truth or falsity of the simpler statements of which it is composed. In effect, what we need is a general procedure for proving or disproving any assertion within the statement logic. With such a procedure, it would be possible, say, given that "It is raining" and "It is hot" are true, to determine the truth or falsity of a statement like "If it is raining, then it is hot or it is *not* hot" by strictly mechanical means.

    One way to accomplish this is to translate statements into Venn diagrams. Recall that in any problem situation the set of logical possibilities includes (at any one level of analysis) all possible conditions under which statements about the situation are meaningful. For example, the set of logical possibilities for any statement about names of the months is {January, February, $\cdots$, December}. Similarly, the set of logical possibilities for statements about the sum obtained in rolling two dice is the set {2, 3, 4, $\cdots$, 12}; the set of logical possibilities for statements about the weather is the set of possible weather conditions (at some level of analysis).

    We begin by representing the set of logical possibilities for a given (or implied) collection of statements by the universal set, U, in a Venn diagram.

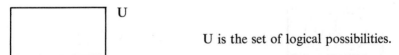

U is the set of logical possibilities.

    For any statement, its *truth set* is the set of all logical possibilities that make the statement true. For example, the statement "It is a month with 30 days" has {April, June, September, November} as its truth set. The statement "The sum of the two dice is an even number" has {2, 4, 6, 8, 10, 12} as its truth set. We cannot list the exact truth set of "It is precipitating" because there are too many different possibilities. But we can represent the truth set of any statement as a set in the Venn diagram.

P is the truth set of the statement.

The truth set of the *negation* of a statement includes all logical possibilities which make the original statement false. For example, the truth set of "It is *not* a month with 30 days" is {January, February, March, May, July, August, October, December}. The truth set of "The sum of the two dice is *not* an even number" is {3, 5, 7, 9, 11}. From this, we see that the truth set of the negation of a statement is the *complement* of the truth set of the original statement. Again, using Venn diagrams, if a statement has truth set P, then its negation has truth set $\bar{P}$. Thus

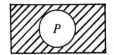

The *complement* of P is the truth set of the *negation* of the statement.

It is possible, of course, to consider the truth set of more than one statement in the same Venn diagram. To see what is involved, let "p" and "q" represent an arbitrary pair of statements, each with its own truth set. A logical possibility will be in the truth set of the compound conjunctive statement, "p and q" if and only if it is in the truth set of "p" *and*, also, in the truth set of "q." This implies that the truth set of "p and q" is the *intersection* of the truth set of "p" and the truth set of "q." Similarly, a logical possibility will be in the truth set of the disjunction, "p or q," if and only if it makes "p" true *or* makes "q" true. Equivalently, the truth set of "p or q" is the *union* of the truth set of "p" and the truth set of "q." For example, consider the statements "It is a month with 30 days" with truth set {April, June, September, November} and "It starts with J" with truth set {January, June, July}. The truth set of "It is a month with 30 days *and* starts with J" is {June}, which is the intersection of the two truth sets. The truth set of "It is a month with 30 days *or* starts with J" is {January, April, June, July, September, November}, which is the union of the two truth sets. You may wish to construct similar examples using statements about the sum of two dice.

With these relationships in mind, it is a simple matter to represent truth sets of compound statements with Venn diagrams. If statement p[14] has truth set P, and statement q has truth set Q, then "p and q" has truth set $P \cap Q$

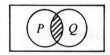

and "p or q" has truth set $P \cup Q$.

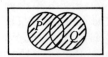

To determine the truth set of a compound conditional statement, "If p then q," it is perhaps simplest to use the equivalent form "Not p or q." That is, if p has truth set P and q has truth set Q, then "If p then q" has truth set $\bar{P} \cup Q$. For

---

[14] From now on in this chapter quotation marks will be used only with compound statements (e.g., "p and q") where there is a possibility of confusion.

example, the statement, "If it is a month with 30 days, then it starts with J" has truth set {January, February, March, May, June, July, August, October, December}. We represent this by the Venn diagram

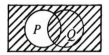

More generally, we can determine the truth or falsity of an arbitrary statement (called a "conjecture") given the truth of an arbitrary set of assumptions or premises. To do this, we proceed as follows:

1. Determine the truth set (or represent it on a Venn diagram) for each of the premises.
2. Determine the truth set for all of the premises taken together. That is, find the *intersection* of all of the sets found in step 1. Call this truth set the *premise set intersection.*
3. Determine the truth set of the conjecture; call this the *conjecture set.*
4. a. If the premise set intersection is a subset of the conjecture set, then the conjecture must be true whenever the premises are.

b. If the premise set intersection lies entirely outside of the conjecture set, then the conjecture must be false whenever the premises are true.

c. If the premise set intersection is partly inside and partly outside the conjecture set, then the premises tell us nothing about the conjecture. That is, when the premises are true, we cannot tell without further information whether the conjecture is true or false.[15]

Let us consider some examples to see how this procedure works in practice. Example (1):

Premises: *p*   It is a month with 30 days.
           q   It is a month beginning with J.
Conjecture: It is a month with 30 days or beginning with J.

Letting P and Q be the truth sets of premises p and q, respectively, the procedure gives us

1. $P = $ {April, June, September, November}
   $Q = $ {January, June, July}
2. Premise set intersection $= P \cap Q = $ {June}
3. Conjecture set $= P \cup Q$ (because the conjecture set is of the form "*p* or *q*") $= $ {January, April, June, July, September, November}
4. Because the premise set intersection is a subset of the conjecture set, the conjecture must be true whenever the premises are.

---

[15] This procedure parallels an alternative procedure involving what are called "Truth Tables." For an introduction compatible with the present discussion, the reader is referred to J. G. Kemeny, J. L. Snell, and G. L. Thompson, *Introduction to Finite Mathematics*, Englewood Cliffs, N.J., Prentice-Hall, 1957.

Let us repeat the example using a Venn diagram. Let P and Q again be the truth sets of premises p and q.

1. Draw overlapping circles to represent P and Q.
2. The premise set intersection, $P \cap Q$, is the cross-hatched area in the diagram.
3. Because the conjecture is of the form "p or q" the conjecture set is $P \cup Q$, which is the entire shaded area.
4. Because the cross-hatched area is a subset of the total shaded area, the conjecture must be true whenever the premises are.

Example (2):

    Premises: p It is a month with 30 days.
               q It is a month beginning with J.
    Conjecture: It is a month with 30 days and does not begin with J.

We will use the Venn diagram procedure and leave it to the reader to verify the results by determining the actual truth sets for each statement.

1. Draw overlapping sets to represent P and Q.
2. The premise set intersection $P \cap Q$ is the dotted area in the diagram.
3. Because the conjecture is of the form "p and not q" the conjecture set is $P \cap \bar{Q}$, which is the cross-hatched area in the diagram.
4. Because the dotted area is entirely outside the cross-hatched area, the conjecture must be false whenever the premises are true.

Taken collectively, Examples (1) and (2) illustrate the fact that the same premises (e.g., p, q) can be used to determine the truth or falsity of an arbitrary number of conjectures formed from the premises together with one or more of the logical symbols—$\sim$ (not), $\cdot$ (and), $\vee$ (or), $\rightarrow$ (implies). Try to determine the truth or falsity of the following: (1) $\sim p \vee q$, (2) $\sim(\sim p \cdot q)$, (3) $\sim(\sim p \vee \sim q)$.

Example (3): Here, we consider a case where Venn diagrams *must* be used because it is impossible to list all of the logical possibilities.

    Premises: a If it is raining, then it is wet.
               b It is wet or it is dry.
               c It is not dry.
    Conjecture: d It is raining.

We proceed as follows:

1. Let R be the truth set of "It is raining," let W be the truth set of "It is wet," and let D be the truth set of "It is dry." We represent them as

overlapping sets in our Venn diagram (because we do not know where they overlap and where they do not).

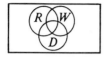

The truth set of premise a is $\bar{R} \cup W$, which is shaded horizontally below. The truth set of premise b is $W \cup D$, which is shaded vertically. The truth set of premise c is $\bar{D}$, which is shaded diagonally.

2. The premise set intersection is the area shaded in all three directions.
3. The conjecture set is set R.
4. Because the premise set intersection is partly inside and partly outside the conjecture set, the truth or falsity of the premises tell us nothing definite about the truth or falsity of the conjecture.

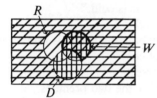

Dark outline indicates set intersection.

This issue, of course, could be settled by adding premises to those we already have. For example, if we are informed that "It is raining or not wet" (which translates into sets as $R \cup \bar{W}$) is true, then we could state with certainty that the conjecture "It is raining" is true. On the other hand, if "It is not raining or it is dry" is found to be true, then the conjecture would necessarily be false.

Finally, we should point out that Venn diagrams also provide an easy way to determine when two or more statements are equivalent. Two assertions are said to be equivalent when they are true for exactly the same set of logical possibilities. Hence, we only need to find the truth sets corresponding to the two statements to be compared. If the truth sets are identical, the statements are equivalent; otherwise, they are not. For example, consider the two statements (1) "a or (b and c)" and (2) "(a or b) and (a or c)" and let A, B, and C be the truth sets corresponding to a, b, and c, respectively. Then the truth set of statement (1) corresponds to the entire shaded area in the left diagram and the truth set of statement (2) to the doubly shaded area in the right diagram. Because both areas are

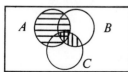
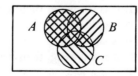

identical, the statements are equivalent.

EXERCISES 3–8

S–1.   On a Venn diagram let U represent the set of all logical possibilities. Let C be the truth set of "It is cloudy," and let W be the truth set of "It is wet." (Be sure that C and W overlap.) Shade the truth set of each of the following statements.

    a. It is cloudy and wet.
    b. It is cloudy or wet.
    c. It is not cloudy.
    d. If it is wet, then it is cloudy.

[S–2.] Work the following problem using the process described above.

    Premises: It is raining or it is cold.
                 It is not cold.
    Conjecture: It is raining.

    Procedure: 1. On a Venn diagram let U represent all logical possibilities.
                    Let R be the truth set of "It is raining."
                    Let C be the truth set of "It is cold."
                    (Be sure sets R and C overlap.)
                    Find the truth sets corresponding to "It is raining or it is cold" and to "It is not cold."
             2. Find the premise set intersection. That is, find the intersection of the truth sets for the two premises.
             3. Find the conjecture set.
             4. What can you say about the conjecture?

[S–3.] Repeat the above process for each of the following.

    a.   Premises: It is raining or it is cold.
                       It is not cold.
      Conjecture: It is not raining.
    b.   Premises: It is raining or it is cold.
                       It is not cold.
      Conjecture: It is raining and windy.
    c.   Premises: If it is snowing, then it is cold.
                       It is snowing.
      Conjecture: It is cold.
    d.   Premises: If it is raining, then it is wet.
                       It is raining.
      Conjecture: It is not wet.
    e.   Premises: If it is snowing, then it is cold.
                       It is cold.
      Conjecture: It is snowing.
    f.   Premises: If it is snowing, then it is wet.
                       It is not wet.
      Conjecture: It is not snowing.
    g.   Premises: If it is raining, then it is windy.
                       It is raining or it is cold.
                       It is not cold.
      Conjecture: It is windy.

[S–4.]   Let p and q be statements. Use Venn diagrams to show that each of the following pairs of statements are equivalent.

a. $\sim(p \text{ or } q)$ and $\sim p$ and $\sim q$
b. $\sim(p \text{ and } q)$ and $\sim p$ or $\sim q$
c. $\sim(p \rightarrow q)$ and p and $\sim q$

9. THE QUANTIFIERS "SOME" AND "ALL"

Unfortunately, the simplicity of the statement logic is not matched by its universality. The statement logic is not sufficient to describe all the various kinds of deductive patterns (inference rules) used by mathematicians. (It is even less adequate for describing the kind of logic used in everyday life.) For example, given the premises

"All integers correspond to rational numbers" and
"5 is an integer,"

then

"5 corresponds to a rational number"

is certainly a valid conclusion. Yet, the validity of this argument cannot be established within the statement logic. Each of the three statements involved is different, and though they are apparently related, it is impossible to represent these relationships within the statement logic.[16] The statement logic is sufficient only for the analysis of statements in terms of their component statements. It provides no machinery for other kinds of analysis.

The applicability of the statement logic can be greatly enlarged by making two simple additions. These additions involve appending the words *all* or *some* to arbitrary statements as a means of adding finer distinctions in meaning.[17] The statement "Men are animals," for example, is indivisible in the statement logic. But, by taking on one of the words "all" or "some," we can construct two clearly related statements with quite distinct meanings: "All men are animals" and "Some men are animals." In addition to the obvious connotative differences, the two *quantified* statements have very different logical implications. The former (quantified) statement means that anything which is a member of the set of men is also a member of the set of animals; the latter (quantified) statement means that there is some (at least one) member of the set of men which is also a member of the set of animals.

Venn diagrams can also be used successfully to analyze the validity of *syllogistic* arguments of the sort given above. To see what is involved, let A

[16] In the statement logic, each statement would be represented by a distinct sign, say p, q, and r. Hence, using only the statement logic we cannot prove that r follows logically from p and q.

[17] In mathematical circles, these words are referred to as the *universal* and *existential* quantifiers, respectively. Quantifiers are basic to a more powerful form of logic, called the Predicate Logic. For a readable introduction, consult R. R. Stoll, *Sets, Logic, and Axiomatic Theories*, San Francisco: Freeman, 1961.

correspond to the set of integers and *B* to the set of rational numbers. The premise "All integers correspond to rational numbers," can be represented by showing set *A* entirely within set *B*. Then, "5 is an integer" can be represented by a logical possibility (marked *x*) within set *A*.

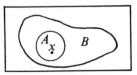

The resulting Venn diagram shows clearly that 5 (marked by *x*) is also in set *B* (i.e., is a rational number).

For additional practice, return to the syllogisms we have used in previous sections and formulate them in terms of the present machinery. For example, in the "Socrates" syllogism, the premise "All men are mortal" may be represented by including the set of all men entirely in the set of all beings that are mortal. The minor premise, "Socrates is a man," may then be represented by an *x* within the set of men. By constructing the Venn diagram described, you will easily see why Socrates (represented by *x*) must also be an element of the set of all beings that are mortal—hence, the conclusion, "Socrates is mortal."

EXERCISES 3–9

S–1.   Give two related statements with distinct meanings for each of the following statements.

    a.   Babies are cute.
    b.   Multiples of 4 are even.
    c.   My debts are large.

S–2.   Use Venn diagrams to test the validity of each of the following deductions:

    a.   All dogs have tails.
        Fido is a dog.
        Therefore, Fido has a tail.
    b.   All prime numbers are natural numbers.
        All natural numbers are whole numbers.
        Therefore, all prime numbers are whole numbers.
    c.   All star scouts can swim.
        John can swim.
        Therefore, John is a star scout. (Be careful.)
    d.   All men get tired.
        Mr. Smith is a man.
        Therefore, Mr. Smith gets tired.

E–3.   Let *A* correspond to the set of even numbers and *B* correspond to the set of "distinguished" numbers. How would we represent each of the following in Venn diagrams?

    a.   All distinguished numbers are even.
    b.   Some distinguished numbers are even.

   c. All even numbers are distinguished.
   d. No even numbers are distinguished.

E–4. The negation of a quantified statement is determined as follows. Replace "all" by "some" (or "some" by "all") and negate the rest of the statement. For example, the negation of "All men are animals" is "Some men are not animals," and the negation of "Some men are beasts" is "All men are not beasts."

    Write the negation of each of the following statements.

   a. All politicians are dishonest.
   b. All men are liars.
   c. Some dogs have three legs.
   d. Some men are 7 feet tall.
   e. All men are mortal.
   f. Some zips are zaps.
   g. Some girls are not nice.

10. MORE ON QUANTIFIERS (OPTIONAL)

Four different kinds of meaning can result from using the terms "some," "all," and "not." Each kind of meaning will be illustrated by simple algebraic statements (e.g., $x + 3 = 7$). The four basic meanings can each be stated in two equivalent forms and we shall see how this can be done. In addition to describing each of these forms we shall introduce, by example, the symbols: $\exists\, x$ (there exists an x), $\in$ (is a member of), N (set of natural numbers), $\forall\, x$ (for all x), $\sim$ (not).

    First, consider the statement, "There exists a natural number denoted by x such that $x + 3 = 7$"—symbolically, $\exists\, x \in N \mid x + 3 = 7$.[18] Exactly the same meaning is implied by the statement, "Not all natural numbers are such that $x + 3 \ne 7$"—symbolically, $\sim \forall\, x \in N \mid x + 3 \ne 7$. Letting set N represent the natural numbers and set S the solution set of $x + 3 = 7$ (i.e., $\{4\}$), Venn diagrams may be used to see that these statements are equivalent.

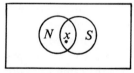

    The first statement guarantees that there is a (some) number which is both a natural number (i.e., in N) and a member of the solution set of $x + 3 = 7$ (i.e., in S). The second says that not all natural numbers in N are outside of the solution set (i.e., in the complement of S), which amounts to precisely the same thing.

    Second, consider, "There is *no* natural number x[19] such that $x + 7 = 3$"— symbolically, $\sim \exists\, x \in N \mid x + 7 = 3$. Any natural number when added to 7 will give something greater than 7 and certainly cannot be equal to 3. Equality could only be guaranteed by selecting from the negative integers, which is not allowed here. An alternative way of stating the same thing is, "For all natural

---

[18] This statement simply means that there is a natural number (i.e., 1, 2, 3, . . .) denoted by x such that the open sentence $x + 3 = 7$ holds (namely, $x = 4$).

[19] This is a less awkward way of saying "It is not true that there is some . . . ."

numbers x, $x + 7 \neq 3$"—symbolically, $\forall~x \in N \mid x + 7 \neq 3$. Letting N correspond to the set of natural numbers and S to the solution set of $x + 7 = 3$, we have two disjoint sets.

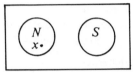

The first statement indicates this directly because it says, in effect, that there is no natural number in the solution set. The second does it indirectly by saying that every natural number lies outside the solution set of $x + 7 = 3$.

Third, consider, "There is a natural number x such that $x + 3 \neq 7$"—symbolically, $\exists\, x \in N \mid x + 3 \neq 7$. The equivalent statement here is, "Not all natural numbers x, are such that $x + 3 = 7$"—symbolically, $\sim \forall~x \in N \mid x + 3 = 7$. The common meaning can be represented by two sets in which the set representing the natural numbers must not be entirely contained in the solution set of the equation $x + 3 = 7$. In particular, there must be an x in N–S. The reader should satisfy himself that this is true.

Finally, consider the statement, "All natural numbers x are such that $x + 3 = 3 + x$"—symbolically $\forall~x \in N \mid x + 3 = 3 + x$. The alternative is, "There is no natural number x such that $x + 3 \neq 3 + x$"—symbolically, $\sim \exists~x \in N \mid x + 3 \neq 3 + x$. Here again the intended meaning may be represented by two sets, this time with the set of natural numbers included in the solution set of the equation. Verify this.

Frequently in mathematics (and sometimes in everyday reasoning), it is important to obtain the denial of a statement involving quantifiers. To obtain the denial of such a statement, we need only to change the quantifier from $\forall x$ to $\exists x$ or from $\exists x$ to $\forall x$ and negate the statement following the quantifier. For example, consider the statement "All natural numbers x are such that $x + 9 = 9$," which may be written $\forall~x \in N \mid x + 9 = 9$. The denial of this statement (i.e., $\sim[\forall~x \in N \mid x + 9 = 9]$) is $\exists\, x \in N \mid x + 9 \neq 9$ or "There exists a natural number x such that $x + 9 \neq 9$." As an example of the denial of an existence statement, consider the assertion "There exists a natural number x such that $x < 1$," which may be written $\exists\, x \in N \mid x < 1$. The denial is $\forall~x \in N \mid x \nless 1$ or "All natural numbers x are such that $x \nless 1$," which is the same as saying "All natural numbers x are greater than or equal to ($\geq$) 1.

The interested reader may show through the use of Venn diagrams, or otherwise, that the denial of any statement of one of these four types is also a statement of one of these types (i.e., that there are only four different kinds of meaning).

EXERCISES 3–10

S–1. Write each of the following statements symbolically.

    a. All natural numbers x are such that $x + x \neq x$.

    b. There is no natural number x such that $x + x = 0$.

    c.   There is a natural number x such that $x + 5 = 9$.

    d.   Not all natural numbers x are such that $3x = 12$.

S–2.   For each of the statements in Problem S–1, write (both in English and symbolically) another statement which is equivalent to it.

S–3.   Demonstrate the equivalence between the statements you have written in Problem S–2 and the corresponding statements in Problem S–1 by drawing Venn diagrams.

S–4.   Give the denial of each of the following statements. State the denial both in English and symbolically.

    a.   All men are mortal.

    b.   All natural numbers x are such that $x + 1 = x$.

    c.   There is a dog that does not bark.

    d.   There is a natural number x such that $x > x$.

4

Algebraic Systems, Theories, and Relationships Between Systems

In Chapter 2 we discussed the basic mathematical ideas of sets, relations, and operations, and the special cases of equivalence relations, order relations, and binary operations. We now turn to some additional notions without which mathematics as we know it would not exist: algebraic systems and embodiments, properties of systems and algebraic theories, and relationships between systems. Many examples of algebraic systems exist in the real world, but the average person is not aware of them. And yet, they are extremely important. We feel that elementary school children should be introduced to the basic ideas as soon as they can grasp them. They are so basic and generally applicable, in fact, that they may help to shape the very way in which the child views the world in which he lives.

1. ALGEBRAIC SYSTEMS AND CONCRETE EMBODIMENTS

An algebraic system consists of one or more basic sets together with one or more operations and/or relations and/or distinguished elements of the basic sets.[1] Rather than worry about what this all means in an abstract sense, let us consider a simple example—the system:

(1) where the basic set is $\{A, B, C\}$, where A, B, and C are "undefined" elements;

---

[1] By capitalizing on certain logical equivalences it is possible to reduce the characterizing elements to one basic set and one or more relations.

(2) where *A* is distinguished in the sense that it serves as an "identity" (for those who are familiar with modern algebra); and

(3) whose defining (binary) operation is

$$\circ = \{(A,\ A) \to A,\ (A,\ B) \to B,\ (B,\ A) \to B,\ (A,\ C) \to C,$$
$$(C,\ A) \to C,\ (B,\ B) \to C,\ (C,\ C) \to B,\ (B,\ C) \to A,\ (C,\ B) \to A\},$$

where the arrows point toward the (third) elements (e.g., *A*) associated with the initial pairs (e.g., (*A*, *A*)). This is a system in which the distinguished element *A* "maps" every element it is paired with into itself (i.e., (*A*, *A*) → *A*, (*A*, *B*) → *B*, (*B*, *A*) → *B*, (*A*, *C*) → *C*, (*C*, *A*) → *C*). When *B* is combined with *B*, the result is *C* and when *C* is combined with *C*, the result is *B*. Finally, *B* combined with *C* in either order results in *A*.

No meaning is specified for any of the elements *A*, *B*, *C*, or for the operation, ∘. They are simply marks on paper and are said to be "undefined terms."

What may be called a concrete system or *embodiment* of an algebraic system results on assignment of meaning to the undefined elements.

Example (1): In the system just cited, the undefined terms might correspond to certain clockwise rotations of a triangle where: *A* corresponds to a rotation of 0°; *B*, to a rotation of 120°; and *C*, to a rotation of 240°. To be more definite, consider a triangle with one marked vertex in which case the rotations may be represented as in Figure 4–1. In this case, the operation, denoted ∘, would

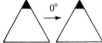 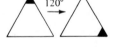 

Figure 4–1

correspond to "followed by." That is, the result of combining two rotations is that single rotation which results in the same action as first doing one rotation and then the other. For example, a rotation of 120° followed by one of 240° results in the same action as a rotation of 0°.

This embodiment, then, involves the set of rotations {0°, 120°, 240°} and the operation "followed by," the basic results of which can be summarized in Table 4–1. To read the table identify the first rotation in a pair with the corresponding row and the second rotation with the corresponding column. Then associate this pair of rotations with that rotation which lies in the same row and column. For example, 120° ∘ 240° → 0° as indicated in the table.

Table 4–1

∘	0°	120°	240°
0°	0°	120°	240°
120°	120°	240°	0°
240°	240°	0°	120°

The remainder of this section is devoted to embodiments, because they are generally of more interest than abstract systems in the elementary school. It must be emphasized, however, that many young children are inclined to view abstract systems as interesting games and find them fascinating. In fact, there is reason to believe that certain abstract systems are easier for children to learn than their corresponding embodiments. After gaining some confidence in working with *concrete* embodiments in the classroom, the teacher may want to get a "feel" for the way children (say, fifth-graders) deal with *abstract systems*. To construct an abstract system from a given embodiment, all the teacher needs to do is to replace the elements of the basic sets and the operators and relations with meaningless *symbols*. In this case, for example, we would simply reverse the above assignments and let A correspond go 0°, B to 120°, C to 240° and ∘ to the operation, "followed by." Any one set of symbols would be as good as any other, but it is generally a good idea to select symbols that are easy to remember and have some intuitive appeal.

In introducing such embodiments to elementary school children it is generally helpful to use concrete materials (like triangles) and have the children manipulate them physically. After they become familiar with the individual rotations, they may be asked to find rotations that are equivalent to pairs of given ones. Ultimately, the children may be asked to do such things as to construct a "multiplication" table and to perform computations involving more than two rotations, e.g., (120° ∘ 0°) ∘240°.

Example (2): "Clock arithmetic" provides another embodiment of the *same* abstract system. In this case, we let the elements A, B, and C of the abstract system correspond to clockwise rotations of 0, 1, and 2 positions, respectively, on the face of a (one-handed) "clock." A clockwise rotation of 2 positions is shown in Figure 4–2A.

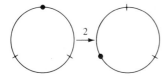

Figure 4–2A

The binary operation, which we call "clock followed by" and denote ⊕, associates to each pair of rotations that single rotation which is equivalent to the first followed by the second. For example, a rotation of two positions followed by another rotation of two positions is equivalent to a rotation of one position as indicated in Figure 4–2B. We may denote this $2 \oplus 2 = 1$.

Figure 4–2B

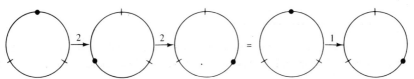

As in Example (1), we can define the binary operation by Table 4–2.

Table 4–2

⊕	0	1	2
0	0	1	2
1	1	2	0
2	2	0	1

Compare this table with that of Example (1). Do you notice any parallels? Would knowing one of these tables help you to learn the other? Why do you suppose this might be so? Think about these questions. The investment of your time now will be returned with interest later on when we discuss these and related matters more fully.

Example (3): We now consider another "clock arithmetic," one that is *not* an embodiment of the original abstract system. In this case, we increase the number of positions on the face of the clock to five. The number of allowable rotations is similarly increased: clockwise rotations of 0, 1, 2, 3, and 4.

The binary operation of "clock followed by," ⊕, is changed only in the number of rotations we need to consider. Test yourself to see if you can fill in each blank. Figure 4–3 should help you to carry out the first computation.

$$2 \oplus 2 = \underline{\qquad},$$
$$2 \oplus 3 = \underline{\qquad}, \text{ and}$$
$$3 \oplus \underline{\qquad} = 2.$$

Figure 4–3

Work the problems by first performing one rotation and then the other or by undoing a rotation as the problem requires. Check your answers against Table 4–3.

Table 4–3

⊕	0	1	2	3	4
0	0	1	2	3	4
1	1	2	3	4	0
2	2	3	4	0	1
3	3	4	0	1	2
4	4	0	1	2	3

Two things should be clear from Examples (1), (2), and (3). First, there are any number of so called "clock arithmetics." In fact, there is a clock arithmetic of the sort described above corresponding to each natural number. The clock arithmetics corresponding to the numbers 1 and 2, of course, would be quite simple, involving as they would, only one and two rotations, respectively. On the other hand, the number of rotations may be increased indefinitely (just as the natural numbers) by considering finer and finer divisions on the face of the clock.

The second point, as illustrated by Examples (1) and (2), is that an algebraic system may have any number of embodiments. Instead of rotations of triangles and three-rotation clock arithmetic, for example, one might consider the binary operation, "followed by," this time defined on the three cyclical permutations of an ordered set, $\langle a, b, c \rangle$:

No-cycle, denoted $\langle a, b, c \rangle \rightarrow \langle a, b, c \rangle$
One-cycle, denoted $\langle a, b, c \rangle \rightarrow \langle c, a, b \rangle$
Two-cycle, denoted $\langle a, b, c \rangle \rightarrow \langle b, c, a \rangle$

This embodiment is explored more fully in the exercises. Similar embodiments may be constructed to parallel the five-rotation clock arithmetic (as well as others which might be constructed) and you should try to do this. *Hint:* Consider symmetry rotations (multiples of 72°) defined on a regular pentagon.

EXERCISES 4–1

S–1 . Complete the following table for the embodiment of clock arithmetic with four positions 0, 1, 2, 3 on the face of the clock. Refer to Examples (2) and (3) of this section.

$\oplus$	0	1	2	3
0	0	1	2	3
1	—	2	3	—
2	—	—	—	1
3	3	—	—	—

[S–2.] Tables are used extensively in this chapter to represent the results of combining operations. The table below defines a binary operation represented by $*$ on the set A, B, C. For example, according to the table, $A * C = B$. (Notice that the first element, A, is selected from the column on the left, and the second element, B, from the row on the top.)

$*$	A	B	C
A	C	A	B
B	B	C	A
C	A	B	C

Use the table to fill in the following:

a. $B * C =$ _____
b. $C * B =$ _____
c. $C *$ _____ $= C$
d. _____ $* B = A$

S–3. Devise an embodiment to parallel the five-position clock arithmetic (Example (3)). That is,

    a. Identify the basic set
    b. Identify the operation
    c. Give a table (which defines the operation)

Hint: Consider rotations (multiples of 72°) defined on a regular pentagon.

[S–4.] Consider the tables below for the embodiments of Examples (1) and (2) of this section, and list all the similarities you can find. *Hint:* Look at corresponding entries, rows, columns, and diagonals.

∘	0°	120°	240°
0°	0°	120°	240°
120°	120°	240°	0°
240°	240°	0°	120°

⊕	0	1	2
0	0	1	2
1	1	2	0
2	2	0	1

[S–5.] Consider the following three rearrangements of an *ordered* set $\langle a, b, c \rangle$.

    N: (no cycle) $\langle a, b, c \rangle \rightarrow \langle a, b, c \rangle$
    $\emptyset$: (one cycle) $\langle a, b, c \rangle \rightarrow \langle c, a, b \rangle$
    T: (two cycle) $\langle a, b, c \rangle \rightarrow \langle b, c, a \rangle$

    a. Describe verbally what the rearrangement $\emptyset$ does. What does $\emptyset$ do to the set $\langle b, c, a \rangle$? to $\langle c, a, b \rangle$?
    b. Let $*$ represent the operation "followed by." Consider $T * \emptyset$, i.e., the rearrangement T "followed by" the rearrangement $\emptyset$. T takes $\langle a, b, c \rangle$ into $\langle b, c, a \rangle$, and $\emptyset$ in turn takes $\langle b, c, a \rangle$ into $\langle a, b, c \rangle$. Thus $T * \emptyset$ takes $\langle a, b, c \rangle$ into $\langle a, b, c \rangle$. Looking at the definitions we see that the rearrangement N does the same thing; so $T * \emptyset = N$. Compute $\emptyset * T$, and $T * T$.
    c. Complete the table for this embodiment.

$*$	N	$\emptyset$	T
N	—	$\emptyset$	—
$\emptyset$	—	—	N
T	—	N	—

M–6. What might you ask a child to do to convince himself that in the clock arithmetic of Example (3), $4 \oplus 3 = 2$ and not 7?

[M–7.] What concrete examples of the system defined in Problem [S–5] can you think of which might be used in the classroom?

For further thought and study: Examine several textbook series for the elementary grades to see how "clock" or "modular" arithmetic is treated. What are some reasons for teaching this subject in the elementary school?

1.1 Some Two-Element Embodiments

Example (4): We now consider an embodiment with only two elements in its basic set. This time we concern ourselves with rotations of a rectangle that leave the position of the rectangle unchanged. In this case, the only rotations that do this (except for the mark which is not part of the rectangle) are rotations of 0° and 180° (see Figure 4–4). Thus, the basic set of the embodiment may be denoted

Figure 4–4

{0°, 180°}. Once again, the binary operation is "followed by," this time applied to the two rotations. This operation is defined by Table 4–4. Notice, in this case, that either element (rotation) followed by itself is equivalent to a rotation of 0° (i.e., $0° \circ 0° = 0°$ and $180° \circ 180° = 0°$).

Table 4–4

∘	0°	180°
0°	0°	180°
180°	180°	0°

Example (5): Another example of a two-element embodiment involves flipping a switch (see Figure 4–5). In this embodiment the basic elements are "flip,"

Figure 4–5

denoted 1, and "no flip," denoted 0. The binary operation is "followed by," this time denoted $*$. The basic results of $*$ are summarized in Table 4–5. Compare this table with that of Example (4). You should now be able to detect similarities between the tables of Examples (1) and (2). If you cannot, ask someone to help you.

Table 4–5

*	0	1
0	0	1
1	1	0

Example (6): In the examples considered so far, the elements in the basic sets of the embodiments have all been operators of one sort or another (e.g., rotations, flips, and so on).[2] This is not necessary, however. Consider, for example, the equivalence class of even integers, E, (i.e., $\{X \mid X$ is an integer and X is divisible by 2$\}$) and the equivalence class of odd integers, O, (i.e., $\{X \mid X$ is an integer and X is *not* divisible by 2$\}$). In addition, consider four rules of arithmetic with which you may be familiar: even $\oplus$ even = even; odd $\oplus$ odd = even; and even $\oplus$ odd = odd; odd $\oplus$ even = odd. The first rule may be interpreted to mean that the sum of any two even numbers is necessarily even and, similarly, for the other rules (e.g., $6(E) + 8(E) = 14(E)$). These two equivalence classes (i.e., E and O) may be thought of as elements in the basic set of an embodiment in which the four rules: $E \oplus E = E$, $O \oplus O = E$, $E \oplus O = O$, $O \oplus E = O$ define a binary operation (see Table 4–6).

Table 4–6

$\oplus$	E	O
E	E	O
O	O	E

Again, compare this table with those of Examples (4) and (5). What can you say about them?

EXERCISES 4–1.1

[S–1.] Make a table to represent the algebraic system consisting of the set $\{-1, 1\}$ and the operation of ordinary multiplication.

[S–2.] Consider the tables below for the embodiments in Examples (4), (5), and (6) of this section. In what ways are they basically the same?

$\circ$	$0°$	$180°$
$0°$	$0°$	$180°$
$180°$	$180°$	$0°$

$*$	0	1
0	0	1
1	1	0

$\oplus$	E	O
E	E	O
O	O	E

E–3. Construct a table for the embodiment of clock arithmetic with six positions on the face of the clock $\{0, 1, 2, 3, 4, 5\}$.

*S–4. Suppose the set P_0, P_1, P_2, P_3, P_4, P_5 is the set of all rearrangements of the ordered set of letters $\{a, b, c\}$, defined below:

$P_0: \langle a, b, c \rangle \rightarrow \langle a, b, c \rangle$
$P_1: \langle a, b, c \rangle \rightarrow \langle c, a, b \rangle$
$P_2: \langle a, b, c \rangle \rightarrow \langle b, c, a \rangle$
$P_3: \langle a, b, c \rangle \rightarrow \langle a, c, b \rangle$
$P_4: \langle a, b, c \rangle \rightarrow \langle b, a, c \rangle$
$P_5: \langle a, b, c \rangle \rightarrow \langle c, b, a \rangle$

[2] There is a good reason for this. It can be proved mathematically that all embodiments (indeed, abstract systems) of the sort considered here are *isomorphic* (see discussion below) to some embodiment involving simple permutations (rearrangements) on the elements in an ordered set. Permutations, of course, are operators.

a. P_3 can be stated verbally: Leave the letter in position one where it is and interchange the letters in the second and third positions. For example, P_3 takes $\langle b, c, a \rangle$ into $\langle b, a, c \rangle$. Describe verbally what P_4 does to any ordered set of three letters. What does P_4 do to $\langle a, c, b \rangle$? To $\langle c, a, b \rangle$?

b. Let the operation $*$ be "followed by." Consider $P_3 * P_4$. P_3 takes the arrangement $\langle a, b, c \rangle$ into $\langle a, c, b \rangle$. Then P_4 takes $\langle a, c, b \rangle$ into $\langle c, a, b \rangle$. Thus $P_3 * P_4$ takes $\langle a, b, c \rangle$ into $\langle c, a, b \rangle$. But P_1 does this, too; so $P_3 * P_4 = P_1$. Compute $P_2 * P_3$, $P_4 * P_1$, and $P_5 * P_2$.

c. Complete the table for this embodiment.

$*$	P_0	P_1	P_2	P_3	P_4	P_5
P_0	P_0	P_1	P_2	—	—	—
P_1	—	P_2	P_0	P_5	P_3	—
P_2	—	—	P_1	—	P_5	P_3
P_3	P_3	P_4	P_5	—	P_1	—
P_4	P_4	—	P_3	—	P_0	P_1
P_5	P_5	P_3	—	P_1	P_2	—

1.2 Some Four-Element Embodiments

Example (7): The next embodiment may be "characterized" in terms of the basic set, $\{0°, 90°, 180°, 270°\}$, and the binary operation defined by Table 4–7. To make things definite, the elements in the basic set may be thought of as rotations of a square. Compare Table 4–7 with that of Example (4), in which we considered two rotations of a rectangle.

Table 4–7

$\circ$	0°	90°	180°	270°
0°	0°	90°	180°	270°
90°	90°	180°	270°	0°
180°	180°	270°	0°	90°
270°	270°	0°	90°	180°

Example (8): Another four-element embodiment can be constructed by adding an additional switch to the display of Example (5) and considering *pairs* of flips. Paralleling our previous discussion, we represent each pair of flips as a pair consisting of the numerals 1 and 0 (see Figure 4–6). (The first member of

Figure 4–6

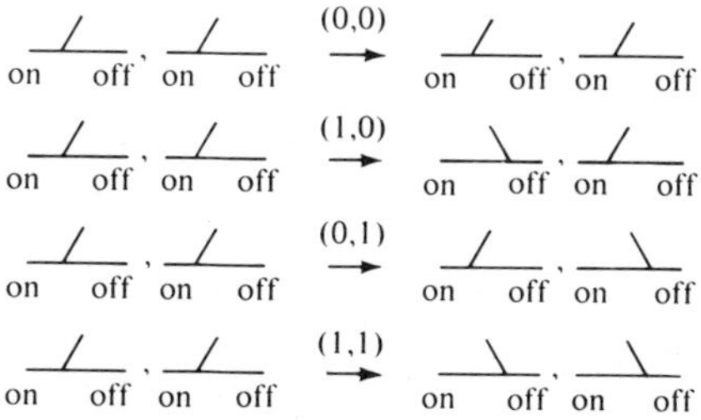

each pair indicates whether or not the first switch is flipped, and the second member indicates whether or not the second switch is flipped.) There are, then, four possible combinations of flips: (0, 0), (1, 0), (0, 1), and (1, 1). These four pairs constitute the elements in the basic set of our embodiment.

Any one pair of flips followed by (∗) any other pair of flips has precisely the same effect as one of the above four pairs. For example, (0, 1) ∗ (1, 0) = (1, 1) (see Figure 4–7).

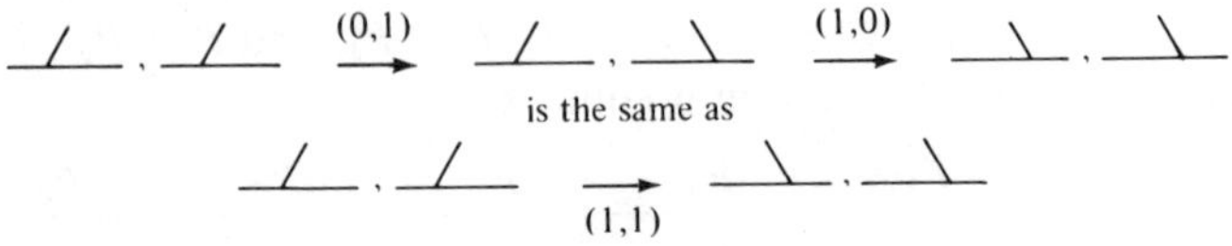

Figure 4–7

Verify the other results summarized in Table 4–8.

Table 4–8

∗	(0, 0)	(1, 0)	(0, 1)	(1, 1)
(0, 0)	(0, 0)	(1, 0)	(0, 1)	(1, 1)
(1, 0)	(1, 0)	(0, 0)	(1, 1)	(0, 1)
(0, 1)	(0, 1)	(1, 1)	(0, 0)	(1, 0)
(1, 1)	(1, 1)	(0, 1)	(1, 0)	(0, 0)

EXERCISES 4–1.2

E–1. Use "clock arithmetic" to construct a *multiplication* table for the set {1, 2, 3, 4} when there are five positions on the clock face, i.e., {0, 1, 2, 3, 4}. Multiplication $\otimes$ will be defined as repeated addition $\oplus$, as indicated by the following examples:

$$4 \otimes 3 = 3 \oplus 3 \oplus 3 \oplus 3 = 2$$
$$2 \otimes 3 = 3 \oplus 3 = 1$$
$$3 \otimes 2 = 2 \oplus 2 \oplus 2 = 1$$
$$3 \otimes 1 = 1 \oplus 1 \oplus 1 = 3$$

Complete the table below using Table 4–3 to facilitate your computation.

$\otimes$	1	2	3	4
1	—	—	—	—
2	—	—	1	—
3	3	1	—	—
4	—	—	2	—

S–2. For the switch flipping embodiment of Example (8), compute:

    a. (0, 1) ∗ (0, 1) =
    b. [(1, 1) ∗ (1, 0)] ∗ (1, 1) =
    c. (1, 1) ∗ [(1, 0) ∗ (1, 1)] =
    d. {[(1, 0) ∗ (0, 1)] ∗ (0, 1)} ∗ (1, 1) =

E–3.   Consider the set of all chains formed by linking together in any order the letters a and b; e.g., $a, b, ab, ba, aa, bb, aab, \ldots, abbab, \ldots, baababaaa, \ldots,$ and so on.

    a.   How many elements are there in this set?
    b.   If "addition" is defined as simply joining two chains together, then is the sum of two chains always in the set?
    c.   Is $abbab$ "$+$" $bab = bab$ "$+$" $abbab$?

[M–4.]   Describe a concrete display or game which would help your pupils learn about the switch flipping embodiments of Examples (5) and (8).

1.3 Some Infinite Embodiments

Example (9): So far all of the embodiments have involved only a finite number of elements. As one might guess, this is not a necessary restriction.

The set of natural numbers, $N = \{1, 2, 3, \cdots\}$ together with the usual operation of addition, constitutes an infinite algebraic system. (For our purposes, we may think of this system as an embodiment.) Can you think of other infinite embodiments? *Hint:* Consider other arithmetic operations. (See also Chapters 5 through 9.)

Example (10): To keep things as simple as possible, we have considered only embodiments involving *single, binary* operations that are *everywhere defined.* None of these restrictions is necessary.

For example, consider the set of natural numbers, $N = \{1, 2, 3, \cdots\}$ together with the operations of addition, subtraction, multiplication, and division. These are all binary operations as before, but there are now four of them defined on the same basic set. The operations of addition and multiplication always give natural numbers as a result. But subtraction and division of natural numbers is not always allowed. That is, $2 - 5$, $3 \div 4$, $4 - 5$, and $7 \div 3$ are *not defined in the system of natural numbers.*

As an example of an operation that is not binary, consider the set of positive rational (fractional) numbers together with the operations of addition, multiplication, and reciprocal. Addition and multiplication are still binary operations, but the reciprocal operation is unary (i.e., one-ary). It maps $\frac{2}{1}$, $\frac{3}{4}$, and $\frac{7}{8}$ into $\frac{1}{2}$, $\frac{4}{3}$, and $\frac{8}{7}$, respectively.

With the help of your instructor, see how many different kinds of embodiments you can come up with.

2.   PROPERTIES OF EMBODIMENTS

In the previous section, we displayed each embodiment directly in terms of the elements and operations involved in it. The embodiments have been defined intentionally, that is, in terms of their internal or intrinsic nature.

Rather than actually display the elements and operations involved, we might instead simply *describe* the embodiment (i.e., describe its *properties*). To see what is involved, consider the following examples of properties.

1. For all pairs of elements in the basic set, there is a unique third element equivalent to the first followed by the second. Symbolically: if a and b are any two elements of the basic set, then $a \oplus b$ is also in the basic set. This is called the *closure* property.

2. For every triple of elements in the basic set, the result of the first followed by the second followed by the third is equivalent to the first followed by the result of the second followed by the third. Symbolically: Given a, b, c; $(a \oplus b) \oplus c = a \oplus (b \oplus c)$. This is called the *associative* property.

3. There exists a unique element, denoted e, such that $e \oplus a = a \oplus e = a$ for every element a (of the embodiment). e is called an *identity* for the operation $\oplus$.

4. For every element a, one can find a unique element a' such that, $a \oplus a' = a' \oplus a = e$ (where e is the identity defined above). a' is called the *inverse* of a relative to the operation $\oplus$.

5. For every pair of elements, call them a and b, $a \oplus b = b \oplus a$. This is called the *commutative* property.

6. There exists at least one element, call it g, such that for all a, there is a
$$\overbrace{}^{n \text{ times}}$$
number n with $a = g \oplus g \oplus \cdots \oplus g$. (In other words, every element can be generated from g.) g is called a *generator*.

7. Sometimes there is a unique generator, often denoted by 1, such that for
$$\overbrace{}^{n \text{ times}}$$
all a, there is a number n with $a = 1 \oplus 1 \oplus \cdots \oplus 1$. (In other words, every element can be generated from 1 and only from 1).

Test some of the embodiments in the previous section to see which of the above properties hold (i.e., determine which properties they *have*). It will turn out that *almost all* of the properties hold in *all* of the embodiments. (As you might suspect, there are systems—embodiments—where this is not true but we defer discussion of such systems until Part III.) Property 6 holds in all but one of the illustrative embodiments. Which one? Property 7 does *not* hold in some of the embodiments in which property 6 holds. Can you find them? Does property 6 hold whenever property 7 does? (This shows that property 7 is more specialized than property 6.)

Mathematical theories are concerned more with *properties of abstract systems* than with systems themselves. The same is true of scientific theories, only here the basic entities being described are complex embodiments in the real world. Loosely speaking, for example, psychological theories are concerned with people and events, the relationships between them, and the operations governing their interaction. Sociological theories substitute groups for individuals. Physical theories are concerned mainly with inanimate elements, like atoms.

To help see why this is so, we first make two observations. First, most abstract systems of interest to mathematicians, and to people generally for that matter, are extremely complex. This is even more true of embodiments (e.g., a rotation of 90° is more "complex" than the symbol "90°"). Try, for example, to actually *display* the set of all numbers. Even more crucial, there are deep relationships both within and between systems which cannot easily (if at all) be displayed directly. It is much simpler to describe them.

The second point is that it is equally impossible to list *all* of the properties of any but the simplest of systems. The list would go on indefinitely.

Fortunately, mathematicians have found a way out of this apparent dilemma. It is usually possible to identify a relatively small finite number of properties from which all of the others may be logically deduced.[3] Such properties are called *axioms* (as indicated in Chapter 3), and together they constitute what is called an *axiom system*. The properties which can be derived from the axioms in a given axiom system are called *theorems*. *Axiomatic theories* include both the axioms and the theorems.

We do not mean to imply that teachers should necessarily attempt to systematically teach axiomatics in the elementary school. The main thing is that teachers become generally aware of what axiomatic theories are, and that they provide their pupils with frequent opportunities to *describe* them.[4] This can usually be accomplished by simply asking them to make "true" statements about the system. Gradually they will come to realize that certain properties are more basic than others in that the latter always hold whenever the former do. There is no harm, of course, in introducing new terms such as *axiom, theorem,* and *axiomatic theory* or in pointing out some of the relationships involved. Young children frequently like this sort of thing. But, a little usually goes a long way in the elementary school and teachers should not attempt to force things. There is time enough for that in high school and college.

EXERCISES 4–2

Note: All examples referred to in the following problems appear in Sections 1, 1.1, and 1.2.

S–1.   Show that the associative property (property 2) holds for the given elements of the following embodiments:

  a.   The rotations 1, 3, and 4 of Example (3).
  b.   The rotations 180°, 90°, and 270° of Example (7).
  *c.   The permutations P_1, P_4 and P_2 of Problem S–4 of Exercises 4–1.1.

---

[3] Actually, this is not quite true in view of Gödel's (1931) famous incompleteness theorem. There will always be some properties which can neither be proved nor disproved in any axiomatization of a system as complex as arithmetic. For a thorough discussion of this and other questions raised in this section, see Chapters 4 and 5 in J. M. Scandura, *Mathematics and Structural Learning*, Englewood Cliffs, N.J., Prentice-Hall, to be published .

[4] A certain amount of this has been done with some success. See R. B. Davis, *Discovery in Mathematics*, Reading, Mass., Addison Wesley, 1964, and P. Suppes, "Mathematical Logic for the School," *Arithmetic Teacher*, **9**, 1962, pp. 396–399.

S–2. Find the unique element e (identity element) mentioned in property 3 for each of the following embodiments:

    a. Example (1)
    b. Example (3)
    c. Example (8)

S–3. For each of the following, find the unique element a' (inverse element) referred to in property 4.

    a. $a = 120°$ (Example (1))
    b. $a = 3$ (Example (3))
    c. $a = (0, 1)$ (Example (8))

S–4. Find a generator (see property 6) for Example (3) and verify that each element in the embodiment can be generated by it. Is the generator you found unique? That is, is there any other element that will serve as a generator?

S–5. In Example (8) can you find a *single* element that will serve as a generator? For example $(1, 1)$ is *not* a generator, since $(1, 1) * (1, 1) = (0, 0)$, and $((1, 1) * (1, 1)) * (1, 1) = (1, 1)$. That is, it is not possible to generate $(1, 0)$ or $(0, 1)$ from $(1, 1)$.

S–6. a. What elements are generated by the natural number 2 with the operation of ordinary addition?
        b. What elements are generated by the natural number 2 with the operation of ordinary multiplication?

S–7. Consider the set a, b, c, d, e, on which an operation $*$ is defined by the following table:

$*$	a	b	c	d	e
a	b	a	d	c	a
b	a	b	c	d	e
c	d	c	b	a	d
d	c	d	a	b	c
e	a	e	d	c	b

Show whether or not this embodiment possesses

    a. closure property (property 1)
    b. associative property (property 2) *Hint:* Consider $(e * c) * d$
    c. identity element (property 3)
    d. commutative property (property 5)

S–8. Consider the embodiments of Examples (1), (3), and (8). Show whether or not these embodiments possess the commutative property (property 5).

S–9. Consider the basic set whose elements are the sets $\{\ \}$, $\{a\}$, $\{b\}$, and $\{a, b\}$. *Note:* $\{\ \}$ denotes the empty set. Let $\cup$ denote the operation of union.

a. Complete the following table:

∪	{ }	{a}	{b}	{a, b}
{ }	—	—	{b}	—
{a}	—	{a}	—	—
{b}	—	—	—	{a, b}
{a, b}	—	—	—	—

b. Show that this embodiment does or does not possess a unique generator (property 7).

[E–10.]   a.   Using the letters a, b, and c as a basic set, construct an embodiment
(i.e., construct a table) which has closure, commutativity, and an identity element
(properties 1, 5, 3).

∗b.   How many such embodiments is it possible to construct?

[E–11.]   Have you detected a regularity in the various tables which enables you
to determine whether or not the commutative property holds? (*Hint:* Draw a
diagonal through the table from upper left to lower right.)

E–12.   Property 1 is called the *closure property*. For example, the set of natural
numbers {1, 2, 3, ...} is closed under multiplication, since the product of any
two natural numbers is a unique natural number. This set is not closed under
division, however, since 2 ÷ 3 is *not* a natural number.

a.   Is the set of natural numbers closed under addition?
b.   Is the set of natural numbers closed under subtraction?
c.   Is the set {0, 1} closed under addition?
d.   Is the set {0, 1} closed under multiplication?

E–13.   Why is it often more difficult to verify that a property holds for a given
finite embodiment than to prove that it does not hold?

M–14.   What activities would you assign children to enable them to detect the
following regularities in a clock arithmetic table?

a.   commutative property (property 5)
b.   associative property (property 2)
c.   existence of an identity element (property 3)

M–15.   The most obvious concrete display of clock arithmetic is a clock. Since
there is no "0" on a normal clock face, which element will serve as the identity
element for addition (property 3)?

2.1 Concrete Examples of Commutative Nonassociative Systems (Optional)[5]

In Section 1 of this chapter we saw several examples of embodiments which satisfied most of the properties in Section 2. None of the examples, however, satisfied

---

[5] Portions of this section are reprinted from an article with the same title by the author
in *Mathematics Teacher*, **59**, December, 1966, by permission of the National Council of Teachers of Mathematics.

the commutative property (property 5) without also satisfying the associative property (property 2). Indeed, there are few examples of *commutative nonassociative* systems (embodiments) that are readily comprehended and meaningful to elementary school children. Some examples of such systems are outlined in this section.

Consider first the set of possible *moves* an elevator may make starting from the basement (zero) floor and ending at some (not necessarily different) floor in the building. Single moves that take the elevator to the top, all the way back to the bottom and, then, up to the initial floor or further are replaced by the shortest possible equivalent move. For example, a move of four floors in a building having two floors above the basement is equivalent to a move of zero floors. The operation can be defined by the following display.

Second Moves

$\otimes$	0	1	2	3
0	0	1	2	3
1	1	2	1	0
2	2	1	0	1
3	3	0	1	2

Moves from Basement

If at least one of the moves is 0 and the other m, $0 \leq m \leq 3$, the product $0 \otimes m = m \otimes 0 = m$. If both are non-zero, then the associated third move is the *shortest* move (from the basement) which takes the elevator to the same floor as that obtained by making the first move followed by a move of length equivalent to the second while moving in a cyclic manner (i.e., up until the top is reached, then down, and so on).[6] The elevator, however, does not necessarily reach this same floor from the same direction.

This simple system has the following properties. First, it is closed. Any two moves are equivalent to some single move. Second, the system has an identity (i.e., a move of zero floors). Third, each move has an inverse (i.e., a move that brings the elevator back to the basement). (Notice that a move of two floors is its own inverse.) Fourth, the system is commutative. The order of making two moves is immaterial. In the illustrative two-story building, for example, moves of two and three floors, in either order, are equivalent to a single move of one floor. This system, however, is *not* associative—a move of $1 \otimes (2 \otimes 3) = 1 \otimes 1 = 2$ is not the same as a move of $(1 \otimes 2) \otimes 3 = 1 \otimes 3 = 0$. In effect, the elevator system has all the properties of what is called an *Abelian* group except associativity.[7]

As a second, but more analytic, example of a commutative nonassociative system, consider the set of non-negative real numbers and the operation $a * b = |a - b|$, where $|x|$ denotes the absolute or numerical value of x. (For example,

---

[6] It is a simple matter to generalize this example to the case where the building has n floors above the basement.

[7] *Note to the Instructor:* By redefining the operation so that "the associated third move is the *shortest* ..." applies to *all* pairs of moves, one destroys the identity property but preserves commutativity and nonassociativity. In this case, $3 \otimes 0 = 0 \otimes 3 = 1$.

$|4| = |^-4| = 4$.) This system is closed (i.e., $|x| \geq 0$ for all x), has an identity (i.e., zero), has inverses (i.e., the inverse of a is a for all $a \geq 0$), and is commutative (i.e., $|a - b| = |b - a|$ for all a, $b \geq 0$). The system, however is *not* associative [e.g., $(1 * 3) * 7 = 2 * 7 = 5$, whereas $1 * (3 * 7) = 1 * 4 = 3$].

Taking absolute values lends commutativity to the nonassociative (and noncommutative) operation of subtraction. Any situation in which consideration is given solely to the magnitude of the difference between two magnitudes can serve as a model for this system or one closely related to it. Try to think of some concrete models.

A final concrete example, which is related to the absolute-value system but not a model for it, is provided by a betting situation in which two opponents start with the same amount of money. On any given play, a positive number is taken to represent a gain for one opponent and a negative number a gain for the other. Consider the set of all possible gains and losses (e.g., the integers between $\pm n$ when whole units are bet on each play). The binary operation is simply "followed by." The system has an identity (i.e., zero), is commutative (i.e., $a + b = b + a$ for all a, b such that $^-n \leq a + b \leq n$), and has inverses (i.e., ^-a is the inverse of a for all a such that $^-n \leq a \leq n$). But the system is *not* associative. In this case, the associative property does not hold because the system is *not closed*. Thus, if $n = 6$, $3 \oplus (4 \oplus {}^-5) = 3 \oplus {}^-1 = 2 \neq (3 \oplus 4) \oplus {}^-5$ which is undefined (it is impossible to win $3 \oplus 4$ units; the maximum gain is 6).[8]

3.   RELATIONSHIPS BETWEEN SYSTEMS AND EMBODIMENTS

As noted previously, many of the embodiments described in Section 1 bear certain relationships to one another. In the following four sections, we shall try to make the nature of these relationships more explicit. Particular attention is given to four relations between systems (or embodiments): generalization, isomorphism, embedding, and homomorphism. These relations are extremely basic in all mathematics, from the most elementary levels. Only by being aware of such relations is it possible to learn mathematics as a unified and coherent body of knowledge. Without such an awareness, mathematics tends to become a morass of discrete bits of information that are hard to remember and even harder to apply correctly. Is it any wonder that many elementary school students learn early to fear and sometimes "hate" mathematics? By becoming familiar with these ideas, the teacher will be in a far better position to avoid such disastrous and long-lived results.

3.1 Generalization

In this section, we consider the relation of *generalization* and its inverse of *restriction*. Rather than give a formal definition (which is more complex than the idea itself), we simply give some examples. The embodiment of Example (3), for

---

[8] This example is actually more complex than indicated above, because after one player loses some money he has less to bet on the next play. Perhaps the reader will want to consider some of the implications of this fact.

example, is a *generalization* of Example (2). (Conversely, the latter is a restriction of the former.) You will recall that both are clock arithmetics but that the first embodiment has only *three* elements (movements), whereas the second has *five*. The elements are of the same general type, however, as are the binary operations (we called both "followed by"). More important, all of the movements (0, 1, 2) of the first embodiment—Example (2)—can be generated by repeated applications of one movement (i.e., 1). Thus,

$$1 = 1,$$
$$1 \oplus 1 = 2, \text{ and}$$
$$1 \oplus 1 \oplus 1 = 0.$$

(Notice in this case that movements of 0, 1, and 2 positions correspond to rotations of 0°, 120°, and 240°, respectively.) The movements (0, 1, 2, 3, 4) of the second embodiment—Example (3)—may be similarly generated, again by a movement of one position. This time, however, the one movement corresponds to a rotation of $360°/5 = 72°$. Thus: $1 = 1, 1 \oplus 1 = 2, 1 \oplus 1 \oplus 1 = 3, 1 \oplus 1 \oplus 1 = 4, 1 \oplus 1 \oplus 1 \oplus 1 \oplus 1 = 0.$

The other clock arithmetics, referred to in discussing Example (3), are also related by generalization to the embodiments of Examples (2) and (3). In each case, the movement of zero positions (0) can be generated by a movement of one (1) in the respective clock arithmetic by applying the binary operation $n - 1$ times where n is the number of movements. That is, $0 = 1 \oplus 1 \oplus \ldots \oplus 1$, where the operation $\oplus$ is repeated $n - 1$ times (but where there are n 1's). In a clock arithmetic with six movements, for example, $0 = 1 \oplus 1 \oplus 1 \oplus 1 \oplus 1 \oplus 1$. This type of similarity is characteristic of many situations where one embodiment is a generalization of another.

The switch-flipping embodiments of Examples (5) and (8) illustrate another kind of generalization. In this case, the four-element embodiment of Example (8) was obtained from the embodiment of Example (5) by adding a second switch. Here, however, the elements of the two embodiments are not of the same type. In one case, they are simple flips and, in the other, they are *pairs* of flips. More to the point, the elements in the second embodiment (Example (8)) may *not* be generated by a single element (pair of flips) whereas the first can. Thus, flip (1) generates both flip (1) itself and no flip (0) (e.g., flip (1) followed by flip (1) is equivalent to no flip (0), denoted, $1 * 1 = 0$). To generate the elements of the other embodiment (Example (8)), two elements (pairs of flips) are required. The pairs denoted (1, 0) and (0, 1) are adequate for this purpose. Thus, $(1, 1) = (1, 0) \oplus (0, 1)$ and $(0, 0) = (1, 0) \oplus (1, 0) = (0, 1) \oplus (0, 1)$. (Of course, (1, 0) and (0, 1) generate themselves.)

As you may have guessed, it is possible to generalize flipping embodiments of this sort still further by simply adding more switches. A set of three switches, for example, leads to an eight-element embodiment.

Represent each element by a triple (e.g., (1, 1, 0)) and convince yourself that indeed there are exactly eight different elements. Can you construct the "addition" table for this embodiment? How many elements are needed to generate all of the elements in the embodiment? Can you identify a set of generators?

Finally, we simply point out that an embodiment may be generalized in both of these ways simultaneously. You may wish to consider how one might go about generalizing the embodiment of Example (2) to an embodiment consisting of 25 *pairs* of movements, together with the binary operation, "followed by." *Hint:* Consider *pairs* of movements of the embodiment of Example (3). How many generators are required to generate all 25 pairs? Identify a set of generators that will work. (This set is *not* unique.) Can you identify two generating pairs such that there is only one non-zero element in each pair? How many replications of each of these generators is required before the no-move pair is obtained?

EXERCISES 4–3.1

S–1. Identify several examples of generalizations—restrictions relating embodiments in Examples (1)–(8) in sections 1, 1.1, and 1.2.

S–2. Give a generalization of each of the following:

a. clock arithmetic with six elements
b. rotations of a square through 0°, 90°, 180°, 270°
c. flips of a single switch
d. flips of a pair of switches

S–3. If we generalize the switch-flipping embodiments of Examples (5) and (8) (Section 1) to include three switches, list the eight elements (i.e., triples) of the resulting embodiment $((0, 0, 0), (0, 0, 1), (0, 1, 0),$ etc.). Construct a table for the embodiment using the operation "followed by." (Use Table 4–8 as a model.)

∗S–4.   a.   Generalize the embodiment of Example (1), to an embodiment consisting of nine *pairs* of movements, together with the binary operation "followed by." *Hint:* Consider *pairs* of movements of the embodiment of Example (1), for example, (120°, 240°) or (0°, 120°).
b. How many generators are required to generate all nine pairs? Identify a set of generators. (This set is not unique.)
c. Identify two generators (two pairs) such that there is only one non-zero element in each pair.
d. How many replications of each of these generators is required before the identity, (0°, 0°), is obtained?

∗S–5.   Clock arithmetic can be generalized to include any number, n, of positions on the clock face. In this case, assuming the smallest number on the clock face is 0, the largest number would be $n - 1$. For this n-position clock arithmetic,

a. What is $(n - 1) \oplus 1$?
b. What is $(n - 1) \oplus 2$?
c. What is $(n - 1) \oplus (n - 1)$?
d. Without actually constructing a table for this embodiment, can you tell what the entries of the fifth row would be? (Assume $n \geq 5$.)

[M–6.] Suppose you have your class construct (using cardboard, etc.) physical representations of some of the embodiments discussed in this chapter. What are some activities the class could do, or what are some questions you could ask to lead the students to

    a. detect regularities *within* embodiments?
    b. detect regularities *between* two or more different embodiments (i.e., similarities)?

3.2 Isomorphisms

In Section 1, we pointed out that Examples (1) and (2) are embodiments of the same abstract system. We also suggested that there is a close relationship between these embodiments. In this section, we make the nature of that relationship clear and show how other pairs of embodiments may be related in the same general way.

    We first note that the two embodiments have the same number of elements; the elements may be paired in one-to-one fashion. Consider the following correspondence: $0° \leftrightarrow 0$, $120° \leftrightarrow 1$, $240° \leftrightarrow 2$. (See Examples (1) and (2) for the meaning of these symbols; the "$\leftrightarrow$," of course, means "corresponds to.") This *particular* one-to-one correspondence has a most interesting property. As we shall say, the correspondence "preserves the structure of the two embodiments." This means that the two operations (one in each embodiment) have corresponding "effects" in the two embodiments. More precisely, given any pair of rotations in the first embodiment (e.g., 120°, 240°), the result (0°) of following the first rotation (120°) by the second (240°), corresponds, in the second embodiment, to the result (0) of following the movement (1), corresponding to the first (120°), by the movement (2), corresponding to the second (240°). This can be seen more simply in schematic form.[9]

$$120° \circ 240° = 0°$$
$$\updownarrow \qquad \updownarrow \qquad \updownarrow$$
$$1 \ \oplus \ 2 \ = 0$$

We can check the other pairs by comparing corresponding entries in the "addition" tables of the two embodiments (see Table 4–9). (The elements considered above are encircled for ready comparison.)

Table 4–9

$\circ$	0°	120°	240°		$\oplus$	0	1	2
0°	0°	120°	240°		0	0	1	2
120°	120°	240°	0°	$\leftrightarrow$	1	1	2	0
240°	240°	0°	120°		2	2	0	1

    Check at least one or two additional pairs of entries. Try the pair (240°, 240°). Find the corresponding pair. Does the so-called "sum" (of 240° and 240°)

---

[9] Alternatively, by letting f denote the correspondence, we can represent this structure preserving relationship symbolically as, $f(120° \circ 240°) = f(120°) \oplus f(240°)$.

correspond to the "sum" of the elements in the corresponding pair? What about the pair (120°, 0°)?

When two embodiments are related in this way, we say that they are *isomorphic* and that the one-to-one correspondence between the elements of the embodiments is an *isomorphism*. The notion of isomorphism provides a precise way of talking about embodiments of the same system, because any two isomorphic embodiments are, by definition, embodiments of the same abstract system.

Another example of an isomorphism is provided by the embodiments of Examples (4) and (5). In this case, we introduce the particular correspondence: $0° \leftrightarrow 0$ (no flip), $180° \leftrightarrow 1$ (flip). Examination of Table 4–10 indicates that this correspondence preserves the structures imposed by the operations, $\circ$ and $*$. For example, we note that

$$180° \circ 0° = 180°$$
$$\updownarrow \quad \updownarrow \;\; = \;\; \updownarrow$$
$$1 \quad * \; 0 \;\; = \;\; 1.$$

Table 4–10

$\circ$	$\boxed{0°}$	$180°$
$0°$	$0°$	$180°$
$\boxed{180°}$	$\boxed{180°}$	$0°$

$\leftrightarrow$

$*$	$\boxed{0}$	1
0	0	1
$\boxed{1}$	$\boxed{1}$	0

Other pairs of isomorphic embodiments are given in the examples of the first section. Can you find them? There are four different isomorphisms altogether. (*Note:* One embodiment may enter into any number of isomorphisms.) Are the four element embodiments of Examples (7) and (8) isomorphic? Why or why not?

EXERCISES 4–3.2

S–1. Complete the following schematic forms for the isomorphism

$$0° \rightarrow 0$$
$$120° \rightarrow 1$$
$$240° \rightarrow 2$$

between the embodiments of Examples (1) and (2).

a. $240° \circ 240° = $ ___
   $\quad \updownarrow \qquad \updownarrow \qquad \updownarrow$
   ___ $\oplus$ ___ $=$ ___

b. $120° \circ 0° = $ ___
   $\quad \updownarrow \qquad \updownarrow \qquad \updownarrow$
   ___ $\oplus$ ___ $=$ ___

c. $240° \circ $ ___ $= $ ___
   $\quad \updownarrow \qquad \updownarrow \qquad \updownarrow$
   ___ $\oplus$ 1 $=$ ___

S-2.   Consider the following correspondence between the elements in the embodiments of Examples (7) and (8).

$$0° \rightarrow (0, 0)$$
$$90° \rightarrow (1, 0)$$
$$180° \rightarrow (0, 1)$$
$$270° \rightarrow (1, 1)$$

Complete the following schematic forms:

a. $\quad 0° \circ \quad 90° = \underline{\quad}$
$\qquad \updownarrow \qquad \updownarrow$
$\qquad \underline{\quad} * \underline{\quad} = \underline{\quad}$

b. $\quad \underline{\quad} \circ \underline{\quad} = \underline{\quad}$
$\qquad \updownarrow \qquad \updownarrow$
$\quad (0, 1) * (0, 1) = \underline{\quad}$

c. $\quad 90° \circ \underline{\quad} = \underline{\quad}$
$\qquad \updownarrow \qquad \updownarrow$
$\qquad \underline{\quad} * (0, 1) = \underline{\quad}$

d. $\quad 270° \circ \underline{\quad} = \underline{\quad}$
$\qquad \updownarrow \qquad \updownarrow$
$\qquad \underline{\quad} * (1, 1) = \underline{\quad}$

e.   Do you think the given correspondence is an isomorphism? Explain.

S-3.   The embodiments of Examples (1) and (4) are not isomorphic. Explain why.

S-4.   The first table below refers to the clock arithmetic table of Example (2) and the second table refers to the table of Problem [S-5] in Exercises 4–1.

$\oplus$	0	1	2
0	0	1	2
1	1	2	0
2	2	0	1

$*$	T	$\emptyset$	N
T	$\emptyset$	N	T
$\emptyset$	N	T	$\emptyset$
N	T	$\emptyset$	N

a.   Is the correspondence $0 \leftrightarrow T$, $1 \leftrightarrow \emptyset$, $2 \leftrightarrow N$ an isomorphism? Explain why or why not.

b.   Is the correspondence $0 \leftrightarrow N$, $1 \leftrightarrow \emptyset$, $2 \leftrightarrow T$ an isomorphism? Explain why or why not.

S-5.   The first table below refers to the familiar clock arithmetic with four rotations; the second table refers to an abstract system with an operation $*$ defined by the table. These two embodiments are in fact isomorphic. Find a one-to-one correspondence between these embodiments and show that it preserves the structure. (*Hint:* The correspondence $0 \leftrightarrow a$, $1 \leftrightarrow b$, $2 \leftrightarrow c$, $3 \leftrightarrow d$ does *not* preserve the structure.)

$\oplus$	0	1	2	3
0	0	1	2	3
1	1	2	3	0
2	2	3	0	1
3	3	0	1	2

$*$	a	b	c	d
a	a	b	c	d
b	b	a	d	c
c	c	d	b	a
d	d	c	a	b

*E–6. We have represented the correspondence between isomorphic embodiments by using double-headed arrows between corresponding elements, e.g., $0° \leftrightarrow 0$, $120° \leftrightarrow 1$, $240° \leftrightarrow 2$, for the embodiments of Examples (1) and (2). We could also express this regularity as a function: $f(0°) = 0, f(120°) = 1, f(240°) = 2$. To check that the function preserves the structure we must show that $f(a \circ b) = f(a) \oplus f(b)$ for all a and b in the first embodiment.

    a.   Check that $f(0° \circ 240°) = f(0°) \oplus f(240°)$, $f(120° \circ 120°) = f(120°) \oplus f(120°)$, and $f(120° \circ 240°) = f(120°) \oplus f(240°)$.

    b.   What other methods, besides arrows and function notation, might you have your children use to represent a one-to-one correspondence?

*[E–7.]   We can set up a one-to-one correspondence between the natural numbers $\{1, 2, 3, 4, \ldots\}$ and the positive fractions with denominator equal to 1 $\{\frac{1}{1}, \frac{2}{1}, \frac{3}{1}, \frac{4}{1}, \ldots\}$ as follows: $1 \leftrightarrow \frac{1}{1}, 2 \leftrightarrow \frac{2}{1}, 3 \leftrightarrow \frac{3}{1}, \ldots$. We cannot check all possible pairs of natural numbers to demonstrate that this correspondence preserves multiplication (because there are infinitely many pairs). We must use deductive reasoning. We proceed as follows:

$$m \leftrightarrow \frac{m}{1}$$

$$n \leftrightarrow \frac{n}{1}$$

$$m \cdot n \leftrightarrow \frac{m \cdot n}{1} = \frac{m}{1} \cdot \frac{n}{1}$$

That is, $m \cdot n$ corresponds to the "correct" fraction, namely, the one obtained by multiplying $m/1 \cdot n/1$.

    Follow the pattern to show (deduce) the fact that this correspondence also preserves the "addition structure."

M–8.   How would you lead your students to discover that the embodiments of Examples (1) and (2) are isomorphic? What visual aids might you use?

[M–9.]   How would you have your students demonstrate that they understand the idea of an isomorphism?

3.3 Embeddings

One embodiment is often isomorphic with a part (subembodiment) of a second embodiment. In this case, we say that the first embodiment is isomorphically *embedded* in the second.

    For example, the embodiment of Example (5) can be embedded in the embodiment of Example (7). That is, the embodiment of Example (5), with basic set $\{0, 1\}$ and operation $*$ is isomorphic with a subembodiment of Example (7), which has basic set $\{0°, 90°, 180°, 270°\}$ and operation $\circ$. The operation on the subembodiment is still $\circ$ (with restricted range of application), but the basic set of the subembodiment contains only the elements $0°$ and $180°$. The one-to-one correspondence is given by: $0 \leftrightarrow 0°$, $1 \leftrightarrow 180°$. Since there are only four entries in the "addition" table ($*$) of Example (5), we verify directly that this one-to-one correspondence is an isomorphism.

1. $0 * 0 = 0$
    $\updownarrow \quad \updownarrow \quad \updownarrow$
    $0° \circ 0° = 0°$

2. $1 * 1 = 0$
    $\updownarrow \quad \updownarrow \quad \updownarrow$
    $180° \circ 180° \quad 0°$

3. $0 * 1 = 1$
    $\updownarrow \quad \updownarrow \quad \updownarrow$
    $0° \circ 180° = 180°$

4. $1 * 0 = 1$
    $\updownarrow \quad \updownarrow \quad \updownarrow$
    $180° \circ 0° = 180°$

Table 4–11 clarifies the idea that the embodiment of Example (5) and a subembodiment of Example (7) are isomorphic. (Corresponding elements of the embodiment are encircled for easy reference.)

Table 4–11

*	0	1		∘	(0°)	90°	(180°)	270°
0	0	1		(0°)	(0°)	90°	(180°)	270°
			↔	90°	90°	180°	270°	0°
1	1	0		(180°)	(180°)	270°	(0°)	90°
				270°	270°	0°	90°	180°

In some sense, of course, it seems absurd to think of flips of a switch as being embedded in rotations of a square. The two embodiments have completely distinct meanings. Nevertheless, the embodiment of Example (5) and the sub-embodiment of Example (7) are equivalent, as we say, "up to isomorphism." They both have precisely the same structure and, hence, are embodiments of the same abstract system.

When we use the term "embedding" we normally think of an isomorphism between one embodiment and a subembodiment of another which involves the same general type of element and operation. For example, the embodiment of Example (5) (flips of one switch) does not differ in any essential way from that subembodiment of Example (8) (flips of two switches) where the second switch is always in the 0 position. We define the isomorphism as follows: $0 \leftrightarrow (0, 0)$, $1 \leftrightarrow (1, 0)$.[10] The fact that the one-to-one correspondence preserves the operations can be seen by checking Table 4–12 of the embodiments of Examples (5) and (8).

[10] The fact that the second elements of the ordered pairs (0, 0) and (1, 0) are both 0 indicates that the second switch is kept in a constant position. Hence, in comparing the subembodiment of Example (8) with embodiment (5), we need only consider what happens to the first switch. It is easy to see that they are isomorphic to one another because, in effect, they both involve a single switch.

Table 4–12

*	0	1
0	0	1
1	1	0

*	(0, 0)	(1, 0)	(0, 1)	(1, 1)
(0, 0)	(0, 0)	(1, 0)	(0, 1)	(1, 1)
(1, 0)	(1, 0)	(0, 0)	(1, 1)	(0, 1)
(0, 1)	(0, 1)	(1, 1)	(0, 0)	(1, 0)
(1, 1)	(1, 1)	(0, 1)	(1, 0)	(0, 0)

We can similarly "embed" the embodiment of Example 4 in the embodiment of Example 7 by introducing the correspondence $0° \leftrightarrow 0°$, $180° \leftrightarrow 180°$. Table 4–13 makes the embedding intuitively obvious. Check your intuition by

Table 4–13

°	0°	180°
0°	0°	180°
180°	180°	0°

°	0°	90°	180°	270°
0°	0°	90°	180°	280°
90°	90°	180°	270°	0°
180°	180°	270°	0°	90°
270°	270°	0°	90°	180°

showing more formally that the correspondence is indeed an embedding. That is, show that it is a one-to-one correspondence that preserves the operation.

In Chapters 7 and 8, we will see that the system of *natural numbers* is embedded in the system of *positive rationals* (fractions) and also in the system of *integers*. Although the integers and positive rationals are defined in terms of ordered pairs of natural numbers, each system contains a subsystem which is isomorphic to the system of natural numbers. The system of integers and the system of positive rationals, in turn, are embedded in the system of *rationals*. Similarly, we shall see that the system of rationals is embedded in the system of *real numbers*, and the system of reals is embedded in the system of *complex numbers*.

EXERCISES 4–3.3

S–1. Determine whether each of the following correspondences defines an embedding of the first embodiment in the second embodiment.

a. Examples (1) and (3) with correspondence
  $0° \leftrightarrow 0$
  $120° \leftrightarrow 2$
  $240° \leftrightarrow 4$

b. Example (2) and Problem *S–4 of Exercises 4–1.1 with correspondence
  $0 \leftrightarrow P_0$
  $1 \leftrightarrow P_1$
  $2 \leftrightarrow P_2$

c. Example (4) and Problem S–5 of Exercises 4–1 with correspondence
$$0° \leftrightarrow N$$
$$180° \leftrightarrow \emptyset$$

d. Example (4) and Problem *S–4 of Exercises 4–1.1 with correspondence
$$0° \leftrightarrow P_0$$
$$180° \leftrightarrow P_3$$

e. Example (4) and Problem *S–4 of Exercises 4–1.1 with correspondence
$$0° \leftrightarrow P_0$$
$$180° \rightarrow P_1$$

S–2. Find an embedding of clock arithmetic with three elements in clock arithmetic with nine elements. Give the correspondence.

S–3. Find an embedding of the embodiment of Problem S–5 of Exercises 4–1 in the embodiment of Problem *S–4 of Exercises 4–1.1.

For further thought and study: What embeddings can you think of besides those mentioned in the text?

3.4 Homomorphisms (Optional)

Clearly, isomorphic embodiments bear a very close relationship to one another. They not only have the same internal structure, but the elements correspond in one-to-one fashion. To be sure, isomorphism is but a special case of a much broader class of relationships between embodiments called homomorphisms.[11]
  Let $A = \langle \{a, a', \ldots \}, \circ_A \rangle$ and $B = \langle \{b, b', \ldots \}, \circ_B \rangle$, be embodiments such that $\{a, a', \ldots \}$ and $\{b, b', \ldots \}$ are the respective basic sets and $\circ_A$ and $\circ_B$ are corresponding binary operations. A *homomorphism* from A to B is a function (denoted "$\xrightarrow{h}$") from the elements of set A into the elements of set B which "preserves the operations." More specifically, for each pair of elements a and a' in the basic set of A, if $a \xrightarrow{h} b$ and $a' \xrightarrow{h} b'$ where b and b' are elements in the basic set of B, then $(a \circ_A a') \xrightarrow{h} (b \circ_B b')$. (*Note:* $(a \circ_A a') = a''$ is a distinct element of set A, and $(b \circ_B b') = b''$ is a distinct element of set B.)
  Consider the embodiments of Examples (7) and (5). Define the following function (operator) which maps the elements of the former embodiment—Example (7)—into the basic set of the latter—Example (5).

EXAMPLE 7   EXAMPLE 5

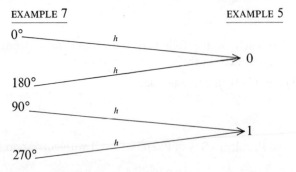

This function may be represented symbolically as:

$$0°, 180° \xrightarrow{\ h\ } 0; \ 90°, 270° \xrightarrow{\ h\ } 1.$$

This function turns out to be a homomorphism. The structures imposed on the two embodiments by the respective binary operations ($\circ$ and $*$) correspond in the required way. For example, consider the two rotations 90° and 180°. In this case, we see that

$$
\begin{array}{ccccc}
90° & \circ & 180° & = & 270° \\
\downarrow h & & \downarrow h & & \downarrow h \\
1 & * & 0 & = & 1,
\end{array}
$$

as required by the definition (of homomorphism).

Checking back, there are 16 entries in the "addition" table of Example (7). In order to be absolutely sure that our function is, indeed, a homomorphism, we still need to do a lot of checking. In fact, there are 15 additional pairs to check (i.e., 16 minus the one above). We show that the required correspondence holds in three of these cases and leave the remainder for the reader.

$$
\begin{array}{lccccc}
1. & 90° & \circ & 90° & = & 180° \\
 & \downarrow h & & \downarrow h & & \downarrow h \\
 & 1 & * & 1 & = & 0
\end{array}
$$

$$
\begin{array}{lccccc}
2. & 0° & \circ & 180° & = & 180° \\
 & \downarrow h & & \downarrow h & & \downarrow h \\
 & 0 & * & 0 & = & 0
\end{array}
$$

$$
\begin{array}{lccccc}
3. & 270° & \circ & 270° & = & 180° \\
 & \downarrow h & & \downarrow h & & \downarrow h \\
 & 1 & * & 1 & = & 0
\end{array}
$$

If you are tempted not to bother checking further because you think it unnecessary, then consider the function: $0°, 90° \to 0; 180°, 270° \to 1$. This function is *not* a homomorphism because

$$
\begin{array}{ccccc}
270° & \circ & 270° & = & 180° \\
\downarrow & & \downarrow & & \downarrow \\
1 & * & 1 & = & 0,
\end{array}
$$

where "$\to$" means "does *not* correspond."

The above homomorphism is an example of what is called a homomorphism *onto*. That is, it is a homomorphism of embodiment 7 *onto* embodiment 5. To be a homomorphism onto, the basic requirement is simply that *every* element in the second embodiment is, as we say, the *image* of some element in the first. *No* elements in the second embodiment are left out, so to speak. Thus, in the above illustration, the second embodiment (the image) has only two elements, 1 and 0, and each of these is the image of at least one element in the first embodiment. In fact, each is the image of exactly two.

A second type of homomorphism is called a *homomorphism into*. The basic requirement in this case is that some element(s) in the second embodiment is (are) *not* the image of an element in the first. For example, consider the function from the embodiment of Example (5) to the embodiment of Example (7) given by $0 \to 0°, 1 \to 180°$. It can be seen that this function defines a homomorphism.

It is a homomorphism *into* (or equivalently, *not* a homomorphism *onto*) because there are elements in the embodiment of Example (7) (90° and 270°) which are not the image of any element in the embodiment of Example (5).

In Sections 3.2 and 3.3 we discussed two special kinds of homomorphisms. An isomorphism is a homomorphism that is both *one-to-one* and *onto*. An embedding is a *one-to-one* homomorphism *into* (i e., *not onto*).

EXERCISES 4–3.4

S–1. Suppose we map the natural numbers onto the even natural numbers as follows:

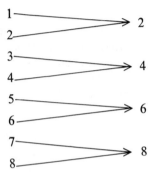

Show that this is not a homomorphism under the operation of addition. (*Hint:* Find two natural numbers where the sums do *not* correspond.)

S–2. a. Map the multiples of 3 into the natural numbers as follows:

$$3 \to 1$$
$$6 \to 2$$
$$9 \to 3$$
$$12 \to 4$$

Convince yourself, by a few computations, that this is a homomorphism under the operation of addition.

      b. Is this a homomorphism onto? Is this an isomorphism (under the operation of addition)?

      c. Show that this correspondence is *not* a homomorphism under the operation of multiplication. (*Hint:* Find two multiples of 3 whose product corresponds to the wrong number.)

[S–3.] Describe the difference between a homomorphism and an isomorphism? If a mapping is a homomorphism, must it also be an isomorphism? If a mapping is an isomorphism, must it also be a homomorphism? Explain.

∗E–4. If two embodiments each have *two* operations defined on them, then a homomorphism must preserve *both* operations. That is, if embodiment A has operations $\oplus$ and $\odot$ while embodiment B has operations $\boxplus$ and $\diamondsuit$, then a function h from A to B must have the property: If

$$a \xrightarrow{\;h\;} b \text{ and } a' \xrightarrow{\;h\;} b'$$

then $a \oplus a' \longrightarrow b \boxplus b'$ *and* $a \odot a' \longrightarrow b \diamondsuit b'$.

Map a four-element clock arithmetic into a two-element clock arithmetic as follows:

$$0 \rightarrow 0$$
$$1 \rightarrow 1$$
$$2 \rightarrow 0$$
$$3 \rightarrow 1$$

Check several pairs of elements of the four-element clock arithmetic to illustrate that this function preserves *both* addition and multiplication.

∗E–5. To show that a function defined on an infinite embodiment is a homomorphism, one must use deduction since it is impossible to check all of the pairs of elements. The usual way is to let a and a' represent arbitrary elements of the first embodiment, and *prove* that if

$$a \xrightarrow{h} b \text{ and } a' \xrightarrow{h} b', \text{ then } a \circ_A a' \longrightarrow b \circ_B b'$$

where $\circ_A$ and $\circ_B$ are the operations on the first and second embodiments, respectively.

Prove (deduce) that the function in Problem S–2 is a homomorphism. (*Hint:* Let $a = 3 \cdot m$ and $a' = 3 \cdot n$ be arbitrary elements of the first embodiment. Find the corresponding elements for a, a' and $a + a'$.)

3.5 Multiple Relationships

As a final note, we point out that some pairs of embodiments may be related in more than one way. For example, we have shown that the embodiments of Examples (5) and (8) are related both by generalization (see Section 3.1) and by embeddedness (see Section 3.3). The embodiment of Example (2) (three-element clock arithmetic) is also related in both of these ways to a six-element clock arithmetic. In Section 3.1, we noted that the six-element clock arithmetic is a generalization of the three-element clock arithmetic. To see that the latter is embedded in the former, we define a one-to-one correspondence between the three-element embodiment of Example (2) and a subembodiment of the six-element clock arithmetic: $0 \leftrightarrow 0$, $1 \leftrightarrow 2$, $2 \leftrightarrow 4$. (See Table 4–14.) Verify for yourself that this correspondence is an embedding.

Table 4–14

$\oplus$	0	1	2
0	0	1	2
1	1	2	0
2	2	0	1

$\xleftrightarrow{h}$

$\oplus$	(0)	1	(2)	3	(4)	5
(0)	(0)	1	(2)	3	(4)	5
1	1	2	3	4	5	0
(2)	(2)	3	(4)	5	(0)	1
3	3	4	5	0	1	2
(4)	(4)	5	(0)	1	(2)	3
5	5	0	1	2	3	4

Look back at the examples of Section 1. What other pairs of related embodiments can you identify? Even if you cannot find any, the search will be instructive.

Three
Number
Systems

The System of Natural Numbers

The system of natural numbers is concerned with (1) the familiar numbers 1, 2, 3, 4, ..., (2) the basic arithmetical operations of addition, subtraction, multiplication, and division, and (3) the properties these operations have and the relationships between them. This is the subject of this chapter.

THE CONCEPT OF NATURAL NUMBER

Number is commonly thought to be the first mathematical idea a child is confronted with. As fundamental as the idea of number is, however, it is based on the even more fundamental ideas of sets, correspondences, and order. One-to-one correspondences between (the elements of) *unordered* sets and one-to-one correspondences between *ordered* sets are most directly involved. The former kind of correspondence (between unordered sets) is used to define *cardinality* (i.e., the numerosity of a set) and the latter (between ordered sets) to define *ordinality* (i.e., the order of an element in an ordered set).

1. CARDINALITY

To say that a child knows what the cardinality of a set is, means simply that the child can sort arbitrary sets into two piles, those having the same cardinality (i.e., the same number of elements) and those not having the same cardinality. Sorting sets in this manner is entirely analogous to sorting elements according to some particular property. Thus, just as the color "red" provides a basis for sorting *objects*, the number "3" provides a basis for sorting *sets*. Those sets having three elements go into one pile and those sets having more or fewer than three elements go into another pile.

A class of sets may be correctly sorted by pairing the elements of each set, in turn, with the elements of some predetermined set having the desired cardinality. If the elements in a set can be paired with the elements in the predetermined test set in a 1–1[1] fashion, then the set has the same cardinal number. If the elements cannot be matched in this manner, then the set does not have the same cardinal number. The predetermined set, singled out for testing, obviously plays a special role and is often called a *canonical*[2] set.

Many children and adults frequently have a canonical set in mind when they think of certain numbers. For example, the young child may think of the number 2 in terms of the set consisting of his two hands. All other sets are compared with this basic set to determine whether or not they have two elements. The child might say that a set has two elements only if he can hold them in his hands, one element per hand. With very young children, the number one might similarly refer to a correspondence between the elements in arbitrary sets and the child himself (e.g., "there is only 'one' for me to play with"). Certain other cardinal numbers may also be defined for the child in terms of one-to-one correspondences with canonical sets. Thus, the hand, with its five fingers, frequently serves for many young children as a canonical set with five elements. The set of 10 fingers serves a similar role for the number 10.

EXERCISES 5–1

S–1.   State the cardinal number of each set.

   a.   the set of primary colors
   b.   the set of leap years between 1899 and 1969
   c.   the set of states in the union
   d.   the set of living ex-Presidents of the United States
   e.   the set of letters of the alphabet appearing in the word "happening"
   f.   the set of printed letters appearing in the printed word "happening"
   g.   the set of strings on a violin
   h.   the set of prime numbers less than 20
   i.   the set of members of your immediate family

M–2.   Suggest a system by which two children who cannot count efficiently can trade toys amicably (i.e., a system to assure that each gives the same number of toys as he receives).

S–3.   Give an example of two sets with different cardinal numbers, as shown by matching.

S–4.   When do two different sets have the same cardinal number?

---

[1] We write "1–1" to mean "one-to-one."
[2] "Canonical" is often used by mathematicians as a synonym for "standard" or "natural."

S–5. Suppose that two boxes each contain some pieces of chalk. Describe a procedure to show whether the sets have the same cardinal number, without counting the pieces of chalk in each box.

[S–6.] Let $S = \{$red, blue, green$\}$ and $T = \{$blue, yellow, white$\}$.

    a.   Find the cardinal number of S.
    b.   Find the cardinal number of T.
    c.   Find the cardinal number of $S \cup T$.
    d.   Find the cardinal number of $S \cap T$.
    e.   Does the cardinal number of $S \cup T$ equal the cardinal number of S plus the cardinal number of T? (Note the cardinal number of $S \cap T$.)

[S–7.] Let $S = \{$hat, ball, table, cup$\}$ and $T = \{$watch, bat, dog, saucer, chair$\}$. Answer the questions in [S–6] above.

[S–8.] Let $S = \{\square, !, \triangle, ?, \#, *\}$ and $T = \{\square, \#, \bigcirc, \cap, !\}$. Answer the questions as in [S–6] above.

[S–9.] Let $S = \{4, 6, 8, 10, 12\}$ and $T = \{14, 16\}$. Answer the questions as in [S–6] above.

[S–10.] On the basis of your answers to problems [S–6] through [S–9], when do you think that the (cardinal number of $S \cup T$) = (cardinal number of S) + (cardinal number of T)? Devise a simple rule which relates the cardinal number of $S \cup T$ to the cardinal numbers of S, T, and $S \cap T$.

[S–11.] Let $S = \{a, b, c, d, e\}$ and $T = \{c, e, f\}$. Draw a Venn diagram illustrating these sets. Answer the questions as in S–6 above.

∗S–12. Do the set of even whole numbers and the set of odd whole numbers have the same cardinal number? Why or why not?

S–13. Does the way in which you match the elements of two sets affect the cardinality of the sets? Explain.

2. ORDINALITY

We only talk about the *ordinal* number of (finite) sets in which the elements are *linearly ordered*. A set is linearly ordered if there is a first, second, etc., element in the set—that is, if a linear order relation has been defined on the set. For example, the elements might be ordered according to height, weight, or time of arrival. A set, whose elements are themselves sets, may also be ordered—for example, according to the number of elements in the constituent sets.

Although the child first observes "order" in the world around him at a very early age, his first formal introduction to order is the chant, "1, 2, 3, ..." Being able to repeat this chant, however, does not necessarily imply that ordinality has been mastered.

A child may be said to know, say, that the ordinal number of an ordered set with four elements is 4 (or that the position of the last element in the ordered set is "fourth") *only if he can pair the elements in the ordered set with some particular (i.e., canonical) ordered set which is known to have four elements.* Normally, subsets of the "chant set," $\{1, 2, 3, 4, \ldots\}$ serve the canonical role in determining the ordinal number of arbitrary ordered sets.[3] For example, the subset $\{1, 2, 3, 4\}$ might serve as the basis for comparison in checking whether a given ordered set has ordinal number 4 or, equivalently, that the last element in the set is the "fourth" (i.e., that it matches the 4 in $\{1, 2, 3, 4\}$). Of course, a child cannot be expected to give the order of the last element (in any set) if he cannot count as high as necessary (i.e., assuming the "chant set" is to serve as his basis for comparison). This is sufficient reason for teaching children how to count on indefinitely.

In everyday life, we are accustomed to referring to the *ordinal number* of particular *elements* in ordered sets (as well as the ordinal number of the sets themselves). Thus, for example, we say that 6 is the third element in the ordered set $\{4, 5, 6, 7, 8, 9, 10, 11\}$. What we really mean when we say this is that the ordered subset $\{4, 5, 6\}$, which consists of the "third" element (6) and all of those which precede it (4 and 5) can be paired one-to-one with the ordered set $\{1, 2, 3\}$. As a second example, consider the ordered set of months (January–December). The ordinal number of this ordered set, of course, is 12, as can easily be seen by comparing it with the ordered set $\{1, 2, \ldots, 12\}$. In addition, January turns out to be the *first* month, August the *eighth*, September the *ninth*, and December the *twelfth*. Why? (*Hint:* Choose an appropriate subset of the set of months and find the corresponding subset of the chant set.)

To summarize, we can speak of cardinal and ordinal numbers of *sets*, and we can speak of the *ordinal* number of a particular *element* in an ordered set. We cannot, however, speak of the *cardinal* number of an element of a set.

EXERCISES 5–2

S–1.   State an order relation that could be defined for each set below. Then give the ordinal number of each set.

a.   the set of primary colors
b.   the set of leap years between 1899 and 1969

---

[3] In some of the more advanced treatments, ordinal numbers are actually defined in terms of the set of all smaller ordinal numbers, whose elements (which are ordinal numbers and, hence, also sets) are ordered in accordance with their inclusiveness. Normally, the ordinal number one is defined as the set containing the empty set. The empty set may be denoted by 0 (i.e., $\{\ \} = 0$). In effect, the ordinal numbers are defined as follows:

$$0 = \{\ \}$$
$$1 = \{0\}$$
$$2 = \{0, \{0\}\} = \{0, 1\}$$
$$3 = \{0, \{0\}, \{0, \{0\}\}\} = \{0, 1, 2\}$$
$$4 = \ldots \qquad\qquad = \{0, 1, 2, 3\}$$
$$n = \ldots \qquad\qquad = \{0, 1, 2, 3, \ldots, n-1\}$$

Once the ordinal numbers are defined in this way (i.e., as *specific* (canonical) *sets*), it is not strictly admissible to talk about the ordinal number of other ordered sets. What must be done is to say that an ordered set is *similar* to a particular ordinal number (which is a set) if the corresponding elements (i.e., first, second, and so on) may be paired in a 1–1 fashion.

For more details, see P. Suppes, *Axiomatic Set Theory*, New York, Van Nostrand Reinhold, 1960, p. 129.

c. the set of states in the union
d. the set of letters of the alphabet appearing in the word "happening"
e. the set of strings on a violin
f. the set of prime numbers less than 20
g. the set of members of your immediate family

S–2.   The set of states of the United States could be ordered either alphabetically or by population, giving two different ordered sets. How do the ordinal numbers of the two ordered sets compare?

S–3.   Each of the following sentences involves an ordinal number but is meaningless unless we know the context (i.e., the set and its order relation). Describe a context within which each statement could apply—that is, give a set and an order relation defined on it.

a. Jeffrey is fourth.
b. Philadelphia is fourth.
c. Toothpaste X is first.

S–4.   Below is a table showing the heights and ages of five children:

	AGE	HEIGHT
Tony	7	4'9"
Mark	3	3'4"
Jane	2	2'6"
Rebecca	5	4'7"
Susan	6	4'5"

a. Set up a one-to-one correspondence that preserves the order between the set of ages and the set of heights.
b. What is the ordinal number of the set formed by using the order relation "height"? "Age"? "Alphabetical order"?
c. What ordinal number is associated with Susan in each instance?
d. Consider now the *unordered* set of five children. Why can't you associate an ordinal number with Susan in this case?

S–5.   Does every set have a cardinal and an ordinal number?

S–6.   If a finite set has a cardinal and an ordinal number, must they be the same?

S–7.   Below is a table showing the averages of six students in geography and history:

	GEOGRAPHY	HISTORY
Mary	85	89
George	91	81
Kathleen	90	95
Sally	82	88
Vince	95	97
Jack	80	83

a. Set up a one-to-one correspondence that preserves the order between the set of geography averages and the set of history averages.

b. What is the ordinal number of the set formed by using the order relation "average in geography"? "Average in history"? "Alphabetical order"?

c. What ordinal number is associated with Kathleen in each instance?

3. DEFINITION OF CARDINAL AND ORDINAL NUMBER

Although we have discussed the problem of determining cardinal and ordinal properties of unordered and ordered sets, respectively, nothing direct has been said about exactly what a cardinal or ordinal number is. A rather simple definition suggests itself, if we recall that a number, either cardinal or ordinal, is a property of a set, and, more particularly, a property common to a number of sets. *Hence, the cardinal number 3 might be defined simply as the set of all sets having numerosity three. Similarly, the ordinal 3 might be taken as the set of all ordered sets consisting of "three" elements.*

Unfortunately, these definitions have certain technical limitations. Although relatively unimportant for our purposes, these limitations are important in more advanced mathematics. Mathematicians prefer to associate each cardinal number with a special set, the canonical set, and then say that any arbitrary set that can be placed in one-to-one correspondence with the canonical set has the same cardinal number. Similarly, ordinal numbers can be defined by associating each with a particular ordered set. Then, as above, arbitrary ordered sets, which can be placed in one-to-one correspondence with the canonical ordered set, are said to have the same ordinal number.

Although ordinal numbers refer to ordered sets, and cardinal numbers to unordered sets, do not mistakenly infer that a given set necessarily has only one kind of number attached to it. Any given set has the property of numerosity and hence, cardinality. Furthermore, as soon as any ordering or arrangement of the set is introduced, the (ordered) set has an ordinal number as well. Thus, two kinds of questions pertaining to number might be asked, for example, about a set of people standing in line. The question, *"How many?" requires a cardinal answer, and, "In what position is the last?" involves ordinality.*

In effect, every set has a cardinal number, but only ordered sets have ordinal numbers. If a given set has both a cardinal and an ordinal number, what can you say about them? (*Hint:* They are *not* identical. Do they correspond?)

Whether the cardinal or ordinal aspect of number is acquired first is an open question that does not seem to have a unique answer. Some children seem to be cardinally oriented, whereas others seem to prefer order. To complicate the question further, a child may first acquire the cardinal aspect of one number and (first acquire) the ordinal aspect of another. Many children, for example, may acquire the cardinal aspect of such small numbers as 1, 2, and sometimes 3, 4, and 5, before the corresponding ordinal aspects are mastered. A child of 4 might be able to match three pennies with three pieces of candy but be unable to follow the instruction, "Pick up the third penny." On the other hand, it does not seem likely that the cardinal aspect of such numbers as 24 or 31 will be mastered until the corresponding ordinal numbers are acquired. Counting is an ordinal operation which children typically use to determine the cardinal numbers of large sets.

Because these ideas are extremely basic to all of mathematics, let us summarize by further operationalizing what it means to have mastered cardinal and ordinal number. Mastery of the cardinal number 3 might be indicated, for example, by a child who can select enough china (ahead of time) to set a table for three people or by a child who "knows" how much candy is needed to give each of three dolls one piece of candy. More generally, complete mastery of cardinal number involves the ability to tell whether or not the elements in two or more arbitrary sets may be placed in 1–1 correspondence—no matter how the sets to be composed are arranged relative to one another. To illustrate this, we can draw on Piaget and the task he originally used to test for number conservation: First, two sets of objects (e.g., dolls and candy) are selected and placed on a table in front of a young child (4–7 years) so that the respective objects "obviously" correspond in a 1–1 fashion.

0   0   0   0   0   0   candy

♀   ♀   ♀   ♀   ♀   ♀   dolls

Next, the child is asked which set contains more objects or if they contain the same. The actual phrasing might go, "Is there enough candy for each doll to have one? Are there enough dolls to eat all the candy?" Then the objects in one or both of the sets are rearranged (see below) so that they no longer look the same and the child is asked the same type of question.

```
           0

0     ♀♀    0
      ♀♀
0     ♀♀    0

           0
```

If the child consistently answers the questions correctly no matter what the arrangement, including the possibility of adding or deleting objects from one or both of the sets (with or without the child's knowledge), he is said to conserve number. It is always fascinating to try this with young children for the first time and the reader who has not done so is urged to try it.

Similarly, ordinality would be indicated where a child can pair the elements in two ordered sets. Here the pairing would not be between arbitrary elements in the two sets but between the two first elements, the two second elements, and so on. As indicated above, the "chant set" would probably serve as the comparison set—but any other ordered set might serve equally well (e.g., "a, b, c, d, . . ."). To test mastery of the ordinal number three, for example, a child might be asked to award ribbons to the first-, second-, and third-place winners in a race.

The idea of *number* cannot be said to be fully operational, however, until the child has observed the close relationship between ordinal and cardinal numbers. The desired integration might be indicated by the child's ability to answer a cardinal question on the basis of an ordinal result, and vice versa. *In particular, a child must eventually learn to recognize that the last number enumerated in counting a set is the cardinal number of the set.*

Recognition of this equivalence might be tested, for example, by having a child count two sets of objects and then asking him whether or not the elements may be paired in a 1–1 fashion. Although this may seem so trivial a task that a child could hardly fail to answer correctly, children frequently do make mistakes. Similarly, many nursery school children are able to determine without counting (i.e., by direct perception) the cardinal number of a display containing up to three or four elements but are nevertheless quite surprised to find out that they get the same number when they actually count the objects—that is, of course, assuming that they are able to count. When they first try to count, young children frequently repeat the same element more than once or skip some elements entirely. In short, they are simply unable to make the required 1–1 pairing. Presumably, experiences provided by such tasks eventually lead young children to recognize the desired equivalence.

EXERCISES 5–3

S–1. Tell whether each of the numbers used in these sentences is an ordinal or a cardinal number.

    a. Today is the nineteenth of September.
    b. The enrollment this semester is 2042.
    c. Your name is number 8 on the waiting list.
    d. The National Safety Council announces that the weekend traffic-death toll is 392.
    e. "First is the worst; second is the same; ...."

S–2. Tell whether each of the numbers used in these sentences is an ordinal or a cardinal number:

    a. There are two main Christmas colors.
    b. Have you read the book, *1984*?
    c. There are 28 letters in the word antidisestablishmentarianism.
    d. I've read up to page 52 in the text.
    e. My rank in class is 27.
    f. Mary is the second oldest in her famly.

S–3. Give an example of an ordered set and state its ordinal number.

S–4. Make up a set whose elements are icons and state its cardinal number.

S–5. Consider set $S = \{1, 2, 3, 4, 5\}$

    a. What is the ordinal number of set S?
    b. What is the cardinal number of set S?
    c. What is the ordinal number of the element 5?

S–6. Consider set $S = \{5, 10, 15, 20, 25, 30\}$. Answer the questions in Problem S–5.

S–7. Consider set $S = \{2, 3, 5, 7, 11, 13, 17, 19\}$. Answer the questions in Problem S–5.

M–8. Greg lines up his blocks as shown:

red   □   □   □   □   □   □

green   □   □   □   □   □   □

When asked how many red blocks there are, he counts and says, "Six." He does the same to find the number of green blocks. Nevertheless he says there are more green blocks than red blocks. Which concept of number—cardinal or ordinal—has he failed to learn?

For further thought and study: Refer to some of the experiments of Piaget on conservation of number in *The Child's Conception of Number*. Some of the results may surprise you. If you know a small child, you may want to try some of these experiments yourself.

4. THE SET OF NATURAL NUMBERS

So far, we have considered *particular* natural numbers (e.g., 3, 4). Only indirect reference has been made to sets of numbers (e.g., in discussing conservation of numbers). This is, of course, as it should be. It is impossible to talk about sets of elements unless the elements themselves are already familiar.

A child cannot be said to have acquired a working knowledge of natural number, however, until he has learned how the various natural numbers are related to one another. For example, it is conceivable that a child might be able to state the *cardinal number* of an arbitrary collection of objects (of reasonable size), to give the *order* of any element in any ordered set, or even to determine the *cardinal number of a set by counting*, but still not know how one number may be generated from another. *In particular, children must come to recognize that the cardinal number of the "successor" set, generated from any given set by adding one more element, can be determined by counting "one up" from the cardinal number of the generating set.* Given a set of eight objects, for example, the number of elements in the successor set, in this case nine, can be determined by simply starting with the cardinal number in the generating set (i.e., eight) and counting up one. Many young children, however, do not immediately make this connection. For a time after they learn to determine the cardinality of a set by counting, they may persist in determining the number of elements in a successor set by counting from the beginning (i.e., "1, 2, 3, . . ."). Although such a procedure certainly leads to the "correct" answer, it is far less efficient than counting up one from the number in the original set.

Counting up can also be used when more than one element at a time is added to the generating set; for example, in adding $4 + 3$, the child starts from

"4" and counts up "5, 6, 7." Suppes and Groen[4] have shown that children who are well practiced in simple addition facts are more likely to use a counting-up procedure to find sums than they are to simply count from the beginning (i.e., starting with 1). In effect, counting up is not only more efficient than counting from the beginning; it also appears to be more natural for young children. It appears to be good pedagogical practice, therefore, to provide young children with ample opportunities to discover the counting-up procedure before getting into addition in any serious way. Counting down, of course, would serve an analogous role in subtraction. We shall return to these ideas in a later section.

The set of numbers generated by starting with 1 and incrementing each successive number by one is called the set of *natural* numbers (usually denoted $N = \{1, 2, 3, 4, \ldots\}$.) The set of numbers generated by starting with *zero* (i.e., $\{0, 1, 2, 3, 4, \ldots\}$) is usually called the set of *whole* numbers (denoted $W = \{0\} \cup N$). Zero, of course, simply refers to the cardinality of the empty set, variously denoted, $\{\ \}, \varnothing$, the "null set." Young children typically find the notions of empty set and zero to be quite natural and they should be introduced along with other sets and numbers from the beginning. In doing this, however, the teacher should be aware that although there are an indeterminant number of *different* sets containing 1, 2, 3, 4, ... elements, there is essentially only *one* empty set. In effect, the (empty) set containing three-eyed people is identical to the (empty) set of six-legged cows.

To summarize, we have introduced the natural numbers and showed how they are related to one another. We have noticed that: (1) the set of natural numbers (N) has a smallest element (i.e., the number 1); (2) the cardinal number of a finite set can be obtained by counting; and (3) the cardinal number of the set formed by adding one element to any generating set is equivalent to the ordinal successor of the generating set.

EXERCISES 5–4

S–1. a. What is the smallest element of N?
　　　b. Give another set that has a smallest element and state the smallest element.
　＊c. Give an ordered set that has no smallest element.

S–2. Find the cardinal number of the following sets:

　　a. $\{r, t, w, n\}$
　　b. $\{8, 9, 10\}$
　　c. $\{22, 23, 24, 25, 26, 27\}$
　　d. $\{ace, king, queen, jack, ten\}$

S–3. If you know that the cardinal number of a set S is 210, and one element is added to the set S, what is the cardinal number of the new resulting set?

[4] P. Suppes and G. Groen, "Some Counting Models for Simple Addition Facts in First-Grade Arithmetic." In J. M. Scandura (ed.), *Research in Mathematics Education*, Washington, D.C.: National Council of Teachers of Mathematics, 1967.

S–4. Which of the following expressions represent a natural number?

    a. $9 - 2$
    b. $14 + 3$
    c. $15 \div 5$
    d. $2 - 9$
    e. $7 \div 3$
    f. $3 \div 7$
    g. 4×6

∗S–5. a. If a and b represent natural numbers, which of the following expressions will *always* represent a natural number?

    (1) $a + b$
    (2) $a - b$
    (3) $a \times b$
    (4) $a \div b$

  b. In the case of those which do not, what conditions can be placed on a and/or b in order that the expressions will represent natural numbers?

SYSTEMS OF NATURAL NUMBERS AND WHOLE NUMBERS

A mathematical system is a set on which one or more operations and/or relations are defined. The systems we consider here each involve an ordered set (i.e., a set with an order relation defined on it), a single binary operation, and an equivalence relation. The first such mathematical system consists of the ordered set of natural numbers (N) together with the binary operation of addition ($+$) and the equals relation ($=$).

5. ADDITION: DEFINITION AND PROPERTIES

The sum, a $+$ b, *of two* natural *numbers,* a *and* b, *may be defined as the cardinal number of elements in the union of any two disjoint*[5] *sets which contain* a *and* b *elements, respectively.* Thus, the sum of 2 and 5 is the number of elements in the union, say, of the two disjoint sets, $\{0, \square\}$ and $\{*, x, p, +, q\}$. The union of these sets $\{0, \square, *, x, p, +, q\}$ has seven elements. So, the sum of 2 and 5 is 7.

According to our definition, however, this sum can also be taken to be the number of elements in the union of the sets $\{c, d\}$ and $\{e, f, g, h, i\}$ or of $\{A, B\}$ and $\{C, D, E, F, G\}$. Be careful to note, however, that the definition does not allow us to take *any* two sets containing two and five elements, respectively. For example, we cannot find the sum of $5 + 2$ by taking the union of the sets $\{A, B\}$ and $\{A, C, D, E, F\}$ since they are *not* disjoint (i.e., A belongs to both sets). Nonetheless, as long as we choose disjoint sets, it makes no difference which sets we use as long as the first contains two elements and the second, five. As mathematicians say, *our operation of addition is well defined.*

Before proceeding, it may be helpful to say a little more about what being "well defined" involves, because the idea is a recurrent one in this chapter. Sup-

---

[5] Two sets are disjoint if they have no elements in common. See Chapter 3, Section 6.2, for the meaning of (set) "union."

pose we have an operation that associates with a pair of numbers, say *a* and *b*, a third number *c*. Suppose that we perform this operation by associating some sets, *A* and *B*, with the numbers. We then operate on these sets to obtain a third set *C* whose cardinal number *c* is the one we are seeking. Now it is clear that any variety of sets would be appropriate for *A* and *B*. In each case, the resulting third set *C* could be different. We would always hope, however, that the cardinal number of *C* would be the same, no matter which sets we chose. As has been noted, the operation of addition has this property with disjoint sets.

There are a number of different ways (compatible with the set definition) in which the sum of two numbers may be obtained.[6] *One procedure views addition as a process of "adding" more elements to a given set.* Joey has three apples, for example, and wants to know how many he will have altogether if he picks two more. *Another involves combining the elements in two given sets.* How many apples will Joey have if Jeanne gives him three apples and Janie gives him two (i.e., assuming that 2-year-old Julie doesn't get into the act and foul things up)?

In the first procedure the first number acts as a state (*i.e., the cardinal number of a set*) *and the second number acts as an* operator (*as an ordinal*) *that transforms the given state into another one* (*the sum*). One way to think of the *state* and *operator* conditions is to consider the *state* as a *set* of *n* elements and the *operator* as the *act* of adding *m* new elements to the basic set one by one until a new *state* (i.e., a set) is obtained with $n + m$ elements. For example, in adding $3 + 4$ by this procedure a child would start at 3 (cardinal) and count up by "ones" until the fourth (ordinal) "one" is reached at which point he would be at the new state, 7. Thus, to obtain the sum, the first number is incremented by 1 as many times as the second number. This *iterative* process is clearly a simple generalization of the relationship we have already noted between a cardinal number and its successor (i.e., the successor may be obtained by taking the next higher ordinal). Because the procedure of counting "one up" is so intimately involved in this iterative process, it would be premature to introduce this approach to addition before a child has developed facility in giving "successors."

The second procedure, although mathematically more symmetrical (i.e., both of the numbers being added act as operators), is apparently less desirable operationally.[7] To obtain the sum via this second approach, a child must count

---

[6] The familiar addition algorithm is based on the basic addition table and the properties of our base 10 numeration system. That the addition algorithm does work can be shown by reverting the process to the underlying set definition or, as seems to be currently more favored, by deducing the algorithms from various properties of the system of natural numbers under addition. We shall consider these properties in turn.

[7] P. Suppes and G. Groen ("Some Counting Models for Simple Addition Facts in First-Grade Arithmetic." *In* J. M. Scandura (ed.), *Research in Mathematics Education*, Washington, D.C., N.C.T.M., 1967) have shown that, after young children have acquired a high degree of competence with simple addition facts (with sums up to and including 5), their performance is best represented by a model which says, in effect, "store the value of the largest number and then increment that number by one as many times as the smaller number." This particular procedure is apparently used more widely by well-practiced students than a number of other possibilities including "counting up to the first number and then continuing on as many more times as the smaller number" and "storing the smaller number and incrementing as many times as the larger number."

The procedures used by Suppes and Groen are similar to those described above but should not be confused with them. The algorithms used by these investigators make no reference to the underlying sets.

up from the beginning. That is, he must form the union of the given disjoint sets, start from zero (cardinal) and count the elements in the union. For example, in adding $4 + 5$ a child would form the union of a "4-set" and a "5-set" and then count the elements in the union.

Although it involves no reference to the basic definition of addition, most children use a third procedure, at least with small numbers. They simply remember the sum. It is natural that they should eventually learn the basic addition facts; this is the most efficient way to add. But, it is important that they know what it means to add before they memorize them. Afterward, it is often too late. Many of the difficulties children have with arithmetic can be traced to such early inadequacies.

Although the two counting procedures lead to the same sum, most young children do not automatically become aware of this fact. Simply providing the child with experience in addition will not necessarily suffice. In our own classroom research we have observed many children who can find sums one way but not the other. Thus, in adding 2 to 5, kindergarten children will often hold up five fingers on one hand and two on the other and then proceed to count each of the seven fingers in turn. For them, counting up 2 from 5 is a completely different process than counting from the beginning; they need a variety of experiences before they will be convinced that the same sum is obtained in both ways. To avoid premature reliance on any single technique, children should be led to see the equivalence between these simple counting procedures.

The system of natural numbers under addition has certain desirable properties but is lacking in others. First, whenever we add two natural numbers, we always get a third natural number. For example, adding 3 to 5 gives 8, also a natural number.

If this seems so obvious as to be unworthy of notice, consider the binary operation of combining *equal*-sized pieces of pie. Adding two pieces of a pie cut into six equal pieces to three more such pieces results in five equal sized pieces (i.e., five pieces, each of which is $\frac{1}{6}$ of a pie). *We start with a number of equal-sized pieces and we get equal-sized pieces.* If, on the other hand, we wish to combine two "fourths" of a pie with three "sixths" of a pie we get five different pieces of pie, but they will *not* be of the same size. In short, the set of all conglomerations of equal-sized pieces of pie is *not closed* under the binary operation of combining such conglomerations. We summarize this property of the natural numbers by saying that the *set of natural numbers under addition is closed.*

Second, in adding three natural numbers, it makes no difference whether the sum of the first two numbers is added to the third or whether the first number is added to the sum of the second and third. For example, if the whole numbers are 2, 6, and 7, then $(2 + 6) + 7 = 8 + 7 = 15$ and $2 + (6 + 7) = 2 + 13 = 15$. That is, $(2 + 6) + 7 = 2 + (6 + 7)$. *This property is called the associative property of addition.*

Third, the order of adding any two natural numbers is immaterial. The sum is the same regardless of how the two numbers are added. For example, $7 + 8 = 15$ and $8 + 7 = 15$. Thus, when we add 8 to 7, the result is the same as adding 7 to 8. This equivalence is represented symbolically by $7 + 8 = 8 + 7$. *The natural numbers are said to satisfy the commutative property of addition.*

Notice, however, that there is no natural number (i.e., 1, 2, 3, ...) which, when added to a given natural number, results in the same natural number. The number 1 is the smallest natural number and adding it to any number transforms that number into its successor. If we include the number 0, so that our basic set is now the set of *whole* numbers, we see that 0 added to any whole number does not change its value (e.g., $0 + 5 = 5$). Hence, adding 0 to the set of natural numbers provides our addition system with what is called an *identity* element.

The reader should satisfy himself that adding 0 to the set of natural numbers does not destroy any of the above properties. The system of *whole* numbers under addition, for example, is closed. To see this, note that (1) the subset of natural numbers, $\{1, 2, 3, \ldots\}$, is closed; (2) the addition of 0 to any natural number gives the same natural number; and (3) the addition of 0 to itself gives 0. This simple argument "proves" that adding 0 to the set of natural numbers produces a new set, called the set of whole numbers, which is closed under addition. Each of the other properties can be shown to hold in the system of whole numbers (under addition) in a similar manner.

Notice that the set of natural numbers and the set of whole numbers each contains a smallest element (1 and 0, respectively). Although this property would seem to be desirable, its presence precludes another desirable property characteristic of other mathematical systems. Thus, for example, there is no natural number, which when added to a given (nonzero) natural number, yields a sum of 0. You cannot add a set of objects to a nonempty set of objects and get the empty set. As we shall see later, there are real-world situations where it is desirable that this restriction no longer hold. We shall soon be confronted with mathematical systems in which for each "number" there exists another number which when added to it sums to zero. In this case, the system is said to have the *inverse* property.

The presence (or absence) of such properties provides the teacher with an excellent opportunity to give her students experience in discovering for themselves. Thus, the teacher might start with $1 + 2 = 3$ and ask what $2 + 1$ equals. After giving a few more examples of the commutative law, she might ask if this is true for all pairs of numbers or just the ones she has picked. There are at least two ways in which children at about this stage (e.g., grades 1 and 2) might go about answering this question. They may simply try out a large number of possibilities and thereby greatly increase the degree of confidence they have in the result. Or they may refer (usually indirectly) to the "meaning" of addition itself. Because the sum of two numbers simply refers to the number of elements in the union of two disjoint sets having the desired numerosities, many children at this stage of development would probably agree that *the order of putting the sets together is immaterial*. The former approach is based on induction and the latter, on a crude form of deduction.

Commutativity is "psychologically" more or less obvious depending on how the addition process is viewed. If we think of addition as simply putting two sets together and counting the number in the resulting set, then the order of putting the sets together should clearly be irrelevant. If, on the other hand, addition involves "counting up," it is not at all immediately obvious that starting with one state (i.e., number) and incrementing by one as many times as the second number should be the same as starting with the second and incrementing as many times as the first. For example, in adding $3 + 5$ it may not be obvious to the child that starting at 3 and incrementing by one 5 times would give the same result as starting at 5 and incrementing by one 3 times.

EXERCISES 5-5

S–1. $5 + 3 = 8$. Ruth has three pebbles and David has five. Describe a situation in which it would be natural to:

    a.   add 3 to 5 by counting up from 5
    b.   add 5 to 3 by counting up from 3
    c.   add 3 to 5 by counting up from the larger
    d.   add 3 to 5 by forming a union and counting the elements of the union

S–2. Hank has three friends who like baseball and Joe has four. Yet when Hank and Joe and their friends get together, there are not enough boys to make up a whole baseball team.

    a.   How could this happen? (Assume their friends are boys.)
    b.   Does it contradict $4 + 5 = 9$?
    c.   Construct another illustration to show the necessity of using only disjoint sets when illustrating a sum.

E–3. If a, b, and c are whole numbers and $a + c = b + c$, is it always true that $a = b$? Explain.

E–4. As explained in the text, a set S is closed under a certain *operation* if we can take any two elements in the set, perform the indicated operation, and have the result be an element of the set S. This must work for *any* and *every* pair of elements that we pick. For example, consider the set of natural numbers. This set is closed under *addition* because if we add any two natural numbers we get a natural number. But this set is not closed under subtraction; if we subtract one natural number from another, we do not necessarily get a natural number. For instance, $2 - 5$ does not result in a natural number. Which of the following sets are closed under the indicated operations?

    a.   the natural numbers under multiplication
    b.   the even numbers under addition
    c.   the odd numbers under addition
    d.   the set $S = \{0, 1\}$ under addition

S–5. Look up "commute" and "associate" in an unabridged dictionary. Can you guess why the commutative and associative properties were so named?

[S–6.] a.   Illustrate the commutative property of addition of natural numbers by three specific instances.
       b.   Illustrate the associative property of addition of natural numbers by three specific instances.

S–7. In the real world, stepping on the gas and putting on the brake are certainly not commutative, for if we perform the actions in the order stated, our car would be at rest. If we perform the actions in the reverse order, however, our car would be moving. Give two other examples of real-world operations which are not commutative.

S–8. State the property that justifies each of the following statements. The letters represent natural numbers.

    a. $4 + 7 = 7 + 4$
    b. $a + (w + t) = (a + w) + t$
    d. $(x + y)$ is a natural number
    d. $t + r = r + t$
    e. $(10 + 1) + 6 = 10 + (1 + 6)$

∗E–9. Some of our previous mathematical relations may be defined for the natural numbers in terms of addition. For example, we can define the order relation $>$ as follows:

$a > b$ if and only if there is a natural number c such that $a = b + c$.

According to this definition $7 > 2$ because $7 = 2 + 5$ (Here $c = 5$), but 3 is not greater than 4 (written $3 \not> 4$) because there is no natural number c such that $3 = 4 + c$.

    a. Prove that $10 > 7$.
    b. What is the relation of $10 + 3$ to $7 + 3$?
    c. Show that if $c > d$, then $c + e > d + e$ for any whole number e. (*Hint:* $c + e > d + e$ if there is an n such that $c + e = d + e + n$. How do you know there is such a number n?)
    d. State this result in ordinary English.

E–10. Define a new operation ∗ on the natural numbers by

$a ∗ b = a \times b - a.$

For example, $7 ∗ 3 = 7 \times 3 - 7 = 21 - 7 = 14$
                 $4 ∗ 6 = 4 \times 6 - 4 = 24 - 4 = 20$

    a. Is the system of natural numbers closed under ∗?
    b. If ∗ a commutative operation?
    c. Is ∗ an associative operation?
  ∗d. Is there a right identity for ∗?
  ∗e. Is there a left identity for ∗?
  ∗f. Would any of these answers change if ∗ were an operation on the set of whole numbers rather than on natural numbers?

For further thought and study: Would you allow a child to use his fingers as counters when learning to add? Why or why not?

6. SUBTRACTION: DEFINITION AND PROPERTIES

Subtraction derives from two types of problem situations, each giving rise to a different way of defining subtraction. In one situation, the problem is to determine the cardinal number of a set after a certain number of elements have been removed. In the second situation, the problem is to compare the size (i.e., the cardinal difference) of two distinct sets. The following examples illustrate these two cases: (1) Suppose David has nine marbles and Joey borrows four of them; how many marbles does David have now? (2) Janie has nine dolls in her collection and Jeanne has six. How many dolls more does Janie have?

Let us use the language of sets to clarify what underlies these two definitions. In Example (1), we have two sets, A and B, such that A is a subset of B.[8] In this case, we can define a new set $(B - A)$ that contains all of the elements of B that are not elements of A. Then *we can define the difference between the number of elements in* B *and the number of elements in* A *to be the number of elements of* $(B - A)$. Using problem (1) as an example:

$$B = \left\{ \begin{matrix} O & O & O & OO \\ O & O & O & O \end{matrix} \right\} \qquad \text{David's marbles.}$$

$$A = \left\{ \begin{matrix} O & OO \\ O & \end{matrix} \right\} \qquad \text{The marbles Joey borrows.}$$

$$(B - A) = \left\{ \begin{matrix} O & O \\ O & O & O \end{matrix} \right\} \qquad \text{The marbles David keeps.}$$

David can determine how many marbles remain in a number of ways. He could view subtraction as the removal, one-by-one, of the marbles which Joey borrows. This is a "counting backwards" approach. He has nine marbles and must give away four. Counting backwards as he gives away one marble at a time, David sees his pile become eight, then seven, then six, and, finally, five. This must be the number of elements in $(B - A)$.

Another way would be for David to first give away his four marbles (e.g., by counting to 4) and then count the remaining ones. Notice that the second method is not as efficient as the first; it does not allow David to determine the number remaining until *after* he gives all of the marbles away.

In the first procedure for subtraction, the larger number, 9, acts as a state and the second number, 4, as an operator telling David how many times to count down. The second procedure for subtraction is similar to the second procedure for addition in that both numbers act as operators. In this case, though, after David counts the fourth marble given to Joey, he starts over again and counts the number of marbles he has remaining.

Problem (2) about the dolls corresponds to *defining subtraction as the inverse of addition*. To see this, let B and A be two sets with the cardinal number of B being larger than the cardinal number of A; in the case of problem (2), nine

---

[8] Recall that A is a subset of B if all of the elements of A are also elements of B.

and six, respectively. Place the elements of A in one-to-one correspondence with the elements of some subset of B as shown below.

How large a set C, disjoint from A, must we join to A in order that the union of the two sets (denoted $A \cup C$) will be in one-to-one correspondence with B? That is, we require that the union $A \cup C$ have the same cardinal number as B. Notice that we must use the idea of one-to-one correspondence and not the previous take-away procedure because A is *not necessarily* a subset of B. In fact, none of Jeanne's dolls belong to Janie and vice versa.

As before, there are a number of ways that the set C can be determined. The cardinal number of A could be regarded as an initial state, and the cardinal number of B as a final state. A child might count from the number of A to the number of B, keeping track of the number of times he counted up. Starting with 6, he would count up by 1 (i.e., increment by 1) three times to reach 9.

This procedure certainly provides a solution, but it fails to use knowledge already acquired. To restate the problem: What number must be *added* to 6 to give 9. The fact that we must find a set C, disjoint from A, such that $A \cup C$ can be put into one-to-one correspondence with B, suggests that the operation of addition lies at the root of the problem. In fact, all that is required is to use the related addition fact in a slightly different way. If a child has mastered the addition fact, $6 + 3 = 9$, then it would appear to be but a small step to determine the number he must add to 6 to get 9, denoted $6 + ? = 9$.

All this may be summarized by saying (defining) $a - b = c$ if and only if $a = b + c$. For example, $9 - 6 = 3$ since $9 = 6 + 3$.

Let us now consider properties of the system of natural numbers under subtraction. Recall first that a restriction was placed on the set which was to be removed (that is, it must be smaller than the first set). Therefore, given any two natural numbers, to speak of their difference is *not* always meaningful. For example, $3 - 5$ is nonsense if we wish a natural-number answer; we cannot take away 5 apples, say, from a set containing only 3. Consequently, the system of natural numbers under subtraction is *not* closed.

Consider next the order in which subtractions may be performed. Take as an example $8 - 5 - 2$. Now, $(8 - 5) - 2 = 3 - 2 = 1$, but $8 - (5 - 2) = 8 - 3 = 5$. Since $1 \neq 5$, we see that the order of performing two (or more) subtractions is crucial. This illustrates that the system of natural numbers is *not* associative under subtraction.

Of course, subtraction (in any number system) is not commutative either. For example, $5 - 3 \neq 3 - 5$ since $5 - 3 = 2$ and $3 - 5$ is undefined.

Does subtraction have an identity? That is, is there a number, say e, such that for any number n, $n - e = e - n = n$? We know that for any number n, $n - 0 = n$. Hence 0 is a good candidate for e. But $0 - n$ is undefined. In such a case, mathematicians say that 0 is only a *right* identity for subtraction (i.e., it only acts as an identity when it is on the right).

Subtraction does have what is called the *inverse* property. Any number n subtracted from itself is 0 (i.e., $n - n = 0$).

EXERCISES 5–6

S–1. By reformulating the problem in terms of addition, show that $13 - 7 = 6$.

M–2. Explain how you would illustrate, using pennies and dimes

$$88 - 24 = 64$$

and

$$23 - 7 = 16$$

S–3. Are the following sets closed with respect to the operation of subtraction (see Exercises 5–5, Problem E–4)?

    a. the natural numbers
    b. the even natural numbers
    c. the odd natural numbers
    d. $S = \{0, 1\}$

[S–4.] Give two examples (not in the text) to illustrate that subtraction of natural numbers is not commutative.

[S–5.] Give two examples (not in the text) to illustrate that subtraction of natural numbers is not associative.

S–6. In each of the following, write the two subtraction equalities that can be derived from the given addition statement:

    a. $7 + 4 = 11$
    b. $20 = 9 + 11$
    c. $13 = 1 + 12$

[S–7.] Consider the set $A = \{a, b, c, d, e\}$ and a subset $B = \{c, e\}$ of A. The set $A - B = \{a, b, d\}$ consists of all the elements in A which are not in B.

    a. For each of the following examples, find $A - B$.

        (1) $A = \{!, \#, ?, *\}, B = \{!, \#\}$
        (2) $A = \{4, 10, 7, 6, 9, 14\}, B = \{4\}$
        (3) $A = \{m, n, o, p, q\}, B = \{ \ \}$
        (4) $A = \{⌐, ⚲, ∩, ⚐\}, B = \{⌐, ⚲, ∩, ⚐\}$

    b. Draw Venn diagrams showing A, B, and $A - B$ for each of the examples in part a.

c. What can you say about $(A - B) \cap B$?

d. What can you say about $(A - B) \cup B$?

e. Find the cardinal numbers of sets A, B, and $A - B$, in each of the examples in part *a*.

f. Formulate a general rule concerning the cardinal numbers of sets A, B, and $A - B$ where A is any set and B is any subset of A.

For further thought and study: Examine a set of Cuisenaire rods. Think about how you could develop the ideas of addition and subtraction using the rods.

7. MULTIPLICATION: DEFINITION AND PROPERTIES

Let us now consider the binary operation of multiplication. In order to define the operation in terms of sets we need, first, to introduce the notion of a Cartesian product of sets. A Cartesian (or cross) product of two sets A and B, denoted $A \times B$, is the set of all ordered pairs (a, b) such that a is an element of set A and b is an element of set B. For example, let A be a set of first names, {Mike, Bill, Fred}, and B be a set of last names, {Smith, Brown}; then the Cartesian product set $A \times B$ would be the set of all combinations of first and last names, {(Mike, Smith), (Bill, Smith), (Fred, Smith), (Mike Brown), (Bill, Brown), (Fred, Brown)}. Another example would be to let $A = \{1, 2\}$ and $B = \{e, f, g, h\}$. In this case, $A \times B = \{1, 2\} \times \{e, f, g, h\} = \{(1, e), (1, f), (1, g), (1, h), (2, e), (2, f), (2, g), (2, h)\}$.

By counting the number of elements in the Cartesian product set of the first example we find that there are six elements. So the Cartesian product of a set containing two elements and a set containing three elements is a set with six elements. Similarly, in the second example, the Cartesian product of a set containing two elements with a set containing four elements is a set with eight elements. With this in mind, then, we *define the product of two natural numbers* a *and* b, *denoted* a $\times$ b, *as the number of elements in the Cartesian product of two sets which contain* a *and* b *elements*, respectively. Thus, for example, the product of 3 and 4 can be defined as the number of elements in the cross product, say, of the sets {#, *, $\triangle$} and {#, *, $\triangle$, $\square$}. The Cartesian product

$$\begin{Bmatrix} (\#, \#), (\#, *), (\#, \triangle), (\#, \square), (*, \#), (*, *) \\ (*, \triangle), (*, \square), (\triangle, \#), (\triangle, *), (\triangle, \triangle), (\triangle, \square) \end{Bmatrix}$$

has 12 elements, so $3 \times 4 = 12$.

According to our definition we could use any other pair of sets as long as they have three and four elements, respectively. This raises the question of whether we would always get the same product, that is, whether the operation is (as we say) well defined. It is not difficult to prove that this is so, but the proof is inessential for our purposes and we restrict ourselves to giving an example. Suppose we had chosen the sets {A, B, C} and {W, X, Y, Z}. Would the product still be the same (i.e., 12)? You should convince yourself that it would be by actually forming the Cartesian product of the two sets.

In our discussion of multiplication, it will sometimes be convenient to display Cartesian product sets as rectangular arrays rather than as lists of ordered pairs. For example, the Cartesian product $\{1, 2\} \times \{e, f, g, h\}$ can be represented by an array with two rows and four columns as follows:

	e	f	g	h
1	$(1, e)$	$(1, f)$	$(1, g)$	$(1, h)$
2	$(2, e)$	$(2, f)$	$(2, g)$	$(2, h)$

Because this (2×4) array has the same number of elements as any other 2×4 array, it could be replaced by a simple array consisting entirely of dots. More

```
.     .     .     .

.     .     .     .
```

generally, we can think of the product $a \times b$ simply as the number of elements in any dot array having a rows and b columns.

The Cartesian product approach to multiplication has certain advantages and should be presented along with the more familiar definition of multiplication as *repeated addition*. (In that case, you will recall the product $a \times b$ is defined as $\overbrace{b + b + \ldots + b}^{a \text{ times}}$. For example, $3 \times 4 = 4 + 4 + 4$.) For one thing, providing students with experience using Cartesian products may enable them to apply multiplication to physical situations where it may not otherwise be apparent that this is appropriate. To illustrate this point, consider the following problem: Julie has three new blouses, a white one, a yellow one, and a green one. She also has two new skirts, a red one and a blue one. How many times can Julie go to school wearing a different outfit? This problem lends itself quite naturally to the Cartesian product approach. If we let the letters W, Y, and G represent the white, yellow, and green blouses, respectively, and R and B, the red and blue skirts, respectively, then we can display the solution to this problem as follows:

		Skirts	
		R	B
	W	WR	WB
Blouses	Y	YR	YB
	G	GR	GB

Thus, the number of different combinations of blouses and skirts can be represented by the Cartesian product of the set of three blouses with the set of two skirts. Hence, our definition of multiplication applies and the total number of outfits is $3 \times 2 = 6$.

The system of natural numbers under multiplication has certain properties that will prove useful when we consider the multiplication algorithm in the next chapter. First, we note that the definition of multiplication applies to all pairs of natural numbers (e.g., 6 and 7) and that the product is always a natural number (e.g., 42). Hence, *the set of natural numbers is closed under multiplication.*

Second, given any three natural numbers, it makes no difference whether the product of the first two numbers is multiplied by the third or whether the first number is multiplied by the product of the second and third. For example, 3×4 ($= 12$) times $2 = 24$, which is the same as 3 times the product 4×2. Symbolically this is written $(3 \times 4) \times 2 = 3 \times (4 \times 2)$. In terms of Cartesian products, this property is probably easiest to illustrate in terms of three-dimensional arrays. Thus, in our example, both $(3 \times 4) \times 2$ and $3 \times (4 \times 2)$ can be represented by the array:

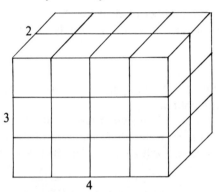

The former (i.e., $(3 \times 4) \times 2$) corresponds to first determining the number of cubes in the "front (vertical) slice" (i.e., 3×4) and then multiplying by the number of "slices" (i.e., 2). The second (i.e., $3 \times (4 \times 2)$) corresponds to multiplying the number of "horizontal slices" (i.e., 3) by the number of cubes in the "top slice" (i.e., 8). Obviously, both methods give the total number of cubes in the (same) array and so are equivalent. *This property is called the associative property of multiplication.*

Third, the order of multiplying any two natural numbers is immaterial; the product is always the same. For example, $3 \times 5 = 15$ gives the same result as 5×3. Symbolically, we say $3 \times 5 = 5 \times 3$. This can be seen directly by noting that 3×5 refers to an array having 3 rows and 5 columns and 5×3, to an array having 5 rows and 3 columns. Thus, the arrays below contain the same number of dots since each can be obtained from the other by a simple rotation.

This property is called the commutative property of multiplication.

Fourth, the number 1 has the property that when any natural number is multiplied by 1, the product is equal to the natural number (e.g., $1 \times 6 = 6 \times$

$1 = 6$). Think about how you might lead second-graders to discover this property for themselves using arrays. *This property is called the identity property of multication and 1 is called the multiplicative identity.*

So far, we have considered only the natural numbers. If we consider the product of 0 with any number a, then $a \times 0 = 0 \times a = 0$. For example, the Cartesian product of a set with five elements with the empty set (i.e., set with no elements) is again the empty set. Since there are no elements in the empty set, it is impossible to construct a single ordered pair which contains an element of the empty set. Thus $\{\ \} \times \{a, b, c, d, e\} = \{\ \}$ the empty set; so the product of 0×5 is 0. (Note: (, a), (, b), and so on are *not* ordered pairs; each pair of parentheses contains only one element).

In effect, the set of *whole* numbers has all of the properties of the natural numbers under multiplication with the added property *that* $0 \times a = a \times 0 = 0$. *This property is called the multiplicative property of* 0.

Before leaving the topic of multiplication, we should note that multiplication is commonly symbolized in at least three ways:

1. $a \times b$ (e.g., 4×3, 2×1)
2. $a \cdot b$ (e.g., $7 \cdot 4$, $3 \cdot 5$)
3. ab, where at least one of the symbols a and b is a letter (e.g., xy, $3r$, AB, $4m$).

EXERCISES 5–7

S–1. Form the Cartesian products $A \times B$ of the following sets:

    a. $A = \{\Box, \triangle, \bigcirc\}$ and $B = \{a, b, c, d\}$
    b. $A = \{1, 2, 3, 4, 5\}$ and $B = \{r, s\}$
    c. $A = \{\cap, \cup\}$ and $B = \{\times, +, \div, -\}$
    d. $A = \{!\}$ and $B = \{w, x, y\}$
    e. $A = \{w, m, t\}$ and $B = \{10, 9\}$

S–2. Find the cardinality of A, B and $A \times B$ for each of the examples in Problem S–1.

S–3. If $A = \{1, 2\}$ and $B = \{a, b\}$, then $A \times B = \{(1, a), (1, b), (2, a), (2, b)\}$. However, $B \times A$ is the set of all ordered pairs such that the first element of the ordered pair is an element of B and the second element of the ordered pair is an element of A. Thus $B \times A = \{(a, 1), (a, 2), (b, 1), (b, 2)\}$. So $A \times B \neq B \times A$ because they do not contain the same ordered pairs. For each of the following examples, form $A \times B$ and then $B \times A$.

    a. $A = \{a, b, c\}$, $B = \{1, 2\}$
    b. $A = \{1, 2\}$, $B = \{1, 2\}$
    c. $A = \{w, x, y\}$, $B = \{y, w, x\}$

∗E–4. Suppose that A and B are sets such that $A \times B = B \times A$. What do you now know about A and B?

*[S–5.] Suppose that set A can be put into one-to-one correspondence with $A \times B$. What can you say about set B?

S–6. Are the following sets closed with respect to the operation of multiplication (see Exercises 5–5, Problem E–4)?

    a.   the natural numbers
    b.   the even natural numbers
    c.   the odd natural numbers
    d.   the set $S = \{0, 1\}$
  *e.   the set $T = \{4t + 1 \mid t \in N\}$

S–7. State the property that justifies each of the following statements. (The letters represent natural numbers.)

    a. $w \times y = y \times w$
    b. $(t \times r)$ is a natural number
    c. $(4 \times 6) \times 2 = 4 \times (6 \times 2)$
    d. $a \times (b \times c) = (a \times b) \times c$
    e. $3 \times 9 = 9 \times 3$

S–8. What is a multiplicative identity? Does the set of natural numbers contain the multiplicative identity?

E–9. If a is any whole number, then $a \times 0 = 0$ and $0 \times a = 0$. What can be said about a and b if $a \times b = 0$?

E–10. Are there any pairs of whole numbers a and b such that

$$a \times b = 1 \quad \text{and}$$

$$b \times a = 1?$$

(Such pairs are called multiplicative inverses.)

E–11. Suppose that for natural numbers a, b, and c we know that $a \times c = b \times c$.

    a. Is it necessarily true that $a = b$?
    b. Can the same conclusion be drawn if a, b, and c are whole numbers?

M–12. Each grade level in an elementary school (grades 1–6) is divided into three sections, A, B, and C. The Cartesian product of the set of grade levels and the set of sections is then the set of all classes (e.g., $1A$, $1B$, and so on in the school). Write a multiplication problem arising from this situation.

8. DIVISION: DEFINITION AND PROPERTIES

The quotient of the natural numbers a *and* b, *denoted* a ÷ b, *is defined only if there exists a natural number* c *such that* b × c = a. *In that case, we say* a ÷ b = c *or, equivalently, that the open statement (i.e., equation)* a ÷ b = □ *has the solution c*. To say that $a \div b$ is not defined, is simply another way of saying that $a \div b = \square$ has no solution (or that the "solution set" is empty).[9]

This definition can be interpreted in terms of rectangular arrays in a manner analogous to that used with multiplication as follows: If a rectangular array with *a* elements has *b* rows, then the number of columns of the array is given by $a \div b = c$. For example, if we are given a rectangular array of 20 elements and 4 rows, then the number of columns must be $20 \div 4 = 5$.

$$20 \div 4 = 5$$

$$
4 \quad
\left.
\begin{array}{ccccc}
x & x & x & x & x \\
x & x & x & x & x \\
x & x & x & x & x \\
x & x & x & x & x
\end{array}
\right\} 20
$$

Recall that in multiplication we were given the number of rows and columns and were required to find the number of elements in the array. *In division, on the other hand, we are given the number of elements and the number of rows (or columns) and must determine the number of columns (or rows).*

Obviously, the two operations are closely related. In fact, every division problem can be reformulated in terms of multiplication and vice versa. In particular, the solution set of the open statement (i.e., equation) $a \div b = \square$ is exactly the same as the solution set of $b \times \square = a$. If $a = 20$ and $b = 5$, for example, then both $20 \div 5 = \square$ and $5 \times \square = 20$ have the same solution, 4. Because of this close relationship, division is called the inverse operation of multiplication.

As indicated above, the quotient, $a \div b$ is not always defined. Ten pennies, for example, cannot be fairly distributed among three people nor can 14 cookies be placed into equal piles of 4. Equivalently, there are no natural numbers which will make the open statements, $10 \div 3 = \square$ and $14 \div 4 = \square$, true. Therefore, the natural numbers are not *closed* under division.

We can, however, define a generalization of the division operation, which is closed over the natural numbers. The motivation for this operation (defined below) can be seen by considering the problem of constructing the largest rectangular array having a given number of rows from a certain number of elements. For example, suppose we have 23 objects and want to arrange them so as to form the largest possible rectangular array with five rows. As the array below shows, the largest such array has four columns and there are three elements left over.

$$
\begin{array}{c}
\hspace{3.5em} 4 \\
5 \quad
\begin{array}{cccc}
o & o & o & o \\
o & o & o & o \\
o & o & o & o \\
o & o & o & o \\
o & o & o & o
\end{array}
\hspace{2em}
\begin{array}{l}
\text{left over} \\
o \\
o \quad o
\end{array}
\end{array}
$$

---

[9] The solution set of a numerical equation is the set of numbers which, when substituted for the unknowns, make the equation true.

We can also express this result by the equation, 23 (total number) = 5 (rows) × 4 (columns) + 3 (remainder). Notice that 3 < 5. It is important to observe that we can always do this. That is, given any two natural numbers, n and d (where n is the total number of objects and d, the divisor), we can always find two whole numbers □ and △ (where □ is the quotient and △ is the remainder) such that

$$n = d \times \square + \triangle \text{ where } \triangle < d.$$

This property of the natural numbers is known as the *Euclidean property* and defines the promised generalization of the division operation.

We have already noted that the system of natural numbers is not closed under division. In fact, the system of natural numbers under division has relatively few interesting properties. For one thing, division is not associative,[10] because

$$(12 \div 6) \div 2 \neq 12 \div (6 \div 2)$$

$$(2) \div 2 = 1 \neq 12 \div (3) = 4$$

Similarly, division is not commutative, because

$$4 \div 8 \neq 8 \div 4$$

($4 \div 8$ is not a natural number and $8 \div 4$ is the number 2).

The natural number 1, however, does have a special property under division. Since $n \div 1 = n$ (e.g., $7 \div 1 = 7$) for any natural number n, we see that 1 is a candidate to be an identity for division. Even this property is limited, though, since $1 \div n$ is *not* defined in the system of natural numbers under division (except for the special case when $n = 1$). The number 1 is said to be (only) a *right identity* for division.

Up to this point we have talked only about the natural numbers. When we include 0 and discuss the whole numbers, we must consider two additional situations. Is there a whole number that will make the following statements true?

(1) $0 \div b = \triangle$ (i.e., $b \times \triangle = 0$)

(2) $a \div 0 = \square$ (i.e., $0 \times \square = a$)

The first case can be answered in the affirmative. Because zero represents the empty set, any partition of the empty set is again the empty set. In other words, the result of dividing zero by any number (except 0) is always zero (e.g., $0 \div 4 = 0$).

To see what is involved in the second situation, suppose $a = 6$. In this case, we want to find some natural number which will make the equivalent sentences $6 \div 0 = \square$ and $0 \times \square = 6$ true. From our earlier discussion of multiplication, we know that $0 \times \square = 0$ no matter what number we put in the box. Hence, there is *no* number that satisfies the equation $6 \div 0 = \square$ or, in fact, any equation

---

[10] If you wish to illustrate this to a child (it is not clear that one should necessarily do this), avoid becoming involved with fractions. The fact that $18 \div (6 \div 3) \neq (18 \div 6) \div 3$ demonstrates the point perfectly well in a logical sense. The inclusion of undefined expressions, for example, fractions such as $4 \div 8$, might confuse the child.

of the form $a \div 0 = \square$ where $a \neq 0$. On the other hand, if we let $a = 0$, the equivalent equations $0 \div 0 = \square$ and $0 \times \square = 0$ would be satisfied by *any* number whatever. For example $0 \times \boxed{5} = 0$, $0 \times \boxed{17} = 0$, and $0 \times \boxed{0} = 0$. At first glance, this may not appear to pose any serious problems. But if we allowed division by 0, then we would be forced into such absurdities as, say, $5 = 7$, because $0 \div 0$ equals 5 and $0 \div 0$ also equals 7. To avoid such nonsense in arithmetic, division by 0 is simply not allowed.

Before leaving the topic of division, we should also note that division is commonly symbolized in at least three ways:

1. $12 \div 4$ (read: 12 divided by 4)

2. $\dfrac{12}{4}$ (12/4)   (read: 12 divided by 4 or 12 over 4)

3. $4\overline{)12}$ (read: 12 divided by 4 or 4 into 12)

Forms 1 and 3 are seldom used in nonarithmetical courses beyond the eighth grade, however.

EXERCISES 5–8

S–1. Identify those symbols which do *not* name whole numbers:

a. $17 - 10$	e. $18 + 3$	i. 3×0
b. $17 - 7$	f. $3 \div 18$	j. $0 \div 18$
c. $7 - 10$	g. $18 \div 1$	k. $0 \div 0$
d. $10 - 0$	h. $18 \div 0$	l. 0×0

S–2. Find the quotient ($\square$) and the remainder ($\triangle$) in each case:

     a. $20 = 6 \times \square + \triangle$
     b. $57 = 7 \times \square + \triangle$
     c. $57 = 3 \times \square + \triangle$
     d. $4 = 10 \times \square + \triangle$

S–3. Find all possible pairs of natural numbers s and t such that $14 = 3s + t$.

S–4. Are the following sets closed with respect to the operation of division?

     a. the natural numbers
     b. the even natural numbers
     c. $S = \{0, 1\}$
     d. the set of all natural numbers which, when divided by 3, leave a remainder of 1

[S–5.] Give two examples to illustrate that division of natural numbers is not commutative.

[S–6.] Give two examples to illustrate that division of natural numbers is not associative.

S–7. Loosely speaking, an operation which "reverses" or "does the opposite of" a given operation is called an inverse operation. As explained, multiplication and division are inverse operations. "Opening the window" and "closing the window" offer another example of inverse operations. State the inverse operations of each of the following:

    a.   turn on the television
    b.   tie your shoe
    c.   boil the water to change it to steam.
    d.   burn the newspaper

[S–8.] State two operations (other than the ones given above) and indicate their inverse operations.

9. THE SYSTEM OF WHOLE NUMBERS UNDER THE FOUR OPERATIONS

At least four operations may be defined on the set of whole numbers. Furthermore, it is obvious that there are close relationships between these operations. For example, we have shown that addition and subtraction are related as are multiplication and division.

In elementary mathematics, it is common practice to consider the set of whole numbers, together with the various binary operations, as a total system. This system, of course, has all of the properties indicated above for addition, subtraction, multiplication, and division. But it also has certain other properties that indicate how addition, multiplication, subtraction, and division are interrelated. One of the two most basic of these properties is the *distributive* property of multiplication over addition. This property says that, given any three whole numbers, the first number multiplied times the sum of the second two is equal to the first multiplied by the second plus the first multiplied by the third. Symbolically, this can be represented $a \cdot (b + c) = (a \cdot b) + (a \cdot c)$. For example, $3 \cdot (4 + 2) = (3 \cdot 4) + (3 \cdot 2)$. The other important property is that the product of each whole number with the additive identity (i.e., zero) is zero (i.e., $a \cdot 0 = 0$).

These two properties are closely related to the definition of the binary operation of multiplication in terms of repeated addition. To see this, we first define the product of two numbers, $a \cdot b$, as the sum of b added a times. That is, the product, $a \cdot b$, is defined to be the number of elements in the set formed by taking the union of a disjoint sets, each of which has b elements. If we let $a = 2$ and $b = 4$, and let $\{*, \triangle, \square, \#\}$ and $\{A, B, C, D\}$ be any two disjoint sets having four elements each, then the number of elements in the union, $\{*, \triangle, \square, \#\} \cup \{A, B, C, D\}$, is the product $2 \cdot 4$.

We can think of non-empty disjoint sets as forming (non-empty) arrays. In the example above we could have viewed the sets $\{*, \triangle, \square, \#\}$, and $\{A, B, C, D\}$ in terms of the array

    *   △   □   #

    A   *B*   *C*   *D*

In this manner, we see that the number of elements in the union of a disjoint sets, each having b elements, corresponds to the number of elements in an array having a rows and b columns. This array, of course, has the same number of elements as one formed by taking the cross product of any set containing a elements with any one containing b elements. So the two methods of defining multiplication, as repeated addition and as Cartesian products, are mathematically equivalent.

Even though these two methods for determining a product are mathematically equivalent, however, they may relate to physically different problem situations. We have already noted that the Cartesian product method can be easily applied to combination problems. The repeated addition method (i.e., finding the product by forming unions of disjoint non-empty sets), however, corresponds more naturally to problem situations involving finding the total number of elements in a collection of (disjoint) sets, each having the same number of elements. For example, if we know that a child has four bags of 20 marbles each, and want to determine the total number that he has, we tend to think of multiplying as a short-cut way of finding the sum

$$20 + 20 + 20 + 20$$

We now show that from the definition of multiplication as repeated addition

$$a \cdot b = \overbrace{b + b + \ldots + b}^{a \text{ times}}$$

we can prove the distributive property. To show this, let $b = c + d$. Then

$$a \cdot b = a \cdot (c + d) = \overbrace{(c + d) + \cdots + (c + d)}^{a \text{ times}}.$$

Because we have $a\ (c + d)$'s, we must also have a c's and a d's. But the associative and commutative properties of addition effectively allow us to reassociate and reorder the c's and d's in any way we wish.

Thus

$$\overbrace{(c + d) + \cdots + (c + d)}^{a \text{ times}} = \overbrace{c + \cdots + c}^{a \text{ times}} + \overbrace{d + \cdots + d}^{a \text{ times}}$$

But

$$(1)\ a \cdot (c + d) = (\overbrace{c + c + \cdots + c}^{a \text{ times}}) + (\overbrace{d + d + \cdots + d}^{a \text{ times}})$$

and

$$(\overbrace{c + c + \cdots + c}^{a \text{ times}}) = a \cdot c \quad \text{and} \quad (\overbrace{d + d + \cdots + d}^{a \text{ times}}) = a \cdot d$$

Hence, we obtain the distributive property

$a \cdot (c + d) = (a \cdot c) + (a \cdot d)$ by substitution of equals in (1).

Perhaps this may more easily be seen if we let

$a = 3, c = 4,$ and $d = 5$

Then

$3 \cdot (4 + 5) = (4 + 5) + (4 + 5) + (4 + 5)$

By reassociating and reordering the 4's and 5's, we get

$3 \cdot (4 + 5) = (4 + 4 + 4) + (5 + 5 + 5)$
$$= 3 \cdot 4 + 3 \cdot 5$$

by definition (of multiplication).

Notice that when $b = 0$, the product $a \cdot 0$ is simply the union of the empty set taken a times. If $a = 3$, and $b = 0$, then $a \cdot b$ is the number of elements in $\{ \} \cup \{ \} \cup \{ \}$ which, of course, is the same as the number of elements in $\{ \}$. In short, $a \cdot 0 = 0$. This proves that the multiplicative property of the additive identity also follows from the definition of multiplication as repeated addition.

In addition to the distributive and additive identity properties, all of the properties of multiplication (e.g., commutativity, associativity, closure, etc.) follow from the definition of multiplication as repeated addition. In each case, it is only necessary to represent the given products as sums and to refer to previously demonstrated properties of addition.

Analogous to considering multiplication as repeated addition, we can think of division as repeated subtraction. That is, the quotient of $a \div b$ (where b divides a) may be found by asking how many times b can be subtracted from a. For example, 4 can be subtracted 3 times from 12 and $12 \div 4 = 3$. Repeated subtraction can be performed as indicated below.

```
    12
   − 4     once
  ─────
     8
   − 4     twice
  ─────
     4
   − 4     three times
  ─────
     0
```

Division, as repeated subtraction, may be interpreted in two ways. The first way may be characterized by the question, "How many subsets can be formed from a set of a elements so that there are b elements in each subset?" For example, "How many boxes of candy can be made from 12 candy bars, if each box is to hold four bars?" In partitioning this set of 12, each subset of four (i.e., each box of four candy bars) would correspond to a repetition of subtraction by 4. The solution (i.e., the quotient, 3) is found by counting the number of subsets (or subtractions).

The second type of situation may be characterized by the question, "If a set of a elements is divided evenly into b disjoint subsets, how many elements are in each subset?" For example, the problem might be to determine how many candy bars there are in each of three boxes if there are 15 candy bars altogether (i.e., assuming that each box contains the same number). In this situation, we know how many subtractions are to be made but we do not know the number to be subtracted each time.

Fortunately, the number is equal to the number of times that 3 can be subtracted from 15. For example, 3 can be subtracted 5 times from 15 and 5 is precisely the number which can be subtracted 3 times from 15. We shall not prove this relationship directly, but you can easily convince yourself that it is true by trying it out in a variety of situations (e.g., How many objects are in each of nine sets if there are 36 objects altogether?) Your students may like to try this, too.

In addition to these relationships, there are many others within the system of whole numbers. One of the most basic is the following property between addition and subtraction.

If $a \geq b$ and $c \geq d$, then

$$(a - b) + (c - d) = (a + c) - (b + d)$$

To illustrate this property, let $a = 6$, $b = 4$, $c = 2$, and $d = 1$. We see, then, that

$$(6 - 4) + (2 - 1) = 3$$

and that

$$(6 + 2) - (4 + 1) = 3$$

This relationship is important because we can prove from it that multiplication is distributive over subtraction. That is,

$$a \cdot (b - c) = (a \cdot b) - (a \cdot c)$$

Perhaps even more important, this relationship is needed to justify our use of subtraction algorithms (as we shall see in the next chapter).

In summary, the system of whole numbers under addition, subtraction, multiplication, and division not only has all of the properties that we specified in discussing each operation separately but many other important properties which allow us to interrelate these operations.

EXERCISES 5-9

S–1. How would you use the distributive law for a rapid computation of $79 \cdot 37 + 79 \cdot 63$?

S–2. Tickets to the zoo cost 75¢ for adults and 25¢ for children. Use the distributive law to quickly compute the admission cost for a group of 14 adults and 14 children.

S–3. Explain how the distributive law is involved in the computation

$$
\begin{array}{r}
7 \\
\times 13 \\
\hline
21 \\
7 \\
\hline
91
\end{array}
$$

(*Hint:* $13 = 10 + 3$)

[S–4.] State the property that justifies each of the following statements. (The letters represent natural numbers.)

a. $r \times (s \times t) = (r \times s) \times t$
b. $m \times (n + p) = m \times n + m \times p$
c. $4 + 1 = 1 + 4$
d. $7 \times (6 + 4) = 7 \times 6 + 7 \times 4$
e. $a + h = h + a$
f. $c \times d = d \times c$
g. $(10 + 1) + 5 = 10 + (1 + 5)$

S–5. A convenient way of computing 7×12 mentally is to use the distributive property. Think $7 \times 12 = 7 \times (10 + 2) = 7 \times 10 + 7 \times 2 = 70 + 14 = 84$. Show how the distributive property can be similarly used to compute each of the following:

a. 7×13
b. 8×15
c. 6×18
d. 9×14
e. 8×19

∗E–6. We have shown that the natural numbers possess a distributive property "for multiplication over addition"; i.e., for any natural numbers a, b, and c, $a \times (b + c) = a \times b + a \times c$.

a. Formulate a distributive property "for addition over multiplication."
b. Does it hold for the natural numbers?
c. Why or why not?

10. PRIME NUMBERS, GREATEST COMMON FACTOR, AND LEAST COMMON MULTIPLE

In this section we define a special kind of natural number, called *prime number* and consider two additional and important properties of natural numbers: *greatest common factor* and *least common multiple*. These concepts are needed in Chapter 7 where we discuss the extension of the natural numbers to the positive rationals, particularly where we define the arithmetical operations of addition and subtraction.

A natural number, *a*, is said to be a *factor* of a natural number, *n*, if there exists a natural number *b* such that $a \times b = n$. For example, 3 is a factor of 6, since $3 \times 2 = 6$. Every natural number, *n*, is a factor of itself, since $n \times 1 = n$. This implies, also, that 1 is a factor of every natural number.

When a natural number greater than 1 has no factors except itself and 1, we call it a *prime number*. Thus, 2, 3, 5, 7, 11, 13, 17, 19, 23, and 29 are the prime numbers less than 30. (What are the next two prime numbers?) Natural numbers greater than 1 which are *not prime* we call *composite numbers*. Thus, 4, 6, 8, 9, 21, and 51 are some of the composite numbers.

Breaking down natural numbers into their factors is called *factorization*. The following are some factorizations of 12 and 30:

$$12 = 4 \times 3 \qquad\qquad 30 = 15 \times 2$$

$$12 = 2 \times 6 \qquad\qquad 30 = 5 \times 6$$

$$30 = 10 \times 3$$

In spite of the apparently large number of different ways in which a given natural number may be factored, the natural numbers have an interesting property in this regard. To see what is involved, let us continue "factoring the factors" until all of the factors are prime. The result of factoring a natural number into its *prime factors* is called a *prime factorization*. This gives:

$$12 = 4 \times 3 = (2 \times 2) \times 3 \qquad\qquad 30 = 15 \times 2 = (3 \times 5) \times 2$$

$$12 = 2 \times 6 = 2 \times (3 \times 2) \qquad\qquad 30 = 5 \times 6 = 5 \times (2 \times 3)$$

$$30 = 10 \times 3 = (2 \times 5) \times 3$$

Notice that no matter which factors we start with we get the same prime factors. Thus, 12 has the prime factors, 2, 2, 3; and 30 has the prime factors, 2, 3, 5.

It is natural to wonder whether this is true of all natural numbers and, happily, the answer is yes. *Each natural number has exactly one set of prime factors.*

Stated slightly differently, the prime factorization of each natural number is unique up to the order of the factors. Although we cannot prove this here, you can convince yourself that it is true by testing some additional examples. For example, find at least two (non-prime) factorizations of 36. Break each factorization down further into a prime factorization. Are the prime factors unique? What are the prime factors?

A scheme that may be used to help children find the prime factors of a natural number is that of using a "factor tree." For example, if we wanted to find the prime factors of 204, we could proceed as follows:

1. Express 204 as the product of two factors (any two will suffice); for example,

$$204 = 4 \times 51$$

2. Express each non-prime factor as the product of two factors and continue this process until all factors are *prime*. We can represent this process in terms of the "tree" diagram:

This diagram gives us the prime factorization, $204 = 2 \times 2 \times 3 \times 17$.

If we had used different factors to begin with, our tree might look like the following:

204
2   102
2   17   6
2   17   3   2

This gives the prime factorization, $204 = 2 \times 17 \times 3 \times 2$, which is exactly the same except for the order of the prime factors.

For some purposes (e.g., adding rational numbers) we will want to know whether two (or more) natural numbers have any factors in common. For example, again consider the numbers 12 and 30. Because the prime factors of 12 are 2, 2, 3, and of 30 are 2, 3, 5, they clearly have factors in common. Thus, 2, 3, and 6 are factors of both. As we shall say, they are *common factors*. In this case, 6 is the largest common factor and we distinguish it with the name, *greatest common factor* (sometimes called greatest common *divisor* or just *GCF*).

In general, a natural number, g, is called the *greatest common factor* of two or more numbers $n_1, n_2, \ldots$ if every common factor of $n_1, n_2, \ldots$ is also a factor of g. In the above example, 6 is the greatest common factor of 12 and 30 because each common factor (i.e., 2, 3, and 6) is also a factor of 6. If the greatest common factor of two numbers is 1, we say that the numbers are *relatively prime*. For example, the numbers 7 and 16 have no common factors other than 1. Hence, the greatest common factor is 1 and the numbers are said to be relatively prime.[11]

We can also use factor trees to find the greatest common factor of any set of natural numbers. We illustrate by finding the common factor of 204 (which we factored above) and some other large number, say 340.

From the two trees, we see that the *common prime factors* are 2 and 17 and that there are two 2's and one 17 common to both trees. Because these are the

---

[11] In mathematical literature, the greatest common factor (GCF) of two numbers n and m is often denoted by (n, m). Thus, $(4, 6) = 2$ and $(12, 16) = 4$. When two numbers, n and m, are relatively prime, $(n, m) = 1$ (e.g., $(2, 3) = 1$, $(10, 13) = 1$, etc.).

only common prime factors of 204 and 340, the common factors which are not prime (i.e., which are composite), must be products of 2, 2 and 17. In this case, the common composite factors are: $2 \times 2 = 4$, $2 \times 17 = 34$, and $2 \times 2 \times 17 = 68$. 68 is the largest of the common factors of 204 and 340 and by definition is the greatest common factor. It is important to notice that $68 = 2 \times 2 \times 17$ is the product of the common prime factors, each one repeated as many times as it appears in both trees.

This approach can easily be generalized to find the greatest common factor of three or more numbers. For example, consider 24, 36, and 60.

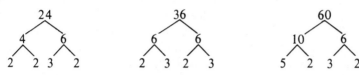

In this case, the only common prime factors of the three numbers are 2 and 3, with the 2 appearing (at least) twice in each tree. Hence, the greatest common factor of 24, 36, and 60 is $2 \times 2 \times 3 = 12$.

Now consider *multiples* of natural numbers, that is, numbers obtained from given natural numbers by taking their product with other natural numbers (e.g., 1, 2, 3, ...). For example, 35 is a multiple of 7 since $7 \times 5 = 35$. Other multiples of 7 are: $1 \times 7 = 7$, $2 \times 7 = 14$, $3 \times 7 = 21$, $4 \times 7 = 28$, and so on.

We are all familiar with the multiples of certain numbers, especially the multiples of 2 and 5: (2, 4, 6, 8, *10*, 12, 14, 16, 18, *20*, ...) and (5, *10*, 15, *20*, 25, ...), respectively. If we extend these lists of multiples far enough, we can easily see that certain multiples are common to both lists. In particular, 10, 20, 30, 40, ... are multiples of both 2 and 5. (We recognize these common multiples as multiples of 10.)

A (natural) number, *m*, is called a *common multiple* of two or more natural numbers if each of them divides *m*. Thus, 18 is a common multiple of 18, 9, 6, 3, 2, and 1. Each divides 18 (with remainder zero). (Of course, every natural number is a common multiple of itself and one—e.g., $7 = 7 \times 1$.)

A natural number *m* is called the *least common multiple* (sometimes denoted LCM) of two or more numbers if every common multiple of these numbers is also a multiple of *m*. Thus, 10 is the least common multiple of 5 and 2, because every common multiple of 5 and 2 is also a multiple of 10. Similarly, 18 is the least common multiple of 6 and 9. However, 18 is *not* the least common multiple of 2 and 3, because 6 is a common multiple of 2 and 3 which is *not* a multiple of 18 (as the definition requires). The notion of least common multiple will be particularly useful in Chapter 7 where we discuss addition of rational numbers.

We can use factor trees to determine least common multiples. For example, to find the least common multiple of 28 and 42, we first find their prime factors.

Next, we observe that if a number, n, is to be divisible by another number, d, then n must include among its prime factors all of the prime factors of d. For example, every multiple of 28 must include among its prime factors 2, 2, and 7. Thus, the multiple, 56, has the prime factorization, $56 = 2 \times 2 \times 2 \times 7$. Hence, if a number is to be divisible by both 28 and 42, it must include among its prime factors all of the prime factors of 28 and 42 (and possibly others). To find the smallest such number (i.e., the least common multiple), the basic idea is to take as few prime factors as possible and still end up with a common multiple. Here we may begin by taking all of the prime factors of 28—namely 2, 2, and 7. (We clearly need all of these to ensure that the multiple is divisible by 28.) The prime factors of 42 are 3, 2, and 7, so the least common multiple must also include 3, 2, and 7 among its prime factors. We have already included 2 and 7 (in fact, 2, 2, and 7) and need only add 3 to the list of factors. This gives $2 \times 2 \times 7 \times 3 = 84$ as the least common multiple of 28 and 42.

The same general approach can easily be extended to find the least common multiple of three or more natural numbers. To illustrate, consider finding the least common multiple of 12, 18, and 30.

As before, the least common multiple must include among its prime factors all of the prime factors of 12, 18, and 30. We begin by selecting all of the prime factors of 12 (i.e., 2, 2, and 3). There is an additional 3 among the prime factors of 18 (i.e., a 3 over and above the factors of 12) so we add it to our list. Moving on to the factors of 30, we see that we must also include the factor 5. In this way, we obtain 2, 2, 3, 3, and 5 as the prime factors of the least common multiple. Therefore, the least common multiple is $180 = 2 \times 2 \times 3 \times 3 \times 5$. One way to check this is to show that the prime factors of 12 (2, 2, and 3), 18 (2, 3, and 3), and 30 (3, 2, and 5) have all been included and that there are no extra prime factors included. (You may also wish to check this by showing that 12, 18, and 30 each divides 180 and that any other multiple of these numbers is divisible by 180.)

Another way to find the least common multiple is to start with the greatest common factor and add all of the other prime factors from each number. Thus, in the above example we would have:

$$180 \; = \; \underline{2 \times 3} \quad \times \; 2 \; \times \; 3 \; \times \; 5$$

least common multiple	greatest common factor (divisor)	from 12	from 18	from 30

In effect, the least common multiple is the product of the greatest common factor together with all of the unique prime factors.

Notice, in particular, that if the greatest common factor is 1, then the numbers are relatively prime so that the least common multiple is the product of the numbers themselves. For example, the least common multiple of 6 and 25 (which are relatively prime) is

$$150 = 6 \times 25 = \quad \underline{1} \quad \times \underline{2 \times 3} \times \underline{5 \times 5}$$

greatest   from   from
common   6   25
factor

EXERCISES 5–10

S–1. Which of the following are prime numbers?

a.	271	f.	23
b.	728	g.	143
c.	17	h.	229
d.	5	i.	5462
e.	19	j.	307

S–2. Which of the following are composite numbers?

a.	143	f.	31
b.	137	g.	7
c.	38	h.	217
d.	161	i.	263
e.	1394	j.	6,083,959

S–3. Factor each of the following into its prime factors:

a.	98	f.	173
b.	40	g.	726
c.	72	h.	371
d.	360	i.	713
e.	150	j.	2592

S–4. Write at least five elements in the set of multiples of:

a.	6	c.	14
b.	11	d.	18

S–5. Find the greatest common factor of each of the following:

a.	30, 72	f.	35, 60
b.	12, 18	g.	36, 108
c.	24, 60	h.	50, 60
d.	75, 110	i.	180, 168
e.	40, 96	j.	28, 140, 100

S–6. Find the least common multiple of each of the following:

a.	2, 3	f.	12, 9
b.	18, 27	g.	22, 48
c.	18, 21	h.	24, 60
d.	18, 30	i.	180, 168
e.	9, 21	j.	16, 24, 40

S–7. Complete the table below for the GCF and LCM of the given numbers:

	GCF	LCM
a. 7, 9		
b. 6, 8		
c. 5, 25		
d. 12, 15		
e. 18, 22		
f. 126, 140		
g. 220, 600		
h. 9, 12, 15		

11. SYSTEMS NOT BASED ON NUMBERS (OPTIONAL)

As we saw in Chapter 4, many kinds of mathematical systems have nothing to do with numbers. Nevertheless, many of these systems share properties in common with the arithmetical systems we have just considered. Consider, for example, the system of rotations of a square. Recall from Chapter 4, Section 1.2, that in this system the basic set consisted of the 0° rotation (no movement), 90° rotation, 180° rotation, and 270° rotation. The binary operation defined on this set of rotations was called "followed by" (e.g., 90° "followed by" 180° = 270°).

This particular system is *closed, associative, commutative,* and has an *identity* (0°). It also has the *inverse* property. That is, to each rotation there corresponds another rotation such that the first followed by the second is equivalent to the identity (0°) rotation. Test yourself to see if you can show that these properties indeed do hold in the system.

In addition, the above rotation system has the property

$$A \odot (B \odot C) = (A \odot C) \odot B$$

where $\odot$ means "followed by" and A, B, and C refer to arbitrary rotations. Although this property is different from any of those mentioned above, it is *not* independent of them. It may, in fact, be deduced logically from the commutative and associative properties listed above. Satisfy yourself that this is true. You may also find it interesting to identify other "new" properties of this system and to try to demonstrate whether or not they follow logically from the ones listed above.

EXERCISES 5–11

S–1. Mr. Baker does not like to carry heavy change in his pocket and always exchanges five pennies for a nickel so that he never has more than four pennies. Suppose we construct a system in which the elements are the number of pennies in Mr. Baker's pocket (including 0). The operation $A \oplus B$ means: If Mr. Baker has A pennies and gets B more, then $A \oplus B$ is the number of pennies he keeps. For example:

$$3 \oplus 1 = 4$$
$$3 \oplus 4 = 2$$
$$4 \oplus 1 = 0$$

a. Compute $3 \oplus 2$
$$2 \oplus 4$$
$$1 \oplus 1$$
$$4 \oplus 4$$

b. Is the system closed under the operation $\oplus$?
c. Is the operation $\oplus$ commutative?
d. Is the operation $\oplus$ associative?
e. Is there an identity?
f. Does 3 have an inverse (i.e., is there a number B such that $3 \oplus B = 0$)?
g. Does every element have an inverse?

∗S–2. Consider the mathematical system which consists of a set S, whose elements are points on a line, and a relation called "between." The relation is expressed symbolically as (abc), and is read "b is between a and c." Intuition would tell us that if (abc), then (cba). On the basis of intuition, state whether each of the following is true or false:

a. if (abc) and (bcd), then (abd)
b. if (abc) and (abd), then (cad)
c. if (abd) and (acd), then (bdc)
d. if (abc) and (bcd), then (acd)
e. if (abc) and (acd), then (abd)
f. if (abc) and (acd), then (bcd)

[S–3.] Consider the set $S = \{\#, !, \$\}$ and the binary operation $\oplus$ as defined in the table below:

$\oplus$	#	!	$
#	#	!	$
!	!	$	#
$	$	#	!

To find $\$ \oplus !$, for example, locate $\$$ in the column under $\oplus$; move your finger to the right until it is under the column headed by !. Read the answer in that space. Thus $\$ \oplus ! = \#$. Similarly, $\$ \oplus \# = \$$.

a. Is this system closed with respect to $\oplus$? Why or why not?
b. Is the operation $\oplus$ commutative?
c. Is there an identity? If so, what is it?
d. Does # have an inverse? If so, what is it?
e. Does ! have an inverse? If so, what is it?
f. Does $ have an inverse? If so, what is it?

[S–4.] Consider set $S = \{\$, !, \$, ¢\}$ and the binary operation $\otimes$ as defined in the table below.

$\otimes$	#	!	$	¢
#	#	#	#	#
!	#	!	$	¢
$	#	$	#	!
¢	#	¢	!	!

Answer questions a through f as in Problem [S–3] above.

E–5.   Consider an equilateral triangle with vertices numbered as in Figure E. If it is rotated one-third of a revolution (120°) clockwise in the plane, then it will appear as in Figure R. If this is followed by another rotation (one-third of a revolution clockwise), the triangle will then appear as in Figure R^2.

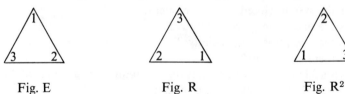

Fig. E   Fig. R   Fig. R^2

Let E denote a rotation of 0° or any rotation which leaves the numbers of the triangle arranged as in Figure E.

Let R denote a rotation of 120° clockwise or any rotation which leaves the numbers of the triangle arranged as in Figure R.

Let R^2 denote 2 rotations of 120° clockwise or any rotation which leaves the numbers of the triangle arranged as in Figure R^2.

Let set $S = \{E, R, R^2\}$ and let the operation ∘ be "followed by."

a.   Cut out an equilateral triangular region from a piece of paper and label it as shown in Figure E. To perform $R^2 \circ R$, position the triangle as in Figure R^2. Follow this by a rotation of 120° clockwise. Do the numbers on the vertices of your triangle appear as in Figure E?

b.   Complete the following table:

∘	E	R	R^2
E	—	—	—
R	—	—	—
R^2	—	E	—

Using the results of part b, answer the following:

c.   Is this system closed with respect to ∘?
d.   Is the operation ∘ commutative?
e.   Does $(E \circ R) \circ R^2 = E \circ (R \circ R^2)$?
f.   Is there an identity? If so, what is it?
g.   Does E have an inverse? If so, what is it?
h.   Does R have an inverse? If so, what is it?
i.   Does R^2 have an inverse? If so, what is it?

Numeration Systems and Computational Algorithms

In the previous chapter we considered the definition of the whole numbers together with the basic arithmetic operations of addition, subtraction, multiplication, and division. We also looked at the basic properties of these operations, such as the commutative law for addition and for multiplication. In this chapter we introduce systems of numeration. Then we show how the properties of the *whole number system* together with those of the *base 10 numeration system* permit us to compute in the usual way.

REPRESENTING NATURAL NUMBERS (SYSTEMS OF NUMERATION)[1]

Once children become familiar with the idea of number in the sense that they can pair up the elements of *equipollent* sets (i.e., sets that have the same number of elements), they will quite naturally recognize the *need for assigning names to numbers*. In fact, the vast majority of children on entering school will already be familiar with the sounds (auditory names) "one," "two," "three," and so on. Unless working with children raised in educationally deprived environments, about all the kindergarten and/or first grade teacher will usually have to do is to

[1] In talking about numeration systems, it is frequently necessary to distinguish between numbers and symbolic representations of numbers (i.e., numerals). Where there is any possibility of confusion, we will adopt the convention of using quotation marks (e.g., "9") with numerals. When we say, for example, that 9 is larger than 2, we will mean the number, nine, is larger than the number, two. On the other hand, when we say "9" is larger than "2," we will mean that the numeral "9" has proportions which make it physically larger than the numeral "2."

make sure that the children learn to pair the proper auditory names with the proper numbers. Many children of this age do not know in counting the elements in a set that the names must be paired in one-to-one fashion with the elements. For example, a child might point to the fifth object in a set, according to his progression of counting, and say "six." The teacher should not become overly concerned when this happens but should simply provide the child with opportunities to count and ask questions that will gradually lead him to discover how to count correctly.

Normally, kindergarten and first-grade children will not be as familiar with *visual* names for numbers. This fact provides the teacher with an excellent opportunity to let the children have a hand in making up their own system for naming numbers. Learning to *represent* classes of abstract ideas (such as the set of natural numbers) with symbol systems is fundamental to all of mathematics and the teacher should take advantage of such opportunities whenever they present themselves.

After the children are familiar with the Hindu-Arabic Base 10 System which is in current use, they will not find the task of naming numbers nearly so interesting. They already know the "correct" way to write (i.e., name) numbers and may be puzzled as to what the teacher wants them to do. In situations like this it is frequently difficult for the learner to recapture preconceptual innocence.

1. CONCRETE AND ICONIC NAMES

It has been important to represent numbers ever since man invented the barter system, in which one set of objects is traded for another. Probably one of the very earliest ways of representing numbers was with certain objects such as small stones or sticks. Rather than having to pair such things as bushels of wheat with goats, cartloads of lumber with cows, or gold bullion with (for) dancing girls, traders could use more easily handled stones or sticks as an intermediary. Suppose, for example, that a herder wants to trade some of his cows for some wheat, one cow for each two bushels of wheat. Then, if the number of cows is at all large (say 25), it would be difficult to effect the trade properly without counting or using some other equivalent way of naming numbers. In this case, assume that the traders determine the appropriate number of bushels of wheat through the intermediate use of stones. First they run the herd into the farmer's pen, putting one stone in an urn every time a cow enters. (None is allowed to leave, of course.) Then they give the herder two bushels of wheat for every stone in the urn.

Later, when materials for writing became more readily available, it may have become still more convenient to make a mark for each object to be traded. This procedure is still used frequently in the form of tally marks (e.g., //), even by research workers. Tally marks represent numbers in an *iconic* (i.e., pictorial) fashion. Each individual mark represents one element in the set being counted.

When asked to "draw a picture" of a set of objects, a child who has learned to conserve the cardinal number involved will typically draw one icon for each object in the set. After repeated experiences of this kind, the kindergarten or nursery school teacher might suggest the use of some sort of standard mark (like the tally mark "/") as a means of representing arbitrary objects. The teacher should not drill her pupils on such activities. The important thing is that the child should be provided with a wide variety of experiences like this during the preschool years as preparation for later learning.

EXERCISES 6–1

S–1. Classify each of the following representations as symbolic, iconic, or verbal:

    a. the Roman numeral III
    b. the holes in a computer punch card
    c. the Roman numeral V
    d. a floor indicator in an elevator consisting of three lights, one of which
        lights up (and stays on) with each successive floor
    e. spoken numerals
    f. holding up seven fingers

S–2. Write five *different* numerals representing four.

E–3. A *numeral* names a *number*. Tell whether each sentence refers to a numeral or to a number:

    a. 3 plus 5 equals 8.
    b. The (number, numeral) on my door is 17.
    c. Write a 6 below the 2.
    d. I ordered four hamburgers.
    e. Multiply 8 by 4.
    f. The product is 32.
    g. The ten's digit in 324 is 2.

2. SYMBOLIC NAMES

Because most children on entering school are already familiar (in varying degrees) with such symbols as "their own name," "1," "2," "*A*," "*B*," etc., it is usually quite easy to get them to make up names for numbers. Usually this can be done simply by presenting a child with a set of objects and asking him to make up a name for the number of objects in the set. After a period of orientation, a typical 5- or 6-year-old might do something like

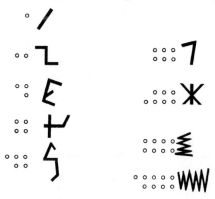

As indicated, children will frequently be able to write reasonable facsimiles of the Hindu-Arabic symbols for the smaller numbers, but will either confuse some of the larger *numerals* (i.e., number names) with each other—as in "7" above—or invent entirely new symbols.

A child must first learn, of course, to attach a different symbol to each number. If he fails to do this (e.g., he writes " 7 " for $\substack{\circ\circ\circ\\\circ\circ\circ}$ and " 7 " for $\substack{\circ\circ\circ\\\circ\circ\circ\circ}$) the teacher might question him with something like, "What number is this (" 7 ") the name of?". If the same name has been assigned to two different numbers, it will be easy to get the child (who "knows" what number is) to recognize his error. The teacher might point out, for example, that although the child is free to make up whatever name he wishes, he must be able to "draw a picture of the number" involved whenever he sees its name. Even where a child has managed to make up a different symbol for each number, the "numeration systems" first considered by the naive child are sure to be inadequate. Unless some short cuts or other memory devices are incorporated into the system a point will soon be reached where it simply becomes impossible for the child to remember the names (let alone the numbers they represent)—no matter how good his memory.

All viable numeration systems necessarily rely on memory devices of one sort or another. We shall consider briefly two such systems, one based on Roman numerals and the other, on the familiar Hindu-Arabic numerals (i.e., 1, 2, ..., 31, ...).

2.1 Roman Numerals[2]

The system of Roman numerals is one of the longest-lived of the early attempts to reduce the number of symbols required to represent numbers. The Roman use of "V" for five, "X" for ten, "L" for fifty, etc. all serve to reduce the number of symbols required to represent numbers. Nonetheless, anyone who has attempted to read the cornerstone of a building knows that Roman numerals can still be quite cumbersome with larger numbers.

To give children more of an appreciation for the economy of their own Hindu-Arabic system, the teacher in the middle elementary grades might have her children attempt to add and subtract, or, even, multiply and divide, with Roman numerals. This should *not* be done, of course, with the idea of teaching the child how to actually compute with Roman numerals. The really instructive part for children would be to have an opportunity to make up algorithms of their own.

It is possible, of course, for the student and the teacher to check the validity of a given procedure (presumably one made up by a child) in any particular case by converting the corresponding Roman numerals to Hindu-Arabic (base 10) numerals, doing the necessary computation, and finally, writing the result in Roman numerals.

[2] As with our own Hindu-Arabic numerals, Roman numerals probably were originally adapted from iconic representations. "I," "II," "III," for example, are certainly suggestive of the numbers they represent.

One procedure for adding Roman numerals (one that children might invent) involves taking away an "I" from the right of one of the numerals when the other numeral has an "I" on the left, writing the resulting number of "I's," and finally writing all of the remaining symbols in the two addends to the left of the "I's." To add IX and XVII, for example, take away one of the two "I's" to the *right* of "XV" in "XVII" (to counter balance the "I" to the left of "X" in "IX"). Next, write the remaining "I," and put the "X" in "IX" and the "XV" in "XVII" to the left of the "I," giving the final result, "XXVI." Checking, we see that IX = 9 and XVII = 17 so that the sum is 9 + 17 = 26. Since 26 = XXVI, the rule works for the numbers considered. Unfortunately, in its simplest form this procedure is insufficient for adding arbitrary Roman numerals. To see why, try to add "IX" and "V."[3]

EXERCISES 6–2.1

E–1. Write each of the following Roman numerals in Hindu-Arabic numerals.

a.	VIII	b.	XXIV	c.	LIX
d.	DCC	e.	CM	f.	XLII

E–2. Write Roman numerals for the following.

a.	28	b.	54	c.	499
d.	594	e.	1776	f.	944

E–3. Add the following numbers without converting from Roman numerals to ordinary numerals.

a. CXII + DCXVI
b. XIX + VI
c. XXXIX + XXVI
d. LVIII + XXVIII

S–4. What goes wrong if one tries to use the procedure described in the text to add IX + IX?

For further thought and study:

(A) There are many ways to represent numbers. In this book only the Roman system and place-value systems are considered. Look up the Egyptian and Babylonian numeration systems in a book such as Eves, *An Introduction to the History of Mathematics*, New York: Holt, Rinehart and Winston, 1964.

(B) Find a procedure (algorithm) for multiplying Roman numerals.

---

[3] The procedure only applies here if we change V to IIIII. Then, IX + IIIII = XIIII = XIV. This is similar to the borrowing technique used in the subtraction algorithm. (See *Subtraction* in this chapter.)

(C) What is the value of teaching Roman numerals in elementary school? Is it worthwhile to teach addition and multiplication of Roman numerals?

(D) How are numbers such as 10,000, 100,000, 1,000,000 represented in Roman numerals?

2.2 The Hindu-Arabic System of Numerals

Our own familiar base 10 numeration system has the dual advantage of requiring only ten different symbols and, even more important, of naming numbers according to a simple pattern. Thus, we use "0," "1," "2," "3," "4," "5," "6," "7," "8," "9" for the first 10 *whole* numbers. The successor of the number 9, then, is represented "10" where the "1" denotes one set of ten elements and the "0" indicates that there are no additional elements. In a similar fashion, "11" refers to *one* set of 10 together with *one* set of *one*; "57," to *five* sets of ten and *seven* sets of one; "346," to *three* sets of 100 (10 × 10), *four* sets of 10, and *six* sets of one.

What is involved, of course, is the idea of *place value. The number represented by one of the basic symbols* (e.g., "5," "2") *depends on the position of that symbol in the number name* (*i.e., numeral*). Thus, "5" refers to the number five (five ones). The "5" in "51" refers to the number fifty (five tens) and the "5" in "4,549" refers to five hundred (five hundreds).

EXERCISES 6–2.2

[E–1.] Even though the number zero is not a natural number, the symbol "0" (or some equivalent) is needed to represent natural numbers in a place-value system. Explain why.

[M–2.]   a.   How would you group a pile of 23 buttons to show the meaning of the numeral 23?

   b.   By adding one button at a time you could illustrate the numerals 24, 25, 26, ... What rearrangement would be necessary when the thirtieth button is added?

[M–3.] Letting girls represent tens and boys represent ones, explain how you could represent the numerals 23, 24, ..., 30, 31.

2.3 Other Place-Value Systems

Other than the fact that human beings have ten fingers, there is nothing magical about using ten symbols to name the numbers. We might just as easily have picked, say, 2, 3, 5, or 12 symbols. The Duodecimal Society, in fact, has long been proposing the use of twelve symbols as a means of simplifying division by certain numbers.[4] Numbers may be represented in the various bases as indicated below.

---

[4] In base 12 arithmetic, division by 3, 6, and 9 is much easier than in base 10. Division by 2, 4, and 8 offers about the same level of difficulty in both systems. Division by 5, however, is more difficult in base 12.

Number	Base 2	Base 3[5]	Base 10	Base 12
zero	0	$a = 0$	0	0
one	1	$b = 1$	1	1
two	10	$c = 2$	2	2
three	11	$ba = 10$	3	3
four	100	$bb = 11$	4	4
five	101	$bc = 12$	5	5
six	110	$ca = 20$	6	6
seven	111	$cb = 21$	7	7
eight	1000	$cc = 22$	8	8
nine	1001	$baa = 100$	9	9
ten	1010	$bab = 101$	10	T
eleven	1011	$bac = 102$	11	E
twelve	1100	$bba = 110$	12	10
thirteen	1101	$bbb = 111$	13	11
. . .				
twenty	10100	$cac = 202$	20	18
twenty-one	10101	$cba = 210$	21	19
. . .				
twenty-four	11000	$cca = 220$	24	20
. . .				
thirty-two	100000	$babc = 1012$	32	28
. . .				
one-hundred forty-four	10010000	$bcbaa = 12100$	144	100

Notice that in each case place value increases by a *common multiple* wherever a symbol is moved one position to the left in a numeral. In base three, for example, the "*c*" in "*bc*" denotes two (sets of) *ones*; the "*c*" in "*cb*" denotes two sets of *three*; and the "*c*" in "*bcbaa*" denotes two sets of *twenty-seven* ($3 \times 3 \times 3$).

This type of regularity is a matter of convenience and simplicity, however, and not of necessity. Numbers may also be represented in what might be called *mixed bases*. Thus, we might *arbitrarily* assign place values as indicated below.

Number	Thirty-six's (3×12)	Twelve's (12×1)	One's
three			3
thirteen		1	1
twenty-seven		2	3
fifty-one	1	1	3
eighty-two	2	0	10

[5] The symbols *a*, *b*, *c* are introduced for the more familiar symbols 0, 1, and 2 to emphasize the fact that the symbols selected for use are quite arbitrary. In electronic computers, for example, which utilize a base 2 system in performing computations, "0" and "1" are represented by *open* and *closed* electrical switches.

Mixed bases of this sort are widely used in measurement. Thus, the example above of mixed-base numerals is highly reminiscent of measuring in terms of yards, feet, and inches, where 12 inches equal 1 foot and 3 feet equal 1 yard. Thus, for example, there is no need to speak of more than 0, 1, or 2 groups of twelve (because three groups of 12 equal one group of 36). To be sure, just about all traditional English systems of measurement (including money) involve mixed-bases. In liquid measure, for example, two cups equal one pint and two pints equal one quart, but four quarts equal one gallon.

Because there is a good deal of confusion in the minds of many teachers as to the importance of teaching different systems of numeration, let us try to set the record straight. It is immaterial whether children ever become familiar with such impractical numeration systems as that involving base 6. Introducing different systems of numeration is worthwhile only to the extent that, first, this helps to elucidate similarities between the various systems which are in current use (e.g., base 10, base 2, mixed bases in measurement, etc.), thereby minimizing the amount that needs to be remembered. It illustrates, for example, the independence from a particular base of the usual arithmetic algorithms. That is, no matter what base is chosen, arithmetic algorithms can be devised for use with that base, and furthermore, they all have the same basic form.

Second, introducing different bases helps to point out regularities which have been overlooked in familiar systems (e.g., noticing how place value increases by a common multiple as one reads to the left in a numeral).

EXERCISES 6–2.3

Note: In the following exercises a subscript will denote the base. For example, 12_8 means 12 in the base 8 numeration system.

S–1. To express 67 in base 5 we must find the number of groups of 25, 5, and 1 contained in 67.

$$67 = 2 \times 25 + 3 \times 5 + 2$$

Therefore, $67_{10} = 232_5$

Express 91 as a base 5 numeral.

S–2. In base 5 we use groupings of 1, 5, 25, 125, etc. What groupings are used in:

    a.   base 2?
    b.   base 3?
    c.   base 8?
    d.   base 12?

S–3. Write the numeral for 100_{10} in:

    a.   base 2.
    b.   base 3.
    c.   base 8.
    d.   base 12.

S–4. What base ten numeral represents the same number as

a. 34_5 b. 34_8 c. 34_{12}

S–5. Consider the mixed base system based on pennies, nickels, quarters, 50-cent pieces, and dollars. A numeral can be represented by considering the minimum number of coins and bills needed to give a specified amount of money. For example:

	$	50 cents	quarters	nickels	pennies
96 cents =		1	1	4	1
243 cents =	2	0	1	3	3

a. Compute the numerals for 57 cents, 185 cents, and 378 cents in this system.
b. Compute the numerals for 57 cents, 185 cents, and 378 cents if dimes instead of quarters are used.

[E–6.] One advantage of place-value systems of numeration is that all numbers can be represented in terms of a fixed number of symbols. What is the total number of symbols needed to represent numbers in:

a. base 2?
b. base 7?
c. base 10?
d. base 12?

E–7. Base 2 numerals generally require fewer symbols than base 10 numerals. In which system does a given number generally have a *longer* numeral?

E–8. Here is another procedure for representing a given number by a numeral having a certain specified base. Divide the given number by the base number. Record the remainder. Divide the quotient by the base number and again record the remainder. Repeat this process until the quotient is 0. The list of remainders in reverse order is the new numeral. As an example, to compute the base 5 numeral for 66:

$$
\begin{array}{rl}
5)\ \underline{66} & \text{remainder} = 1 \\
5)\ \underline{13} & \text{remainder} = 3 \\
5)\ \underline{\ \ 2} & \text{remainder} = 2 \\
\ \ \ 0 &
\end{array}
$$

Therefore $66 = 231_5$

Use this procedure to write the numerals for 103 in:

a. base 5.
b. base 2.
c. base 12.

For further thought and study: Why does the procedure of Problem E–8 work?

THE BASIC COMPUTATIONAL ALGORITHMS

3. ADDITION ALGORITHM[6]

Traditionally, one of the things the good arithmetic teacher has done well has been to lead pupils to an awareness of why the algorithms work. For this reason and because so many excellent treatments are available elsewhere, only the main points are discussed here.

Understanding why an algorithm works amounts to being able to justify (i.e., deduce) the particular steps in the algorithm in terms of the underlying operation. To see why this is important, recall that addition is defined in terms of sets, set operations (i.e., forming unions), and cardinal numbers of sets. The arithmetic algorithms, on the other hand, are rules for operating on numerals, the names of numbers, *not* the numbers themselves. It is not at all immediately obvious why or how such rules correspond to well-defined operations on the numbers. When we "prove" that an algorithm works we demonstrate a close relationship between the manipulation of symbols (i.e., numerals) and the corresponding operations on numbers.

To see what is involved consider the example

```
   2          3
  +4          9
      (carry 1)
   7          2
```

As we say, 3 and 9 make 12. Put down the 2 and carry the 1. Then, 1 and 2 make 3 plus 4 make 7. Put the 7 down in the ten's column.

Since addition involves putting two disjoint sets together we can consider two distinct sets of tally marks

(23 =) ///// ///// ///// ///// ///   and

(49 =) ///// ///// ///// ///// ///// ///// ///// ///// ///// ////,

In the base 10 numeration system, we group elements into bundles (i.e., sets) of 10. For example, ///// ///// makes one bundle of ten which we will denote by ⫿⟋⟋⟋⟋⟋⟋⟍ . Thus, we can represent the numbers in our problem by,

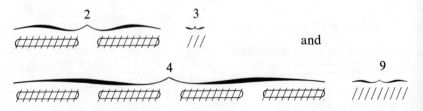

When we add, we put the bundles of ten in one pile and the unbundled tally marks in another as shown below.

6

(1)   ///////// ///////// /////////
      ///////// ///////// /////////

12

///////// ///

Note that *2* ten bundles together with *4* ten bundles give *6* ten bundles and that *3* unbundled tally marks grouped with *9* tally marks give *12* such marks. Putting the 12 tally marks into a bundle of ten plus two

12

(2)   ///////// //

we see that the resulting sum may be denoted by

7

///////// ///////// /////////   2

///////// ///////// ///////// /////////   //

To justify the steps in the algorithm we refer to representations (1) and (2). (1) simply gives the result of combining 2 tens with 4 tens and 3 ones with 9 ones. (2) indicates that $3 + 9$ ones can be represented by 1 ten and 2 ones (hence, justifying carrying the 1) so that there are 7 tens in all (and, of course, only 2 ones).

The addition procedure can also be justified indirectly, without reference to meaning, by assuming the *associative* and *commutative* properties of the system of whole numbers together with the properties of the base 10 numeration system. In particular, whole numbers may be represented as sums consisting of multiples of powers of 10.[7,8] Thus, $23 + 49 = 72$ because:

1. $23 + 49 = (20 + 3) + (40 + 9)$	by renaming 23 and 49 (according to properties of the base ten numeration system)
2. $= 20 + (3 + (40 + 9))$	associative property

---

[7] In general, there is more than one way of representing (i.e., naming) any given number. For example, the number 17 can be represented by "17," "6 + 11," "15 + 2," etc. In the base ten system, the arithmetical algorithms are based on the fact that any number can be written as a sum consisting of multiples of powers of ten. For example, the number seventeen can be denoted by "10 + 7" (or $(1 \cdot 10^1) + (7 \cdot 10^0)$). Similarly, three hundred twenty-seven can be denoted "$(3 \cdot 10^2) + (2 \cdot 10^1) + (7 \cdot 10^0)$."

[8] *Note:* $10^0 = 1$ by *definition*. Intuitive justification for this definition resides in the fact that $10^n \div 10^m = 10^{n-m}$, $n > m$. (For example, $10^5 \div 10^3 = 10^{5-3} = 10^2$.) Hence, it seems reasonable to define $1 = 10^n \div 10^n = 10^{n-n} = 10^0$.

3.	$= 20 + ((40 + 9) + 3)$	commutative property
4.	$= 20 + (40 + (9 + 3))$	associative property
5.	$= (20 + 40) + (9 + 3)$	associative property
6.	$= 60 + 12 = 60 + (10 + 2)$	renaming[9]
7.	$= (60 + 10) + 2$	associative property
8.	$= 70 + 2$	renaming[9]
9.	$= 72$	renaming

Notice that many of the steps above are implicitly assumed in writing the addition problem in the form

$$\begin{array}{r} 23 \\ +49 \\ \hline \end{array}$$

The tens and ones are immediately associated by the positioning of the numerals with respect to one another. The first five steps above were required to rearrange $23 + 49$ in a comparable way. Step 6, then, amounts to renaming $9 + 3$, and "carrying the 1" corresponds to step 7. Adding the tens corresponds to steps 6 and 8, and reading the final sum to step 9.

Proofs of this sort[10] have been given a good deal of attention in courses for elementary school teachers in the past few years. In certain cases, however, this attention has been misguided, and we would like to raise a simple point of caution. The main purpose of proving something is to make that something more acceptable to the learner than it was before. This will only be the case where the assumptions made in constructing the proof are *more* evident (to the learner) than the thing to be proved. It is not clear that most teachers are any more familiar, or trusting, of the associative and commutative properties than they are of the addition algorithm itself. Under these circumstances, a deductive argument based on these laws can hardly be expected to be convincing.

Nonetheless, the commutative and associative laws are extremely basic—more so than the addition algorithm. It may be that the addition algorithm should not be introduced until after the child has become familiar with the commutative and associative properties, but this is an open question. What do you think? Why?

---

[9] Steps 6 and 8 make implicit use of the parallel arguments below.

$20 + 40$	$= 2 \cdot 10 + 4 \cdot 10$	renaming 20 and 40
	$= (2 + 4) \cdot 10$	distributive property
	$= 6 \cdot 10$	renaming $2 + 4$
6.	$= 60$	renaming $6 \cdot 10$
$60 + 10$	$= 6 \cdot 10 + 1 \cdot 10$	renaming 60 and 10
	$= (6 + 1) \cdot 10$	distributive property
	$= 7 \cdot 10$	renaming $6 + 1$
8.	$= 70$	renaming $7 \cdot 10$

[10] It should be emphasized that the above proof demonstrates only that $23 + 49 = 72$. It does *not* prove that the addition algorithm is always valid (which it is when applied to other whole numbers). This would require a somewhat more difficult proof and is beyond the scope of the present discussion.

EXERCISES 6–3

S–1. Use the method of grouping into bundles of ten to illustrate the addition algorithm.

	a.	12	b.	38	c.	17
		+23		+27		29
						+35

S–2. The justification of the addition algorithm (for the special case $23 + 49$) is an example of a deduction. That is, given that the commutative and associative properties hold for addition, we can deduce the fact that $23 + 49 = 72$. Follow the steps in the text to deduce each of the following sums. Give a reason for (justify) each step.

	a.	12	b.	38	c.	47
		+23		+57		+71
		35		95		118

E–3. We have demonstrated the addition algorithm (with carrying) with 2-digit numbers. Can you state the algorithm for 3, 4, and 5 digit numbers? Can you represent the steps in the algorithm iconically? Concretely?

E–4. We can add larger numbers by grouping ten bundles of ten into bundles of one hundred and then proceeding as in the text. Using this method, represent the steps in each of the following.

	a.	234	b.	178	c.	473_8
		+83		+365		$+67_8$

Hint: Part c requires grouping into piles of 8 and $8^2 = 64$.

M–5. a. Explain how you might use bundles of pencils (ten per bundle) and individual pencils to represent the steps in the addition algorithm for $36 + 17$.
   b. What other concrete objects or icons could you use to demonstrate the addition algorithm?

M–6. List two ways you might have your students demonstrate their knowledge of the addition algorithm. Do you have any preferences? Why?

M–7. What examples would you use to see if your students had mastered the addition algorithm? Give at least four test items you would use and explain why you chose these particular ones.

For further thought and study: (A) examine several elementary text book series to see how they introduce the addition algorithms; and (B) see Dienes, *Mathematics in the Primary School*, New York: St. Martin's Press, 1964, to see how multibase blocks can be used to introduce the addition algorithm.

4. SUBTRACTION ALGORITHMS

There are several subtraction algorithms, most of which are designed to circumvent the problem of having to subtract a larger digit from a smaller digit of the same place value $\left(\text{e.g., subtracting 7 from 3 in } \begin{matrix} 83 \\ -47 \end{matrix}\right)$. This situation does not arise, of course, in problems where each digit of the larger number is greater than the corresponding digit in the number to be subtracted. For example, in subtracting 41 from 58 (i.e., 58 − 41) the units digit (8) of 58 is greater than the units digit (1) of 41 and similarly, for the tens digits, 5 and 4, respectively. Simple problems of this sort can be solved by the following procedure: Subtract the units digit (1) of the smaller number from the units digit (8) of the larger number, put the result (7) in the units place of the answer; subtract the tens digit (4) of the smaller number from the tens digit (5) of the larger number, put the result (1) in the tens place of the answer. Do the same for the hundreds digits, thousands digits, and so on. When applied to the problem 58 − 41, this procedure corresponds to removing 1 tally mark from a set of 8, and 4 bundles of ten from a set of 5 bundles of ten, as shown below.

This simple procedure is *not* adequate, however, to deal with such problems as 83 − 47. As it stands, the procedure provides no way to "subtract" the 7 units in 47 from the 3 units in 83. Fortunately, there are many ways to fix things up. We shall consider three which are (or have been) in relatively common use. They are called the "borrowing," "equal additions," and "complement" methods.

The "borrowing" procedure can be represented as follows:

$$
\begin{array}{l}
83 = \quad 8 \text{ tens} + 3 \text{ units} \quad = \quad 7 \text{ tens} + 13 \text{ units} \quad = \quad \overset{7}{\cancel{8}}\,\overset{1}{3} \\
\underline{-47 = -(4 \text{ tens} + 7 \text{ units}) = -(4 \text{ tens} + 7 \text{ units}) = -4\ 7} \\
\qquad\qquad\qquad\qquad\qquad\qquad\qquad\qquad\qquad\qquad\qquad\quad 3\ 6
\end{array}
$$

Stated in words, we "borrow" 1 ten from the tens digit and correspondingly increase the units digit by ten (units) before subtracting. Thus, the *number* 83 is first represented by 7 tens and 13 ones (units) (instead of by 8 tens and 3 ones). We then subtract the 47 as in "simple" subtraction.

In "equal additions," the procedure can be represented:[11]

$$
\begin{array}{l}
83 = \quad 8 \text{ tens} + 3 \text{ units} \quad \overset{+10}{\longrightarrow} \quad 8 \text{ tens} + \textcircled{1}3 \text{ units} \quad = \quad 8\,\overset{1}{3} \\
\underline{-47 = -(4 \text{ tens} + 7 \text{ units}) \overset{+10}{\longrightarrow} -(\textcircled{5} \text{ tens} + 7 \text{ units}) = -\overset{5}{\cancel{4}}\ 7} \\
\qquad\qquad\qquad\qquad\qquad\qquad\qquad\qquad\qquad\qquad\qquad\quad 3\ 6
\end{array}
$$

[11] The $\overset{+10}{\longrightarrow}$ designates an increase of ten to the right.

Here, we add 10 to both numbers but in different ways. The 10 is added to the units place (3) in the larger number (83) and to the tens place (4) in the smaller number (47). The remaining steps are the same as before.

In the "complement method" (which is often used in Europe) the procedure is as follows:

$$83 \rightarrow \quad 83 + 53 \quad = \quad 136$$
$$-47 \rightarrow -(47 + 53) = -100$$
$$\overline{\hspace{3cm}36}$$

The rationale for this procedure is that subtraction problems can be made simpler by first transforming them into equivalent problems in which one needs only to subtract a number like 10, 100, 1000, etc. Thus, by first adding 53, which is called the "complement" of 47, to both numbers, the problem, $83 - 47$, is reduced to subtracting 100 from 136. The complement procedure, of course, will only simplify things for the problem solver who knows (or can quickly find) the needed complements.

We have seen that the "carrying" operation in the addition algorithm can be justified by the fact that it is possible to rename numbers (e.g., 7 tens and 13 ones as 83) or, equivalently, to regroup the elements in the sets used to represent them. In order to justify "borrowing" in the subtraction algorithm, we can rely on this same fact. In this case, however, instead of going from something like 7 tens and 13 ones to 83, we go from 83 to 7 tens and 13 ones. Thus, for example, we can justify borrowing in the problem $83 - 47 = 36$ as indicated below.

The "equal additions" and "complement" methods of subtraction can also be justified in terms of sets of tally marks, but we leave this as an exercise.

Borrowing can also be justified symbolically. For example,

1. $83 - 47 = (80 + 3) - (40 + 7)$ renaming 83 and 47
2. $ = (70 + 13) - (40 + 7)$ renaming $(80 + 3)$ as
                                                    $(70 + 13)$

3.	$= (70 - 40) + (13 - 7)$	the basic property[12] of the system of natural numbers given in Chapter 5
4.	$= 30 + 6$	renaming $(70 - 40)$[13] and $(13 - 7)$
5.	$= 36$	renaming $30 + 6$

In the case of "equal additions," the justification is as follows.

1.	$83 - 47 = (83 - 47) + 0$	Adding 0 to any number does not change the sum
2.	$= (83 - 47) + (10 - 10)$	renaming 0 as $10 - 10$
3.	$= (83 + 10) - (47 + 10)$	applying the basic property[12] in Chapter 5 in reverse direction
4.	$= (80 + 13) - (50 + 7)$	renaming $83 + 10$ and $47 + 10$
5.	$= (80 - 50) + (13 - 7)$	applying the basic property[12] in Chapter 5
6.	$= 30 + 6$	renaming $80 - 50$[14] and $13 - 7$
7.	$= 36$	renaming $30 + 6$

The first three steps justify adding 10 to both 83 and 47. The remaining steps, 4–7, correspond to the equal additions algorithm presented earlier.

The "complement" method of subtraction can be justified in a similar manner. Instead of adding $0 = 10 - 10$, however, we add $0 = $ (the complement) $-$ (the complement). In subtracting $83 - 47$, for example, we add $0 = 53 - 53$. Thus, we get

1.	$83 - 47 = (83 - 47) + 0$	as in steps 1–3 in justification of "equal additions"
2.	$= (83 - 47) + (53 - 53)$	
3.	$= (83 + 53) - (47 + 53)$	
4.	$= 136 - 100$	renaming $83 + 53$ and $47 + 53$
5.	$= 36$	renaming $136 - 100$

The above discussion was brief because ideas sometimes get lost in words. You should make sure that you understand what has been said, however, as you will be called on to deal with this material in your teaching. In particular, you should make up other two-, three-, and four-digit subtraction problems and test yourself to make sure that you can justify the algorithm you employ to solve them.

[12] If $a \geq b$ and $c \geq d$, then $(a - b) + (c - d) = (a + c) - (b + d)$.
[13] See footnote 9.
[14] See footnote 9.

EXERCISES 6–4

S–1.   Use each of the algorithms discussed in this section, i.e., borrowing, equal additions, and complement, to perform each of the following subtractions.

a. $\begin{array}{r} 93 \\ -57 \\ \hline \end{array}$
b. $\begin{array}{r} 476 \\ -239 \\ \hline \end{array}$
c. $\begin{array}{r} 921 \\ -737 \\ \hline \end{array}$

S–2.   Justify the following by reference to the underlying meaning. (*Hint:* Use bundles of ten.)

a. $\begin{array}{r} 53 \\ -38 \\ \hline 15 \end{array}$
b. $\begin{array}{r} 81 \\ -33 \\ \hline 48 \end{array}$

[S–3.]   In the text we presented a deductive proof to justify "borrowing" for the case $83 - 47 = 36$. Follow the deduction to justify the "borrowing" algorithm in each of the following subtraction problems.

a. $\begin{array}{r} 53 \\ -38 \\ \hline \end{array}$
b. $\begin{array}{r} 81 \\ -33 \\ \hline \end{array}$

[S–4.]   Follow the deduction in the text to justify the "equal additions" algorithm in each of the following.

a. $\begin{array}{r} 53 \\ -38 \\ \hline \end{array}$
b. $\begin{array}{r} 81 \\ -33 \\ \hline \end{array}$

[S–5.]   Follow the deduction in the text to justify the "complement" algorithm in each of the following.

a. $\begin{array}{r} 53 \\ -38 \\ \hline \end{array}$
b. $\begin{array}{r} 81 \\ -33 \\ \hline \end{array}$

*E–6.   Use any algorithm you wish to perform the following subtractions in base 8.

a. $\begin{array}{r} 74_8 \\ -16_8 \\ \hline \end{array}$
b. $\begin{array}{r} 34_8 \\ -27_8 \\ \hline \end{array}$

M–7.   Explain how you might use bundles of pencils (ten per bundle) and individual pencils to demonstrate the (a) borrowing, (b) equal additions, and (c) complement algorithms for the subtraction problem $35 - 18$.

M–8.   How would you have your students demonstrate their knowledge of each of the subtraction algorithms? Why?

M–9.   Is it important that elementary school children know all three subtraction algorithms? Why?

For further thought and study:

(A) Examine several elementary school textbook series to see how they introduce the subtraction algorithm(s).

(B) Which subtraction algorithm do you prefer to use? Why? Which seems the most "natural"? Which algorithm were you taught in school? Which would you prefer to teach? Why?

5. MULTIPLICATION ALGORITHM

Let us now see how we might justify the use of a typical multiplication algorithm. Consider the problem of finding the product of 23 and 47.

$$
\begin{array}{r}
23 \\
\times 47 \\
\hline
161 \\
92 \\
\hline
1081
\end{array}
$$

As we say,

1. 7 times 3 is 21, put down 1 and carry the 2.
2. 7 times 2 is 14 and 2 makes 16.
3. 4 times 3 is 12, put down the 2 under the 6 and carry 1.
4. 4 times 2 is 8 and 1 makes 9.
5. Now add 161 and 920. (Note that "92" above denotes 920.)

To justify the steps in the algorithm, we make use of the following array.

	3 columns	20 columns
7 rows	A 7×3	B 7×20
40 rows	C 40×3	D 40×20

By the Cartesian product method, we find that array A represents the product, 7×3, and array B, the product, 7×20. Putting the two arrays together to form a new array with seven columns and 23 rows (i.e., making 161 elements) corresponds to taking the sum of 7×3 and 7×20, steps 1 and 2 in our algorithm. Arrays C and D represent the products 40×3 and 40×20. Putting these two arrays together to form an array with 40 rows and 23 columns (with 920 elements) corresponds to steps 3 and 4 in the algorithm. Finally, putting the two newly formed arrays ($A \cup B$ and $C \cup D$) together to form the array with 47 rows and 23 columns corresponds to taking the sum of 161 and 920 (step 5).

The algorithm can also be justified by reference to the distributive property of multiplication over addition. In order to do this, the product 47×23 must first be represented in an expanded form which is analogous to the array above.

1. $47 \times 23 = (7 + 40) \times 23$ — renaming 47 as $7 + 40$

2. $\qquad = (7 \times 23) + (40 \times 23)$ — distributive property of multiplication over addition

3. $\qquad = [7 \times (3 + 20)]$ $+ [40 \times (3 + 20)]$ — renaming 23 as $3 + 20$ (in both places)

4. $\qquad = [(7 \times 3) + (7 \times 20)]$ $+ [(40 \times 3) + (40 \times 20)]$ — distributive property

The multiplication algorithm corresponds to simplifying the expression on the right of step 4. Thus,

4. $\quad [(7 \times 3) + (7 \times 20)] + [(40 \times 3) + (40 \times 20)]$

$= [21 + (7 \times 20)] + [(40 \times 3) + (40 \times 20)]$ — step 1 in algorithm

$= [21 + 140] + [(40 \times 3) + (40 \times 20)]$ — step 2 in algorithm

$= 161 + [(40 \times 3) + (40 \times 20)]$

$= 161 + [120 + (40 \times 20)]$ — step 3 in algorithm

$= 161 + [120 + 800] = 161 + 920$ — step 4 in algorithm

$= 1081$ — step 5 in algorithm

EXERCISES 6–5

S–1. Use an array to justify each of the following.

a. $\begin{array}{r} 13 \\ \times 8 \\ \hline \end{array}$
b. $\begin{array}{r} 13 \\ \times 12 \\ \hline \end{array}$
c. $\begin{array}{r} 23 \\ \times 21 \\ \hline \end{array}$

S–2. The deductive argument given in the text to justify the multiplication algorithm (for the example 47×23) is based on known properties of addition and multiplication and the base 10 numeration system. Follow this deduction to justify each of the following.

a. $\begin{array}{r} 37 \\ \times 6 \\ \hline \end{array}$
b. $\begin{array}{r} 46 \\ \times 51 \\ \hline \end{array}$
c. $\begin{array}{r} 73 \\ \times 46 \\ \hline \end{array}$
*d. $\begin{array}{r} 173 \\ \times 276 \\ \hline \end{array}$

S–3.   Use the distributive law to simplify each of the following problems.
   a. $(39 \times 71) + (39 \times 29)$
   b. $(41 \times 76) + (59 \times 76)$
   c. $(146 \times 468) + (146 \times 532)$

E–4.   Perform the following base 8 multiplications by generalizing the usual multiplication algorithm. (*Hint:* Use the multiplication facts for base 8.)

   a. 23_8
        $\times 41_8$

   b. 27_8
        $\times 35_8$

   c. 216_8
        $\times 43_8$

E–5.   Find a "short-cut" algorithm to perform the following multiplications. State the algorithm both verbally and symbolically and then perform the multiplications.

   a.   25
        $\times 16$

   b.   273
        $\times 25$

   c.   38
        $\times 25$

   d.   736
        $\times 50$

Hint: $273 \times 25 = 273 \times (100 \div 4)$
          $273 \times 25 = (273 \times 100) \div 4$

*E–6.   a.   There is an underlying rule (regularity) in the following multiplications by 11. Express the rule verbally.

   $23 \times 11 = 253$ *Hint:* $5 = 2 + 3$
   $45 \times 11 = 495$ $9 = 4 + 5$
   $71 \times 11 = 781$ $8 = 7 + 1$
   $52 \times 11 = 572$ $7 = 5 + 2$
   $62 \times 11 = 682$ $8 = 6 + 2$

   b.   Express the rule symbolically.
   c.   Use the rule to perform the following:

      i.   34
           $\times 11$

      ii.   27
            $\times 11$

      iii.   63
             $\times 11$

      iv.   16
            $\times 11$

   d.   What are the restrictions on the use of the rule? Will it work for the following examples? Can you find a generalization which will work?

      i.   93
           $\times 11$

      ii.   56
            $\times 11$

      iii.   77
             $\times 11$

M–7.   What is wrong with each of the following *mis*applications of the multiplication algorithms?

   a.   37
        $\times 18$
        296
         37
        333

   b.   37
        $\times 18$
        2456
          37
        2826

   c.   37
        $\times 18$
        286
         37
        656

M–8.   Do children have to know the multiplication facts to use the multiplication algorithm? Explain.

M–9.   How would you have your students demonstrate that they know and can use the multiplication algorithm? Is it enough to state it verbally? Is it enough to do problems with 2-digit numbers? Explain.

For further thought and study:
  (A) State the multiplication algorithm verbally and symbolically.
  (B) Examine several elementary textbook series to see how they introduce the multiplication algorithm.

6. DIVISION ALGORITHM

To see what is involved in justifying "the" division algorithm (i.e., one of the many possible division algorithms), let us consider $77 \div 3$. Listing all of the steps of the division algorithm in even this simple case would be a tedious task; we shall not do it. Instead, we just represent the *result* of carrying out the usual division algorithm.

$$
\begin{array}{r}
25 \quad \text{r } 2 \\
3 \overline{)77} \\
6 \\
\hline
17 \\
15 \\
\hline
2
\end{array}
$$

(1)
(2)
(3)
(4)
(5)

The major steps involved can be rationalized by thinking in terms of sets of elements as follows.

  • Since there are 2 sets of 3 in 7 (the first digit of 77), we see that there are (at least) 20 sets of 3 in 70.
  • But, 20 sets of 3 only account for 60 (the 6—in step (2)) of the 77 elements.
  • This leaves 17 (step (3)) elements unaccounted for.
  • Next, we notice that there are 5 sets of 3 in the unaccounted for 17 elements.
  • This gives a total of 25 sets of 3 in the original 77.
  • However, the 5 sets of 3 account for only 15 (step (4)) of the remaining 17.
  • Hence, 2 of the original 77 elements remain unaccounted for. As we say, the remainder is 2 (r 2).

Another way to justify this algorithm is in terms of the Euclidean Algorithm (which was discussed in the section on division, Chapter 5) together with other properties of the system of natural numbers. In finding the quotient, $77 \div 3$, recall that what we want are the two unique numbers which make true the number sentence $77 = (\square \times 3) + \triangle$, where $\triangle < 3$.

$77 = 70 + 7 = [(20 \times 3) + 10] + 7$ — applying the Euclidean Algorithm to 7 and 3, we get $7 = (2 \times 3) + 1$, which implies (multiplying by 10) that $70 = (20 \times 3) + 10$

$= (20 \times 3) + (10 + 7)$ — associative property of addition

$= (20 \times 3) + 17$ — renaming $10 + 7$ as 17

$= (20 \times 3) + [(5 \times 3) + 2]$ — applying the Euclidean Algorithm to 17 and 3

$= [(20 \times 3) + (5 \times 3)] + 2$ — associative property of addition

$= [(20 + 5) \times 3] + 2$ — distributive property of multiplication over addition

$= (25 \times 3) + 2$ — renaming $20 + 5$ as 25

EXERCISES 6–6

S–1. Perform each of the following divisions. Justify each step.

    a. $5\overline{)76}$ b. $24\overline{)293}$ c. $34\overline{)3296}$

S–2. Use the Euclidean Algorithm to justify each step of the following.

    a. $5\overline{)76}$ b. $24\overline{)293}$ c. $34\overline{)3296}$

E–3. Simple rules exist for testing divisibility by 2, 3, 4, 5, 6, 8, 9, and 10. These rules are as follows:

- A natural number is divisible by 2 if and only if the last (units) digit is even.
- A natural number is divisible by 3 if and only if the sum of the digits is divisible by 3.
- A natural number is divisible by 4 if and only if the numeral formed by the ten and units digit is divisible by 4. (e.g., 3948 is divisible by 4 since 48 is divisible by 4. 3983 is not divisible by 4 since 83 is not.)
- A natural number is divisible by 5 if and only if it ends in 5 or 0.
- A natural number is divisible by 6 if and only if it is divisible by both 2 and 3.
- A natural number is divisible by 8 if and only if the numeral formed by the hundreds, tens and units digits is divisible by 8. (e.g., 293488 is divisible by 8 since 488 is divisible by 8. 293801 is not divisible by 8 since 801 is not.)
- A natural number is divisible by 9 if and only if the sum of the digits is divisible by 9.
- A natural number is divisible by 10 if and only if the final (units) digit is 0.

Use these rules to test each of the following numbers for divisibility by 2, 3, 4, 5, 6, 8, 9, and 10.

    a. 486
    b. 9600
    c. 72,864
    d. 451,008
    e. 977

M–4. a. How could you illustrate the Euclidean Algorithm by dividing a bag of marbles into a given number of piles?

    b. How could you use pennies and dimes to illustrate the division algorithm (for amounts up to 99 cents)?

M–5. Which of the regularities in Exercise E–3 would you expect elementary school children to detect for themselves? What aids could you use to help them?

For further thought and study:

    (A) Examine several elementary mathematics textbooks to see how the division algorithm is introduced.
    (B) Look into the question of when a number is divisible by 7.

7
The System of Positive Rationals

As we learned in Chapter 5, the natural numbers can be used to characterize either the numerosity of sets (of elements) or the order of elements in ordered sets. The natural numbers are not rich enough, however, to deal with a large number of other situations which arise frequently in everyday life. (1) Suppose, for example, we want to talk about part of an object. In this case, the natural numbers are not adequate. (2) Similarly, no single natural number can adequately represent the difference between two given quantities since both magnitude and direction are involved. It is impossible, for example, to distinguish between having a check for $5.00 and giving an IOU for the same amount. (3) The simultaneous representation of two or more distinct quantities poses an even more general problem. The representation of a balance of $103.47 in one bank, a balance of $53.00 in a second, and a debt of $250.00 in a third, cannot adequately be represented by any one number, natural or otherwise.

Clearly, the first example necessitates the use of fractional numbers, whereas the second involves signed numbers. The last example requires the use of vectors. As we shall see, numbers representing each of these quantities may be defined in terms of natural numbers. In particular, although single natural numbers might not suffice to represent the desired quantities in such situations, two or more may. With this in mind, we turn to the task of showing how such quantities may be viewed as *extensions* of the natural numbers.

In this chapter, we discuss fractional numbers and positive and non-negative rational numbers. Chapter 8 is concerned with integers (i.e., signed numbers) and Chapter 9, with rational numbers, real numbers, vectors, and complex numbers.

FRACTIONS AND POSITIVE RATIONAL NUMBERS

1. FRACTIONS

In Chapter 5, we saw that natural numbers may be defined as properties of sets, namely properties that refer to numerosity (i.e., number of elements). In much the same way, *fractions may be thought of as properties of ordered pairs of sets* because in this case we are concerned with the relative "size" of the two sets. For example, the *pair* of sets $(\{a, b, c\}, \{d, e, f, g\})$ has the *fraction property* of having three elements in the first set of the pair relative to four elements in the second. In effect, we can represent the fraction property of a pair of (nonempty) sets by an ordered pair of natural numbers, with the understanding that we are concerned with the relative size of the first number compared with the second.[1] The first number of the ordered pair, which we call the *numerator*, corresponds to the number of elements in the first set and the second number, which we call the *denominator*, corresponds to the number of elements in the second set. Thus, the fraction property "three to four," of the pair $(\{a, b, c\}, \{d, e, f, g\})$, can be represented by the ordered pair $(3, 4)$. Indeed, this is precisely what we are doing when we write fractions in the usual fractional form (e.g., $\frac{3}{4}$ or 3/4) or when we express the ratio of two sets as an ordered pair of rational numbers (e.g., $3:4$—which is read, "the ratio, 3 to 4"). Whatever form is used, however, two natural numbers are required and the same fraction property is denoted. We usually distinguish the particular form a/b (e.g., $\frac{3}{4}$) with the name, *fraction*. The important thing to notice here is that "$\frac{3}{4}$" (or "$3:4$" or "$(3, 4)$," for that matter) is a *name* for the fraction property, three-fourths, and not the property itself.

The relative size of more than one pair of sets, of course, may be represented by the same fraction. For example, the fraction property of each of the following pairs of sets can be represented by the fraction, $\frac{1}{2}$: $(\{0\}, \{*, \square\})$, $(\{\triangle\}, \{\bigcirc, \square\})$, $(\{a\}, \{a, b\})$, $(\{a\}, \{b, c\})$. In general, we say that two pairs (of nonempty sets) are *fractionally equivalent* (i.e., have the same fraction property) if the elements of the first and second sets of one pair can be put into one-to-one correspondence with the elements of the corresponding sets of the second ordered pair. Notice that each of the above pairs is fractionally equivalent to the others. Thus, for example, the respective sets of the pair $(\{0\}, \{*, \square\})$ can be put into one-to-one correspondence with the sets of the pair $(\{a\}, \{a, b\})$.

Recalling that natural numbers may be thought of as equivalence classes of sets, this suggests that fractions effectively define *equivalence classes of pairs* of nonempty sets. Each such equivalence class, of course, contains all and only those pairs which can be put into one-to-one correspondence. Thus, for example, the equivalence class defined by the fraction $\frac{7}{8}$ includes precisely those pairs of sets such that the first set contains seven elements and the second contains eight.

We shall call any pair of sets belonging to the equivalence class defined by a fraction an *interpretation* of the fraction. Logically speaking, every interpretation of a fraction is equally as good as any other interpretation. Nonetheless, it is of some heuristic interest to distinguish between different types of interpretations. Thus, the interpretations (they are *not* called that) used in most elementary school

---

[1] The qualifying phrase involving relative size is important because ordered pairs of natural numbers may also be used to represent signed numbers. See Chapter 8.

textbooks are of one of two general types. In each type, the basic elements are common to both sets of a given pair. Thus, pairs like ({a}, {b, c}) or ({△}, {○, □}), which were included above as examples, are *not* considered.

Perhaps the most familiar type involves sets whose elements are equal-sized parts of some whole. For example, Figure 7–1 can be thought of as an interpretation of the fraction $\frac{2}{3}$ since the two shaded thirds can be thought of as elements in the first set of a pair (e.g., {◒ , ◖▨}, and the three thirds constituting the whole circle can be thought of as elements in the second (e.g., {◒ , ◖▨, ◗}). (Note that the shaded portions of the circle are common to both sets in the interpretation.)

Figure 7–1

Another common type of interpretation involves pairs of sets with common elements, in which the elements are *wholes* (e.g., whole apples). For example, consider the following interpretation of the fraction $\frac{2}{3}$:

$$(\{\mathbf{apple},\mathbf{apple}\}\{\mathbf{apple},\mathbf{apple},\mathbf{apple}\})$$

(In elementary school textbooks, this interpretation would be displayed more simply as, $\mathbf{apple}\ \mathbf{apple}\ \mathbf{apple}$.)

A less common variant of this type involves interpretations where the elements are themselves sets (with equal numbers of elements). For example, consider the display

$$(\mathbf{apple\ apple})\ (\mathbf{apple\ apple})\ (\mathbf{apple\ apple})$$

which can also be viewed as an interpretation of $\frac{2}{3}$. In this case, the display implicitly consists of the pair

$$(\{(\mathbf{apple\ apple}),(\mathbf{apple\ apple})\}\{(\mathbf{apple\ apple}),(\mathbf{apple\ apple}),(\mathbf{apple\ apple})\})$$

We caution the reader that this is *not* an interpretation of the fraction, $\frac{4}{6}$. The elements in the interpretation above are sets of pairs of apples. Contrast this with the following interpretation of $\frac{4}{6}$

$$\mathbf{apple}\quad\mathbf{apple}\quad\mathbf{apple}\quad\mathbf{apple}\quad\mathbf{apple}\quad\mathbf{apple}$$

In the latter situation, the elements are apples, not *pairs* (or sets) of apples. Strictly speaking, then, "$\frac{4}{6}$" does *not* mean the same thing as "$\frac{2}{3}$"; they have quite different interpretations. There is obviously a close relationship between these two fractions, however, and we consider this question again in the section on *rational* numbers.

As a second example of this type, consider the display

$$(a\ a\ a) \quad (a\ a\ a) \quad (a\ a\ a) \quad (b\ b\ b)$$

which may be thought of as an interpretation of the fraction $\frac{3}{4}$, where the basic elements are triples (i.e., sets containing three elements).

So far, we have viewed fractions in only one way, as pairs of sets. We shall call this a *state-state* view of fractions. In this case, we are simply using fractions to represent the "static" fractional property of an observable display of the sort shown above.

In many situations, it seems more natural to view fractions in terms of *operators*. Consider, for example, a situation in which we are given a whole pie and wish to serve a normal-sized piece to each of five people. We must first divide the pie into some agreed-upon number of equal-sized pieces (say, eight) (we normally do not divide pies into fifths). Second, we must give out the correct number of pieces (i.e., five). In this case, rather than representing the quantity of pie needed, the two numbers involved (i.e., eight and five) refer more naturally to two operations (i.e., actions): cutting the pie into equal sized pieces and serving the required number of pieces. Cutting the pie, of course, corresponds to the denominator of a fraction and serving it corresponds to the numerator.

In effect, the fraction $\frac{5}{8}$ may represent the fractional property of a pair of operators (i.e., cutting into eight pieces and selecting five of them). It should be emphasized, however, that this way of viewing a fraction is logically equivalent in every way to viewing a fraction as representing a property of a pair of sets.[2] The only difference is in the kinds of interpretations we allow. The state-state view is restricted to static displays. The operator-operator view deals with pairs of actions.

As a second example of the latter type, consider the actions involved in dividing an orange into quarters and giving one to each of three children. Can you think of other situations of this type? Any pair of operations will do.

All this leads us to still a third and fourth way of viewing fractions. *Fractions may be used to represent properties of either operator-state pairs or state-operator pairs.* In the first case, the denominator of the fraction represents the number of elements in some set (the state) and the numerator represents the number of elements we are to select. Consider, for example, a situation where we are given a dozen eggs (the state) and wish to take five of them. We can represent this operator-state pair as the fraction $\frac{5}{12}$. (Equivalently, of course, we can view this operator-state pair as an *interpretation* of the fraction $\frac{5}{12}$.)

In the state-operator view, everything is reversed. The numerator of the fraction (representing the fraction property) corresponds to a state, and the denominator, to the operation of dividing the state into equivalent parts. The process of dividing three oranges into four equivalent parts, for example, pro-

---

[2] This is because of the close relationships (as discussed in Chapter 2) between sets, relations, and operations.

vides an interpretation of the fraction $\frac{3}{4}$. Can you find an interpretation of this type for the fraction $\frac{2}{3}$?

After reading this section, you may wonder why we have gone into all of this. Doesn't it complicate what is essentially a very simple idea? One answer that can be given to such a question is that apparently simple ideas often turn out to be the most difficult to fully understand. But this does not really get at the intended question. For elementary school teachers the main point is simply that young children are frequently bothered by some of the (rather subtle) differences discussed above. Only by being explicitly aware of such subtleties can the teacher hope to avoid problems that may otherwise arise in the classroom. In particular, it may otherwise be difficult (if not impossible) to detect some of the problems children have in "understanding" fractions—in knowing where and how to use them.

EXERCISES 7–1

S–1.   What property is common to each of the following ordered pairs of sets? How may this property be represented?

$(\{.\ .\}, \{.\ .\ .\ .\})$ $(\{p, q\}, \{a, b, c, d\})$

$(\{*, \circ\}, \{*, \circ, \square, \infty\})$

S–2.   Which of the following pairs of sets is fractionally equivalent to $(\{a\}, \{x, y, z\})$?

    a. $(\{z\}, \{b, c, d\})$
    b. $(\{x, y, z\}, \{a\})$
    c. $(\{\bigcirc\}, \{\bigcirc\bigcirc\bigcirc\})$
    d. $(\{\triangle, \square\}, \{\diamond, \lozenge\})$

S–3.   How many different fractions are represented by the pairs of ordered sets shown in Exercise S–2? Name the fraction representing each pair.

S–4.   Consider the pairs

    a. What fractions represent x and y, respectively?
    b. What is the difference between the elements of the sets in pair x and in pair y?
    c. Are x and y fractionally equivalent?

S–5.   Write a fraction for which

might be an interpretation. (*Hint:* There are four reasonable answers.)

S–6. Give a concrete interpretation of the fraction $\frac{3}{7}$ as

    a. a state-state pair
    b. an operator-operator pair
    c. an operator-state pair
    d. a state-operator pair

E–7. Although $\frac{5}{3}$ is a legitimate fraction, it is often called an "improper" fraction because its numerator is larger than its denominator. Suggest a plausible interpretation of $\frac{5}{3}$ as

    a. a state-state pair
    b. a state-operator pair
    c. an operator-state pair
    d. an operator-operator pair

M–8. Why are fractions with numerator 1 (e.g., $\frac{1}{2}$, $\frac{1}{3}$, $\frac{1}{4}$) often introduced to elementary school children before fractions with larger numerators?

M–9. Write some informal questions which you could ask a child to determine whether he understands the concept of fraction.

For further thought and study:

    (A) How might Cuisinaire rods be used to develop the notion of fractional equivalence?
    (B) The use of "improper" fractions such as $\frac{4}{3}$ is often prohibited by elementary school teachers and "mixed" numbers, such as $1\frac{1}{3}$ are used instead. Do you agree with this usage? If you would retain both types of numerals, which would you introduce first?
    (C) Investigate the treatment of fractions in a particular elementary textbook series.

        (a) At what grade level are fractions first introduced?
        (b) Which particular fractions or types of fractions are considered earliest, which at later levels?
        (c) What type or types of interpretation are given when fractions are first introduced? Are other types of interpretation used at later levels?
        (d) Does the series present "improper" fractions, "mixed" numbers, or both?

2. POSITIVE RATIONAL NUMBERS

In the previous section, we noted that the fractions $\frac{2}{3}$ and $\frac{4}{6}$ can be interpreted in very similar ways. In particular, we showed that the fraction $\frac{2}{3}$ represents the fractional property of whereas $\frac{4}{6}$ is required to represent the fractional property of the closely related display

The former says, in effect, that there are three pairs of apples and that two of these pairs are shaded. The latter says that there are six apples, four of which are shaded. Nevertheless, the same number of apples is involved in each case. This is not a chance occurrence, of course. Thus, for example, the pairs of interpretations represented in Figure 7–2 correspond in a similar fashion.

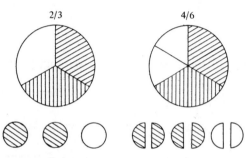

Figure 7–2

In effect, the two fractions represent the same "amount." Furthermore, there are any number of other fractions which also represent this amount. The fractions $\frac{2}{3}$, $\frac{4}{6}$, $\frac{6}{9}$, $\frac{8}{12}$, $\frac{10}{15}$, and $\frac{12}{18}$ are just a few. In fact, any fraction of the form $\frac{2 \times n}{3 \times n}$, where n is a natural number, will do.

This general idea applies to all fractions. Whether we are talking about apples, pieces of pie, or any other particular interpretation, the fractions $\frac{1}{4}$, $\frac{2}{8}$, $\frac{3}{12}$, and so on all represent the same "amount." For illustrative purposes, consider the following interpretations:

If we are primarily interested in representing the quantity of something and not in the particular way that something "looks," then there are any number of different fractions we might use. As we shall say, fractions that represent the same quantity or amount are *equivalent*.

We can formally define this idea of equivalence as follows: Two fractions a/b and c/d are said to be *equivalent*, denoted $a/b \cong c/d$, if and only if $a \times d = b \times c$. For example, the fractions $\frac{2}{3}$ and $\frac{4}{6}$ are equivalent because $2 \times 6 = 3 \times 4$.

As noted in Chapter 2, equivalence relations partition universal sets into exhaustive (i.e., every element is in some subset) and mutually exclusive (i.e., disjoint) subsets, called equivalence classes. Here, the relation, $\cong$, is an equivalence relation which partitions the set of all fractions into equivalence classes. The equivalence classes

$$\left\{\frac{2}{3}, \frac{4}{6}, \frac{6}{9}, \frac{8}{12}, \ldots, \frac{2n}{3n}, \ldots\right\}$$

and

$$\left\{\frac{1}{4}, \frac{2}{8}, \ldots, \frac{n}{4n}, \ldots\right\}$$

are just two examples. Equivalence classes of fractions are called *positive rational numbers*.[3]

EXERCISES 7–2

S–1. Verify that $\frac{4}{6}$, $\frac{6}{9}$, $\frac{8}{12}$, and $\frac{212}{318}$ are all equivalent to $\frac{2}{3}$, but that $\frac{4}{7}$ is not.

S–2. a. Does the set of fractions $\left\{\frac{1}{4}, \frac{2}{8}, \frac{3}{12}, \ldots, \frac{n \cdot 1}{n \cdot 4}, \ldots\right\}$ where n is a natural

number include *all* fractions equivalent to $\frac{1}{4}$? Give a reason for your answer.

   b. Does the set containing all fractions of the form $\frac{m \cdot 2}{m \cdot 6}$ (where m is a

   natural number) include all fractions equivalent to $\frac{2}{6}$? Give a reason for

   your answer.

S–3. Give a precise description of the equivalence class of fractions of the rational number $\frac{28}{12}$.

---

[3] With fractions, the distinction between names and numbers is sometimes important. To see this, consider what we tell a child who asks why we cannot replace $\frac{6}{15}$ by $\frac{2}{5}$ in the statement, "3 divides the numerator of $\frac{6}{15}$" (clearly, 3 does not divide the numerator of $\frac{2}{5}$).

To answer this question, we must distinguish between numbers (i.e., fraction properties) and names of numbers (i.e., fractions). The fractions "$\frac{2}{5}$," "$\frac{6}{15}$," "$\frac{12}{30}$," etc., are all names of a certain (rational) number. This number has many nonfractional names as well (e.g., ".4," "$1 - .6$," and "$\sqrt{.16}$"), and we may use any one of these names in statements about the number. For example, "$\frac{2}{5} = (.5 \times .8)$," "$\frac{6}{15} = (.5 \times .8)$," "$\sqrt{.16} = (.5 \times .8)$," etc., all make the same statement about the number having the names "$\frac{2}{5}$," "$\frac{6}{15}$," "$\sqrt{.16}$," etc.

On the basis of this distinction between a number and its name, our opening statement should read, "3 divides the numerator of '$\frac{6}{15}$,'" because we are making a statement about the name "$\frac{6}{15}$" and not the number it denotes. Hence, there is no reason, *a priori*, to expect this statement to be true when a different name is inserted—whether this name refers to the same number or a different number.

Failure to distinguish between a number and its names also explains the statement, " '.3549724' is larger than '3,' " by UICSM's now famous Eskimo. After all, he measured the names with a ruler.

It is also worth noting that there are relatively few places in mathematics where failure to distinguish between objects and their names results in confusion. The question raised above is, perhaps, a legitimate cause of confusion, whereas few literate adults would share the Eskimo's problem.

It is not suggested that the distinction between a number and its names be thrust formally on elementary school students, but teachers should be aware of it and, by example, should provide their students with an intuitive understanding of this distinction.

S–4. $\frac{2}{5}$ and $\frac{4}{10}$ are equivalent fractions. The following pair of sets has the fractional property $\frac{4}{10}$ but not $\frac{2}{5}$. Show, however, that the elements may be regrouped to yield a different pair of sets with the fractional property $\frac{2}{5}$.

```
     o              x  x
  o     o           x x x
     o              x     x
                    x  x  x
```

[E–5.] $\frac{5}{8}$ is a symbolic representation of a rational number. Give an iconic representation of the same number (see Chapter 1).

E–6. a. Using a/b, c/d, and e/f to represent fractions, state the three conditions which must be satisfied by $\cong$ in order that it be an equivalence relation (refer to Chapter 2, Section 5.1).

b. Show that $\cong$ does satisfy the three properties you listed in *a*.

M–7. What would you expect your students to *do* to demonstrate that they understand the idea of equivalent fractions?

M–8. Make up some questions you could ask your students to test their knowledge of equivalence of fractions.

For further thought and study:

(A) Describe a game one could play with two bags of marbles to develop the notion of equivalence of fractions.

(B) Consulting your favorite elementary textbook series find out

(a) what procedures are given for determining whether two fractions are equivalent (or equal).

(b) whether the series makes use of the idea of equivalence classes of fractions.

3. REPRESENTING POSITIVE RATIONAL NUMBERS

3.1 Fractional Representation of Positive Rationals

In view of the way we have defined (positive) rational numbers, it seems reasonable to represent them in set notation. Thus, we could represent one rational number by

$$\left\{ \frac{1}{1}, \frac{2}{2}, \frac{3}{3}, \frac{4}{4}, \cdots \right\}$$

and another by

$$\left\{ \frac{1}{2}, \frac{2}{4}, \frac{3}{6}, \frac{4}{8}, \cdots \right\}$$

or, if we prefer to use set builder notation, we could represent them by

$$\left\{ \frac{a}{b} \,\middle|\, \frac{a}{b} \cong \frac{1}{1} \right\} \text{ and } \left\{ \frac{c}{d} \,\middle|\, \frac{c}{d} \cong \frac{1}{2} \right\}$$

respectively, where a/b and c/d are fractions.

Fortunately, we do not have to resort to this. Use of such notation would be extremely cumbersome and difficult to work with. The standard approach is to allow any fraction in each equivalence class (a rational number) to represent that rational number. Thus, "$\frac{2}{3}$," "$\frac{4}{6}$," "$\frac{6}{9}$," ..., "$\frac{24}{36}$," and so on are all names for the rational number

$$\left\{ \frac{2}{3}, \frac{4}{6}, \frac{6}{9}, \ldots, \frac{22}{33}, \frac{24}{36}, \frac{26}{39}, \ldots \right\}$$

Frequently we choose that fraction (i.e., member of the equivalence class) which is simplest, in the sense that the numerator and denominator are relatively prime. (Recall from Chapter 5 that two numbers are *relatively prime* if there is no natural number greater than one which divides them both.) Here the fraction "$\frac{2}{3}$" has this property. (Satisfy yourself that this is true.) The fraction "$\frac{1}{2}$" plays the same role with respect to the rational number

$$\left\{ \frac{1}{2}, \frac{2}{4}, \frac{3}{6}, \frac{4}{8}, \ldots \right\}$$

Such fractions are said to be "reduced" to *lowest terms* and may be called *canonical* representations.

For many purposes, canonical representations are not the best way to represent rational numbers. In particular, it is often difficult to determine which of two rational numbers is "larger" when they are both represented in canonical form. Consider, for example, which of the two fractions, $\frac{1}{6}$ or $\frac{2}{15}$, represents the greater amount (i.e., larger rational number). What about $\frac{1}{3}$, $\frac{2}{7}$, and $\frac{5}{16}$?

It is much simpler to order rational numbers as to size when the representing fractions have the same denominator. In this case, one has only to order the (rational) numbers according to the size of the numerators (of the fractions used to represent them). For example,

$$\frac{11}{15} > \frac{8}{15} > \frac{7}{15}$$

Of course, the fractions used to represent a set of rational numbers will not always have the same denominators. Given any (finite) set of fractions, however, it is always possible to find another set (of fractions) which represent the same rational numbers and which also have the same denominator. In fact, there are any number of possible sets of fractions with these properties, but, generally speaking, we will want to find the simplest (i.e., the set in which the fractions have the smallest possible denominator).

The basic approach is to find the *least common multiple* (see Chapter 5) of the denominators of the given fractions and then to find fractions, with this least common multiple as denominator, which are equivalent to the given ones. To order the numbers $\frac{1}{3}$, $\frac{2}{7}$, and $\frac{5}{16}$, for example, we first find the least common multiple, which in this case is $3 \times 7 \times 16 = 336$. Next, we determine the numerators x, y, and z in $\frac{1}{3} \cong x/336$, $\frac{2}{7} \cong y/336$, and $\frac{5}{16} \cong z/336$, respectively.

According to our definition of equivalence (i.e., $a/b \cong c/d$ if and only if $a \cdot d = c \cdot b$) we can find x, y, and z by solving the following equations:

$$3x = 1 \times 336$$

$$7y = 2 \times 336$$

$$16z = 5 \times 336$$

This gives $x = 112$, $y = 96$, $z = 105$. Thus, $\frac{1}{3} \cong \frac{112}{336}$, $\frac{2}{7} \cong \frac{96}{336}$, and $\frac{5}{16} \cong \frac{105}{336}$. The desired ordering follows directly. That is, the rational number (originally denoted by) $\frac{1}{3}$ is greater than the rational $\frac{5}{16}$ is greater than the rational $\frac{2}{7}$ because $\frac{112}{336} > \frac{105}{336} > \frac{96}{336}$.

We need not limit ourselves to least common multiples. As suggested above, any common multiple will do. In the present example, $\frac{1}{3} \cong \frac{224}{672} \cong \frac{336}{1008}$, $\frac{2}{7} \cong \frac{192}{672} \cong \frac{288}{1008}$, and $\frac{5}{16} \cong \frac{210}{672} \cong \frac{315}{1008}$. Hence, the desired ordering could also be obtained by comparing either of the sets

$$\left\{ \frac{224}{672}, \frac{192}{672}, \frac{210}{672} \right\} \text{ or }$$

$$\left\{ \frac{336}{1008}, \frac{288}{1008}, \frac{315}{1008} \right\}$$

There are other possibilities as well, of course.

The notion of least common multiple will come up again in our discussion of the addition and subtraction algorithms.

EXERCISES 7–3.1

S–1. a. List five other names for the rational number $\frac{10}{24}$.
b. Display the number as an equivalence class of fractions.

S–2. Two distinct mathematical ideas are represented by the symbol $\frac{7}{8}$. What are they?

S–3. Order each of the following lists of rational numbers.

a. $\dfrac{2}{11}, \dfrac{12}{11}, \dfrac{7}{11}, \dfrac{19}{11}$

b. $\dfrac{7}{12}, \dfrac{3}{4}, \dfrac{2}{3}$

c. $\dfrac{2}{9}, \dfrac{4}{15}, \dfrac{1}{5}$

E–4. a. A student claims that one fraction is smaller than a second because the denominator of the first is larger, but the numerators are the same. Is he right? How would you react to his method of comparison? (*Hint:* Try some specific examples.)
b. Can you tell anything about the relative size of a/b and c/d if $a > c$ and $b < d$?

c.   Can you tell anything about the relative size of a/b and c/d if $a < c$ and $b < d$?

M–5.   a.   Find the error in this computation.

$$\frac{7 \times \overset{1}{\cancel{3}} + 4}{\underset{3}{\cancel{15}}} = \frac{7 + 4}{3} = \frac{11}{3}$$

How would you explain that it is not correct?

b.   Change part a so that "cancellation" is allowed? What justifies it?

M–6.   How could you make clear to fourth- or fifth-grade pupils the distinction between fractions and rational numbers without using the sophisticated notion that a rational number is an equivalence class of fractions?

E–7.   A common rule for reducing a fraction to lowest terms is, "Divide the numerator and denominator by their greatest common divisor." Is this rule valid? Explain your answer and give an example.

For further thought and study: Refer to the treatment of rational numbers in an elementary textbook series. Is the term "rational number" used? If so, how is it described or defined?

3.2 Decimal Representation of Positive Rationals

There is one additional but very special way of representing (positive) rational numbers. We are all familiar with decimals like .25, .6, 3.72, .0657 and so on. Such decimals are called *terminating decimals*.

Each terminating decimal corresponds precisely to a unique fraction with a denominator that is a power of ten: 10, 100, 1000, ..., 10^n, ... For example, $.25 = \frac{25}{100}$, $.6 = \frac{6}{10}$, $3.72 = \frac{372}{100}$, and $.0657 = \frac{657}{10,000}$. In an important sense, then, terminating decimals are identical to fractions. They simply serve as another type of *canonical* representation of rational numbers. Thus, instead of representing rationals as (canonical) fractions (e.g., $\frac{1}{4}, \frac{2}{5}, \frac{1}{8}$, and $\frac{19}{20}$), we can (in certain cases— see below) represent them as fractions (e.g., $\frac{25}{100}, \frac{4}{10}, \frac{125}{1000}$, and $\frac{95}{100}$), whose denominators are powers of ten, or corresponding decimals (e.g., .25, .4, .125, and .95).[4]

Given a fraction, the procedure traditionally used to find the corresponding decimal representation is to divide the denominator into the numerator.[5] Thus,

---

[4] When we discuss the computational algorithms used with rational numbers, we shall see important advantages in representing rational numbers as terminating decimals. In particular, the algorithms used with the natural numbers carry over directly to rationals, which can be represented as terminating decimals.

[5] Strictly speaking, it may not be possible to divide the numerator by the denominator of a fraction. Denominators and numerators are natural numbers and, as we saw in Chapter 5, division over the natural numbers is not always defined. Methods exist, however, for finding decimal equivalents which are always defined. One such method goes as follows: Multiply the numerator n by 10^k and divide this product by the denominator d, obtaining a quotient q

$\frac{1}{4} = .25$ because 4 divided into 1.0 is .25. Similarly, for the other fractions above. Check them!

Unfortunately, only *certain* rational numbers (and, hence, only certain fractions) can be represented as terminating decimals. Consider, for example, the rational number named by the fraction $\frac{1}{3}$. In this case, 3 divided into 1.0 gives .333 ... where the "..." indicates that the process does not terminate but goes on forever.

To show that there is no terminating decimal that represents the rational number named by $\frac{1}{3}$, we assume first that there is such a decimal and show that this forces us to accept some absurdity (like $1 = 0$). From this, we reason that the assumption that there *is* such a terminating decimal is false. This method of proof is used widely in mathematics and is called an *indirect proof*.

1. If there were a terminating decimal that represented the same rational number as the fraction $\frac{1}{3}$, then there would also be some equivalent fraction of the form $a/10^n$ where $\frac{1}{3} \cong a/10^n$ (and a is a natural number). (Remember that every terminating decimal can be represented as a fraction of the form $a/10^n$.)

2. By our definition of equivalence, this would imply that $1 \cdot 10^n = 3 \cdot a$ or that $a = 10^n/3$.

3. Since a is a natural number, 3 would have to divide 10^n (evenly). But if 3 divided 10^n, then 10^n would have to contain 3 as a (prime) factor.

4. And if 10^n contained 3 as a (prime) factor, then 10 would have to contain 3 as a (prime) factor (because the only prime factors of 10^n are the prime factors of 10, each repeated n times).

5. The only prime factors of 10, however, are 5 and 2 ($10 = 5 \cdot 2$).

6. In effect, the assumption that $\frac{1}{3}$ is equivalent to a fraction of the form $a/10^n$ has led us to an absurdity (i.e., that 10 has 3 as a factor).

7. Hence, there is no fraction of the form $a/10^n$ which is equivalent to $\frac{1}{3}$ and, in turn, there is no terminating decimal which represents the same rational number as $\frac{1}{3}$.

This proof (that there is no terminating decimal that corresponds to $\frac{1}{3}$) says something even more general. The same argument can be repeated with any other fraction, which is *reduced to lowest terms* and whose denominator is not 1. And the only times one will not reach a contradiction will be when the only prime factors of the denominator of the fraction in question are 5 or 2, which are the prime factors of 10. (These factors may be repeated any number of times, of course.) In every other case, a contradiction (absurdity) will be attained. Convince yourself of this by testing the fraction $\frac{2}{7}$, on the one hand, and $\frac{3}{5}$, on the other. All you need do is replace $\frac{1}{3}$ in the above proof by the fraction in question. In the case of $\frac{3}{5}$, of course, no absurdity will be obtained.

These observations lead us to the following general statement: *The only fractions* (reduced to lowest terms) *that correspond to terminating decimals are*

(and a remainder r, which we will ignore). k should be chosen large enough to give as many digits in the quotient as desired. (The more digits in the quotient the greater the degree of accuracy where there is a remainder.) If q does not have at least k digits, add enough zeros to the left to make k digits. Place the decimal point to the left of the last k digits of q. The result is the decimal equivalent of n/d (to some degree of accuracy).

those in which the only prime factors of the denominator of the fraction are 2 and/or 5.[6]

Rational numbers that cannot be represented as terminating decimals can be represented as *non*terminating, *repeating* decimals. For example, we can represent the rational number $\{\frac{1}{3}, \frac{2}{6}, \frac{3}{9}, \ldots\}$ by the repeating decimal, .333 ... (where "..." indicates an infinite sequence of 3's). The rational number $\{\frac{2}{7}, \frac{4}{14}, \frac{6}{21}, \frac{8}{28} \ldots\}$ can be represented similarly by the repeating decimal, .285714285714 ... (where the "..." here indicates that the sequence of digits 285714 is repeated indefinitely).[7] Another type of notation that may be used to indicate repeating decimals is $.\overline{3}$ and $.\overline{285714}$, where the bar is placed over the sequence of digits which repeats.

Furthermore, terminating decimals may also be thought of as repeating decimals in which the repeating digit is always 0. For example, the terminating decimal .205 can also be represented as .205000 ... (or $.20\overline{50}$)[8] (Terminating decimals are preferred in practice, of course, because they are easier to work with.) Hence, *every rational number can be represented as a (nonterminating) repeating decimal.* (It is also true that every repeating decimal represents a rational number.)

---

[6] A general proof of this statement would go something as follows:

1. If a rational number is to be represented by a terminating decimal, then there must be some representing fraction, $a/10^n$ whose denominator is a power of 10. This follows from the definition that every terminating decimal may be represented by a fraction whose denominator is a power of 10.

2. Let b/c be a fraction, reduced to lowest terms, which also represents this rational number. Then, $b/c \cong a/10^n$, because they both represent the same rational number. This implies that $a \times c = b \times 10^n$ or that $a = b \times 10^n/c$ by our previous definition of equivalence of fractions.

3. Now, because b and c (in b/c) are relatively prime, the only way we can find the (natural) number a would be if c divides 10^n.

4–5. But, the prime factors of 10^n are simply the factors of 10 (i.e., 2 and 5) each repeated n times.

6–7. Hence, $a = b \times 10^n/c$ has a solution (i.e., a exists) if and only if the only prime factors of c are 2 and/or 5, each of which may be repeated any number of times. If c has any other prime factors, it could not divide 10 (or 10^n).

[7] The reader who may wonder why .333 ... represents the same rational number as $\frac{1}{3}$ may be convinced by the following argument. (We choose not to call this argument a proof because it makes certain subtle assumptions which may be no more obvious than the thing to be proved. A rigorous proof would require a definition of equivalence for infinite sequences of fractions and is beyond the scope of this book.)

$$
\begin{array}{ll}
10 \times .333 \ldots = 3.333 \ldots & \\
-1 \times .333 \ldots = .333 \ldots & \\
\hline
9 \times .333 \ldots = 3 & \text{subtraction and the distributive property of multiplication} \\
& \text{over subtraction} \\
.333 \ldots = \frac{3}{9} & \text{dividing both sides by 9} \\
.333 \ldots = \frac{1}{3} & \text{since } \frac{3}{9} = \frac{1}{3}.
\end{array}
$$

In much the same way one can show that $\frac{2}{7}$ corresponds to .285714 .... (*Hint:* Consider 10^6.)

[8] In fact, every rational number that can be represented by a terminating decimal can also be represented by a nonterminating decimal whose repeating digit is 9. For example, .20500 ... = .20499 ...; 1.0 = 1.000 ... = .999 ...; .25 = .25000 ... = .24999 ...; .301 = .30100 ... = .300999 ....

Not all nonterminating decimals, however, are rational numbers. As we shall see in the section on real numbers, there are decimals for which no corresponding rational numbers exist.

EXERCISES 7–3.2

S–1. Find the terminating decimal that represents each of the following rational numbers.

a. $\dfrac{3}{4}$ b. $\dfrac{5}{8}$ c. $\dfrac{16}{5}$ d. $\dfrac{6}{3}$

(Be careful! The result must be a decimal.)

S–2. Which of the following numbers have no prime factors other than 5, 2, or 5 and 2?

a.	4	e.	64
b.	14	f.	100
c.	15	g.	101
d.	20	h.	160

S–3. Which of the following rational numbers have terminating decimal representations? (It should not be necessary to compute the representations if you did Problem S–2.)

a. $\dfrac{5}{4}$ e. $\dfrac{57}{64}$

b. $\dfrac{3}{14}$ f. $\dfrac{11}{100}$

c. $\dfrac{2}{15}$ g. $\dfrac{1}{101}$

d. $\dfrac{7}{20}$ h. $\dfrac{49}{160}$

*[S–4.]   a.   Imitate the proof that $\frac{1}{3}$ has no terminating decimal representation to show logically (i.e., deduce) that $\frac{2}{7}$ has no terminating decimal representation.
    b.   Try a similar proof for $\frac{3}{5}$. Where does the proof break down?

S–5. What repeating decimal represents each of the following rational numbers?

a. $\dfrac{1}{6}$ b. $\dfrac{8}{9}$ c. $\dfrac{4}{13}$

S–6. What fraction represents the same rational number as the repeating decimal .555 ...? (*Hint:* See Footnote 7.)

M–7. A student challenges you to find a decimal which clearly will never terminate or become repeating. Can you find one? (Write out enough digits to show a scheme by which it may be continued.)

M–8. How would you explain to a student why these statements are inaccurate?

a. $\frac{1}{3} = .33$

b. Every decimal represents a rational number.

*c. A rational number has many fractional names but only one decimal name.

M–9. Write four test questions designed to determine whether a student understands the decimal representation of rational numbers.

E–10. Find a fractional representation for

a. .2727 . . .

b. .1$\overline{36}$

For further thought and study:

(A) Find a set of multi-base blocks. How could they be used to facilitate the understanding of decimal notation?

(B) Consult an elementary textbook series. Does it make clear the meaning of decimal notation?

(C) Why does the procedure described in the text for finding a decimal representation always produce a terminating or repeating decimal?

THE SYSTEM OF POSITIVE RATIONAL NUMBERS

In discussing the various systems of natural numbers, we defined the four arithmetical operations separately in terms of sets. Then, in each case, we discussed the properties of the system of natural numbers under the *particular* operation in question. After doing this for addition, subtraction, multiplication, and division, we briefly considered "the" system of natural numbers under all four operations. This led us to introduce one additional property, called the distributive principle by which addition and multiplication are related. Finally in Chapter 6, we discussed algorithms for performing the four arithmetical operations.

We followed this approach in order to familiarize ourselves with a number of different number systems, thereby extending the variety of systems at our disposal (also see Chapter 4).

In this chapter, we follow the more usual procedure of talking about the properties of the various number systems under all four operations simultaneously. (Actually, subtraction and division may be defined in terms of addition and multiplication, respectively.) Prior to discussing properties of the System of Positive Rationals, we first define each operation; second, prove that it is well defined; third, show that it has a reasonable physical interpretation; and fourth, discuss algorithms for performing the operation. It is convenient to discuss the algorithms following each definition because they either follow directly from the definitions themselves or are merely extensions of the algorithms described for the natural numbers. The same procedure is followed with the signed numbers (i.e., integers) in Chapter 8.

4. ADDITION

Recall from Chapter 5 that the binary operation of addition over the natural numbers was defined in terms of sets. (You may wish to review this definition before reading on.) In this section, we define addition over the (positive) rational numbers in terms of pairs of natural numbers (i.e., fractions). More particularly, because rational numbers have been defined as equivalence classes of fractions, addition is defined in terms of *representative fractions*.

To *define* the *sum* $r_1 + r_2$ of two rational numbers, we let $\dfrac{x}{y}$ and $\dfrac{z}{w}$ be any fraction representatives of r_1 and r_2, respectively. Then $r_1 + r_2$ is the rational number named by

$$\left(\frac{x}{y} + \frac{z}{w}\right) = \frac{xw + yz}{yw}.^9$$

In effect, this definition states that in order to add two rational numbers, you pick a fraction representative for each rational number and combine them in the way indicated.

This is fine as far as it goes but, of course, we must convince ourselves that this is a reasonable definition. If it is not, then we would be better off without it. Let us first check to see that the expression for the sum actually denotes a rational number (i.e., that it is a fraction). This follows directly from the closure properties of the system of natural numbers. Closure of the natural numbers under multiplication assures us that xw, yz, and yw are natural numbers; closure under addition guarantees that $xw + yz$ is a natural number. Thus, each expression

$$\frac{xw + yz}{yw}$$

with natural number values for x, y, z, and w, is a fraction and, hence, represents a rational number. For example, consider

$$r_1 = \left\{\frac{1}{2}, \frac{2}{4}, \frac{3}{6}, \ldots, \frac{m}{2m}, \ldots\right\} \text{ and}$$

$$r_2 = \left\{\frac{1}{3}, \frac{2}{5}, \frac{3}{9}, \ldots, \frac{n}{3n}, \ldots\right\}$$

Then

$$\frac{3}{6} + \frac{3}{9} = \frac{27 + 18}{54} = \frac{45}{54}$$

which is a fraction representative of a rational number.

We still need to know if defining the sum in this way depends on the particular fractions chosen—that is, whether our definition of addition is *well defined*. It certainly would not do if, everytime we picked a pair of fraction representatives, we obtained a different sum. In effect, we wish to show that no matter which fractions we begin with, the fractions we obtain by applying the defini-

---

⁹ Recall that if a and b are letters used to represent numbers, we can write the product of a and b in three different ways: $a \times b$, $a \cdot b$, and ab. In this chapter we use this last form because it is simplest. Thus, for example, xw stands for the product of x and w, yz stands for the product of y and z, and yw stands for the product of y and w.

tion will always be equivalent. We prove this by showing that, given any two rationals, the sum (i.e., rational) named by combining every pair of representative fractions $(x/y =) ma/nb$ and $(z/w =) nc/nd$, is the same as the sum (i.e., rational) named by combining the pair of *canonical* fractions, a/b and c/d.[10] That is, we prove that

$$\frac{ma}{mb} + \frac{nc}{nd} = \frac{(ma)(nd) + (mb)(nc)}{(mb)(nd)} \cong \frac{ad + bc}{bd} = \frac{a}{b} + \frac{c}{d}$$

1. We begin by noting

$$\frac{(ma)(nd) + (mb)(nc)}{(mb)(nd)} = \frac{(mn)(ad) + (mn)(bc)}{(mn)(bd)}$$
$$= \frac{(mn)[ad + bc]}{(mn)(bd)}$$

by the commutative and associative properties of multiplication and the distributive property of multiplication over addition for natural numbers.

2. But

$$\frac{(mn)[ad + bc]}{(mn)(bd)} \cong \frac{[ad + bc]}{bd}$$

by our definition of equivalence because $(mn)[ad + bc](bd) = (mn)(bd)[ad + bc]$.

3. Therefore

$$\frac{(ma)(nd) + (mb)(nc)}{(mb)(nd)} \cong \frac{ad + bc}{bd}$$

by substitution of equals on the left in step 2.

This proves that the fraction obtained by combining any two fractional representatives of r_1 and r_2 is equivalent to $\dfrac{ad + bc}{bd}$. In short, the resulting fractions all represent the same sum (i.e., rational number) so the operation is well defined.

The following example is illustrative: Let $\frac{3}{4} \cong \frac{6}{8}$ and $\frac{1}{3} \cong \frac{3}{9}$ be the fractional representatives of the rational numbers $\{\frac{3}{4}, \frac{6}{8}, \frac{9}{12}, \ldots, 3n/4n, \ldots\}$ and $\{\frac{1}{3}, \frac{2}{6}, \frac{3}{9}, \ldots, m/3m, \ldots\}$, respectively. By our definition of addition

$$\frac{3}{4} + \frac{1}{3} = \frac{9 + 4}{12} = \frac{13}{12}$$

and

$$\frac{6}{8} + \frac{3}{9} = \frac{54 + 24}{72} = \frac{78}{72}$$

But

$$\frac{13}{12} \cong \frac{78}{72},$$ because $13 \times 72 = 936 = 12 \times 78$. Hence, the two fraction representatives of the sum name the same rational number.

***

[10] a/b and c/d are fractions reduced to lowest terms.

So far, so good. The most important step, however, of showing that our definition is a good one still remains. Does the definition conform to our intuition as to what the sum of two rational numbers should be? That is, does the definition have a sensible interpretation?

We can show that it does. Recall first that all equivalent fractions (of a given rational number) represent the same *amount*. Hence, we would hope that every fraction obtained by applying the above definition would represent the *amount* one would obtain in combining the (separate) amounts corresponding to the two given rational numbers. (We already know that the fraction representatives of the sum all name the same rational number, but we do not know for sure that this rational number represents the amount that we think it should.)

Suppose we are given one-fourth of a pie and (later) one-sixth of the same pie, as shown in Figure 7–3. We would like to express as a fraction (i.e., in terms

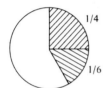

Figure 7–3

of equal-sized pieces) how much pie we have altogether. If our definition is a good one, it should be possible to represent this amount by the fraction

$$\frac{1}{4} + \frac{1}{6} = \frac{6+4}{24} = \frac{10}{24}$$

To see that we can, let us partition the quarter piece into sixths (i.e., $\frac{1}{24}$ of the whole pie) and the sixth of the pie into fourths (i.e., also $\frac{1}{24}$ of the whole pie). In Figure 7–4 we see that we have $6 + 4 = 10$ equal-sized pieces, or $\frac{10}{24}$ of the pie.

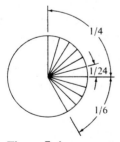

Figure 7–4

This is precisely the same as the fraction representative of the sum

$$\frac{1}{4} + \frac{1}{6} = \frac{6+4}{24} = \frac{10}{24}$$

obtained by the definition. (We usually express fractions in lowest terms ($\frac{10}{24} \cong \frac{5}{12}$) but this is not necessary.)

There are several algorithms for adding rational numbers, all of which are expressed in terms of fractions or decimals. One addition algorithm is given by the definition itself. To add the rationals represented by $\frac{1}{4}$ and $\frac{1}{6}$, we express

$$\frac{1}{4}+\frac{1}{6} \quad \text{as} \quad \frac{6\cdot 1+1\cdot 4}{4\cdot 6}=\frac{10}{24}$$

All the necessary operations have already been defined with respect to the natural numbers.

The usual algorithm for adding rationals, however, involves finding the least common multiple of the denominators of the fraction representatives.[11]

In the present case, for example, we first find the size of the largest piece of pie which will fit into both $\frac{1}{6}$ and $\frac{1}{4}$ of a pie an integral number of times. By finding the least common multiple of 6 and 4, namely 12, and then dividing the pie into 12 equal-sized pieces, we see that three of these $\frac{1}{12}$ pieces fit evenly into $\frac{1}{4}$ of the pie, and two into $\frac{1}{6}$ of the pie, as shown in Figure 7–5.

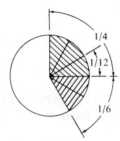

Figure 7–5

From the diagram, we reason that

$$\frac{1}{4}+\frac{1}{6}=\frac{3}{12}+\frac{2}{12}=\frac{3+2}{12}=\frac{5}{12}$$

Now, $\frac{5}{12}\cong\frac{10}{24}$ (because $5\times 24=12\times 10=120$) so "$\frac{5}{12}$" and "$\frac{10}{24}$" are fraction representatives of the same rational number. Although we have not shown that this algorithm is *always* equivalent to the definition, we have shown that it is reasonable to suspect that it is.

If the denominators of two fraction representatives are the same, a particularly simple (but less general) algorithm works. We can represent this algorithm symbolically as

$$\frac{a}{b}+\frac{c}{b}=\frac{a+c}{b}$$

where a/b and c/b are fraction representatives of two rationals. For example

$$\frac{2}{5}+\frac{1}{5}=\frac{3}{5}$$

[11] Review Chapter 5, Section 10, briefly to make sure you can find the least common multiple of two or more numbers before going on.

This algorithm is a special case of the more general algorithm involving the least common multiple, because the least common multiple of the denominators of any two fractions of the form a/b and c/b is b itself. Thus, in the example above, the least common multiple is 5.

We might note parenthetically that in some of our recent experiments we have found that, although less general algorithms are more easily learned, they are sometimes incorrectly applied and, more important, may actually inhibit the learning of more general algorithms when presented later on.[12] Thus, in teaching addition algorithms, it may be advisable to introduce the more general algorithm, involving the least common multiple, before spending too much time on special algorithms as is frequently done in the classroom.

The *addition algorithm for adding decimals* is merely an extension of the algorithm for adding natural numbers. Hence, we can add

$$\begin{array}{r} 3.14 \\ +2.61 \\ \hline 5.75 \end{array}$$

in the same way as we add

$$\begin{array}{r} 314 \\ +261 \\ \hline 575 \end{array}$$

The only additional rule required is that the decimal point is inserted in the sum directly beneath the decimal points in the addends.

We can justify this rule by applying the algorithm for adding fractions with a common denominator. To see this, observe that[13]

$$3.14 + 2.61 = \frac{314}{100} + \frac{261}{100} = \frac{575}{100} = 5.75$$

Another example makes it clear how this algorithm may be extended to decimals like 3.14 and .0261. In this case, the corresponding fractions do not have the same denominator. Here, we can always find equivalent fractions with a common denominator equal to the largest denominator. Thus

$$3.14 + .0261 = \frac{314}{100} + \frac{261}{10,000} = \frac{31,400}{10,000} + \frac{261}{10,000} = \frac{31,661}{10,000} = 3.1661$$

or
$$\begin{array}{r} 3.14 \\ +.0261 \\ \hline 3.1661 \end{array}$$

---

[12] For more details, see J. M. Scandura, E. Woodward, and F. Lee, "Rule Generality and Consistency in Mathematical Learning," *American Educational Research Journal*, **4**, 1967, pp. 303–319; J. M. Scandura and J. H. Durnin, "Extra-Scope Transfer in Learning Mathematical Strategies," *Journal of Educational Psychology*, **59**, 1968, pp. 350–354; and W. G. Roughead, and J. M. Scandura, " 'What Is Learned' in Mathematical Discovery," *Journal of Educational Psychology*, **59**, 1968, pp. 283–289.

[13] Note that $3.14 = 3 + \frac{1}{10} + \frac{4}{100} = \frac{300}{100} + \frac{10}{100} + \frac{4}{100} = \frac{314}{100}$ and

$$2.61 = 2 + \frac{6}{10} + \frac{1}{100} = \frac{200}{100} + \frac{60}{100} + \frac{1}{100} = \frac{261}{100}$$

EXERCISES 7–4

S–1.   Add $\frac{1}{6}$ and $\frac{3}{10}$ in two ways:

    a. by using the definition

    b. by using the least common multiple algorithm

    Check that your answers are equivalent fractions.

S–2.   The simple algorithm $\frac{a}{b} + \frac{c}{b} = \frac{a+c}{b}$ is presented in the text. Is it also true that $\frac{a}{b} + \frac{a}{c} = \frac{a}{b+c}$? Justify your answer.

M–3.   Explain how you could illustrate, using 20 paper clips, that $\frac{1}{4} + \frac{2}{5} = \frac{13}{20}$.

M–4.   Repeat Problem M–3 using a circular disk instead of paper clips.

M–5.   One of your students claims that $\frac{2}{3} + \frac{4}{5} = \frac{6}{8}$.

    a. What algorithm do you think he has been using?

    b. How would you show him that he is wrong?

M–6.   Write four examination questions designed to test whether your students can apply the algorithms for addition of rational numbers.

E–7.   Using the definitions of addition and of equivalence of rational numbers, and any properties of natural numbers you need, show that

$$\frac{a}{b} + \frac{c}{b} = \frac{a+c}{b}.$$

*E–8.   Following steps suggested below, show that the least common multiple (least common denominator) algorithm for adding rational numbers $r_1 = a/b$ and $r_2 = c/d$ actually works.

    Suppose e is the least common multiple of b and d. Then $e = fb$ and $e = gd$ for some natural numbers f and g. Express r_1 and r_2 as fractions with denominators e.

$$\frac{a}{b} \cong \frac{\ }{e}, \quad \frac{c}{d} \cong \frac{\ }{e}$$

Therefore, $\dfrac{a}{b} + \dfrac{c}{d} \cong \dfrac{\ }{e} + \dfrac{\ }{e} = $ _____.

Show that this last fraction is, in fact, equivalent to $\dfrac{ad + bc}{bd}$.

For further thought and study: How can Cuisenaire rods be used to develop understanding of the algorithms for adding rational numbers?

5. SUBTRACTION

Let r_1 and r_2 be two (positive) rational numbers. Then if $r_1 > r_2$,[14] the *difference*, $r_1 - r_2$ may be *defined* (similar to addition) as the rational number named by

$$\left(\frac{x}{y} - \frac{z}{w} \right) = \frac{xw - yz}{yw}$$

[14] If r_1 is not greater than r_2, then the difference $r_1 - r_2 = r_3$ is not defined. In the system of positive rationals "you can't take away as much or more than you have."

where x/y and z/w are arbitrary names of r_1 and r_2, respectively. For example, let $\frac{4}{6}$ and $\frac{6}{10}$ name two rational numbers. Then since $\frac{4}{6} > \frac{6}{10}$, the difference (a rational number) is defined and named by

$$\frac{4}{6} - \frac{6}{10} = \frac{40 - 36}{60} = \frac{4}{60}$$

This operation is well defined (for $r_1 > r_2$) because, if

$$r_1 = \left\{ \frac{a}{b}, \frac{2a}{2b}, \frac{3a}{3b}, \cdots, \frac{ma}{mb}, \cdots \right\}$$

and

$$r_2 = \left\{ \frac{c}{d}, \frac{2c}{2d}, \frac{3c}{3d}, \cdots, \frac{nc}{nd}, \cdots \right\}$$

then

$$\frac{(ma)(nd) - (mb)(nc)}{(mb)(nd)} = \frac{ad - bc}{bd}$$

for each value of m and n. Show this by paralleling the argument in the previous section.

Subtraction may also be defined as the "inverse" of addition: $r_1 - r_2 = r_3$ (where $r_1 > r_2$) if and only if $r_2 + r_3 = r_1$. That is, if the fractions a/b and c/d denote r_1 and r_2, respectively, then the fraction, e/f, denotes r_3 if and only if $c/d + e/f = a/b$. In this case, $\frac{4}{60}$ represents the difference, $\frac{4}{6} - \frac{6}{10}$, because $\frac{6}{10} + \frac{4}{60} = \frac{36 + 4}{60} = \frac{40}{60} \cong \frac{4}{6}$.

The interpretation of subtraction and the subtraction algorithms parallel the interpretation and algorithms for addition. The only difference is that we "take away" or subtract, respectively, instead of "put together" or add. For example, consider the algorithm given by the definition of subtraction as it relates to the pie illustration from the section on addition. Suppose that we are given $\frac{1}{4}$ of a pie and must subtract from it $\frac{1}{6}$ of the pie (see Figure 7–6).

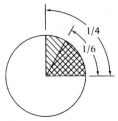

Figure 7–6

We wish to know, in terms of equal sized pieces, how much pie we have left. That is, we hope we can find a fraction which denotes $\frac{1}{4} - \frac{1}{6}$ by applying the definition.

As before, we can partition the $\frac{1}{4}$ piece into sixths and the $\frac{1}{6}$ piece into fourths as shown in Figure 7–7.

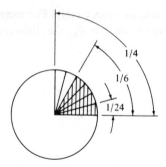

Figure 7–7

Here, $\frac{1}{4}$ of the pie amounts to six of the equal-sized pieces of pie ($\frac{1}{24}$) and $\frac{1}{6}$, to four of these pieces. So, the amount of pie we have left is $6 - 4 = 2$ equal-sized pieces or $\frac{2}{24}$ of the pie. As we had hoped, this fraction is the same as that given by our definition

$$\frac{1}{4} - \frac{1}{6} = \frac{6-4}{24} = \frac{2}{24}$$

Even though everything follows through directly, exactly as in addition, try to justify for yourself the subtraction algorithm, involving the least common multiple

$$\left(\text{e.g.,} \quad \frac{1}{4} - \frac{1}{6} = \frac{3}{12} - \frac{2}{12} = \frac{1}{12} \right)$$

and the special algorithm, $a/b - c/b = (a - c)/b$

$$\left(\text{e.g.,} \quad \frac{3}{5} - \frac{2}{5} = \frac{3-2}{5} = \frac{1}{5} \right)$$

In the process, be sure to convince yourself that these algorithms only make sense when $a/b > c/d$. (Note: In the special algorithm $d = b$.)

The algorithm for subtracting decimals is an equally simple extension of the subtraction algorithm for natural numbers. You can show this also by paralleling the illustrative arguments of the previous section.

EXERCISES 7–5

S–1. Compute $\frac{7}{8} - \frac{7}{10}$ by

    a. using the definition of subtraction
    b. using the least common multiple algorithm

S–2. Using the equivalent fractions, show that

$$\begin{array}{r} 4.30 \\ -1.16 \\ \hline 3.14 \end{array}$$

can be computed just as

$$
\begin{array}{r}
430 \\
-116 \\
\hline 314
\end{array}
$$

(*Hint:* $4.30 = \frac{430}{100}$, $1.16 = \frac{116}{100}$.) Describe how $4.30 - 1.16$ is computed in practice?

S–3.   Show that subtraction is well defined.

E–4.   a.   When is $a/b > c/b$?
   b.   Why is it necessary to require that $a/b > c/b$ in order to define $a/b - c/b$?
   c.   Given our definition of rational number, would $r_1 - r_2$ make sense if $r_1 = r_2$? Why?

E–5.   In Chapter 5 we noted that $c = a - b$ if and only if $a = b + c$. Does this hold true for positive rational numbers as well? Check a few particular cases.

M–6.   How could you use paper clips to illustrate the fact that $\frac{8}{9} - \frac{5}{6} = \frac{1}{18}$?

M–7.   Repeat Problem M–6 using circle diagrams rather than paper clips.

M–8.   What would you have your students do to demonstrate that they understand the algorithm for subtraction of positive rational numbers.

For further thought and study: Check an elementary school textbook series to see whether subtraction of rational numbers is explained independently or in terms of the relation of subtraction to addition.

6. MULTIPLICATION

The *product* $r_1 \times r_2 = r_3$ of two (positive) rational numbers r_1 and r_2, where r_1 is named by x/y and r_2 by z/w, *is defined* to be the rational number named by

$$
\left(\frac{x}{y} \times \frac{z}{w}\right) = \frac{xz}{yw} \quad {}^{15}
$$

That is, in order to multiply two rationals, we choose a pair of representative fractions and form a new fraction by multiplying together their respective numerators and denominators. This new fraction represents the product $r_1 \times r_2 = r_3$. For example, let $r_1 = \{\frac{1}{2}, \frac{2}{4}, \frac{3}{6}, \ldots, m/2m, \ldots\}$ and $r_2 = \{\frac{1}{3}, \frac{2}{6}, \frac{3}{9}, \ldots, n/3n, \ldots\}$. In this case, $r_1 \times r_2 = r_3$ is named by the fraction

$$
\left(\frac{3}{6} \times \frac{3}{9}\right) = \frac{3 \times 3}{6 \times 9} = \frac{9}{54}
$$

We can show that our definition of multiplication is sound in much the same way that we showed this for addition. In fact, we have already implicitly

---

[15] Recall that xz means the product of x and z; yw means the product of y and w.

assumed that our definition always yields a fraction representative of a rational number (i.e., that xz/yw is a fraction). The formal argument directly parallels the discussion in the section on addition and we leave this as an exercise. (*Hint:* Show that the numerator and denominator must be natural numbers.)

Next, we show that our operation is well defined—that is, no matter which pair of fractions we choose to represent r_1 and r_2, we always get a fraction that names the same rational number. As in the section on addition, we do this by showing that every fraction of the form $\dfrac{ma \times nc}{mb \times nd}$ is equivalent to the fraction, $\dfrac{a \times c}{b \times d}$:

1. If $r_1 = \left\{\dfrac{a}{b}, \dfrac{2a}{2b}, \dfrac{3a}{3b}, \ldots, \dfrac{ma}{mb}, \ldots\right\}$ and $r_2 = \left\{\dfrac{c}{d}, \dfrac{2c}{2d}, \dfrac{3c}{3d}, \ldots, \dfrac{nc}{nd}, \ldots\right\}$

   then $\dfrac{ma \times nc}{mb \times nd}$ names $r_1 \times r_2 = r_3$ by the definition of multiplication.

2. Hence, $\dfrac{(mn)(a \times c)}{(mn)(b \times d)}\left(= \dfrac{ma \times nc}{mb \times nd}\right)$ also names r_3 by the associative and commutative properties of multiplication of natural numbers.

3. But $\dfrac{(mn)(a \times c)}{(mn)(b \times d)} \cong \dfrac{a \times c}{b \times d}$ (i.e., both fractions name the same rational) by definition, because $(mn)(a \times c)(b \times d) = (mn)(b \times d)(a \times c)$.

4. Hence, $\dfrac{ma \times nc}{mb \times nd} \cong \dfrac{a \times c}{b \times d}$ by substitution of equals (from step 2 into step 3).

5. Therefore, $\dfrac{ma \times nc}{mb \times nd}$ and $\dfrac{a \times c}{b \times d}$ both name r_3 by definition of equivalent fractions in step 4.

This says that no matter which names (i.e., fractional representatives) of r_1 and r_2 we start with, the fraction we obtain will name the same rational (r_3) as $\dfrac{a \times c}{b \times d}$, where $\dfrac{a}{b}$ and $\dfrac{c}{d}$ are fractional representations of r_1 and r_2, respectively, in lowest terms.

To illustrate, we again consider $r_1 = \left\{\dfrac{1}{2}, \dfrac{2}{4}, \dfrac{3}{6}, \ldots, \dfrac{m}{2m}, \ldots\right\}$ and $r_2 = \left\{\dfrac{1}{3}, \dfrac{2}{6}, \dfrac{3}{9}, \ldots, \dfrac{n}{3n}, \ldots\right\}$. The fractions $\left(\dfrac{3}{6} \times \dfrac{2}{6} = \dfrac{3 \times 2}{6 \times 6} = \right)\dfrac{6}{36}$ and $\left(\dfrac{1}{2} \times \dfrac{3}{9} = \dfrac{1 \times 3}{2 \times 9} = \right)\dfrac{3}{18}$, for example, are both equivalent to $\dfrac{1}{2} \times \dfrac{1}{3} = \dfrac{1 \times 1}{2 \times 3} = \dfrac{1}{6}$, and, hence, name the same rational number.[16]

The product of two rational numbers may be thought of as the result of superimposing two (corresponding) fractional amounts on one another in the

---

[16] That is, $\frac{6}{36} \cong \frac{1}{6}$ and $\frac{3}{18} \cong \frac{1}{6}$, because $6 \times 6 = 36 \times 1$ and $3 \times 6 = 18 \times 1$, respectively. Also, $\frac{6}{36} \cong \frac{3}{18}$, because $6 \times 18 = 36 \times 3$, but this can be shown to follow directly from the former two equivalences by the transitive property of equivalence relations.

sense described below. To be specific, let us represent the rationals named by $\frac{3}{5}$ and $\frac{2}{3}$ as parts of rectangles (see Figure 7–8). The product $\frac{3}{5} \times \frac{2}{3}$, then, is represented by the *cross-hatched* area (six sections) formed by superimposing the two representations as shown in Figure 7–9. Happily, our definition of multiplication gives us precisely what our intuition has led us to suspect—the rational number named by

$$\frac{3}{5} \times \frac{2}{3} = \frac{3 \times 2}{5 \times 3} = \frac{6}{15}$$

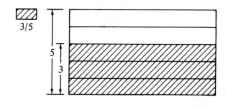

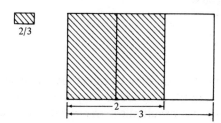

Figure 7–8

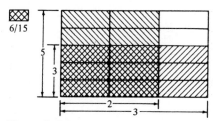

Figure 7–9

This idea can be extended to "improper" fractions as well. Consider the product of the two rational numbers named by $\frac{2}{3} \times \frac{4}{3}$. We can represent the $\frac{2}{3}$ by Figure 7–10 and $\frac{4}{3}$ by Figure 7–11. The product named by $\frac{2}{3} \times \frac{4}{3} = \frac{8}{9}$, then, cor-

Figure 7–10

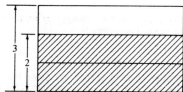

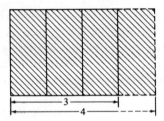

Figure 7–11

responds to the ratio of the cross-hatched area (eight sections) to the area (nine sections) contained entirely in the unit rectangle (see Figure 7–12).[17]

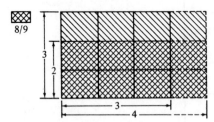

Figure 7–12

The above interpretation of multiplication may be called the "state-state" view because both the initial fractions and the product were represented by fixed objects (i.e., rectangles). Another way to view multiplication considers one number, say, $\frac{3}{5}$, as a state (see Figure 7–13) and the second number, say, $\frac{2}{3}$, as an

Figure 7–13

operator. Here, we partition the state represented by $\frac{3}{5}$, first into thirds (Figure 7–14), and then consider that part of the original shading, included in two of the thirds, giving the cross-hatched area (see Figure 7–15). This gives $\frac{6}{15}$ of the unit rectangle, which, of course, corresponds to what we obtain by using the definition.

A third way to interpret multiplication views both fractions (e.g., $\frac{3}{5}$ and $\frac{2}{3}$)

---

[17] Multiplication, involving *improper* fractions, can also *be thought of* as a sum of products involving natural numbers and proper fractions. For example,

$$\frac{2}{3} \times \frac{4}{3} = \frac{2}{3} \times \left(1 + \frac{1}{3}\right) = \left(\frac{2}{3} \times 1\right) + \left(\frac{2}{3} \times \frac{1}{3}\right)$$

Can you construct an interpretation of the expression $(\frac{2}{3} \times 1) + (\frac{2}{3} \times \frac{1}{3})$? Does the result correspond to what the definitions of multiplication *and* addition say it should? (If not, try again.)

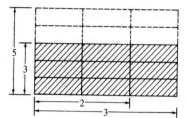

Figure 7–14

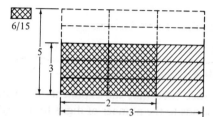

Figure 7–15

as representing operators acting on a given unit state (e.g., the unpartitioned rectangle, ☐).

Is an operator-by-operator view of multiplication also compatible with the definition? Satisfy yourself that it is. (*Hint:* Start with a unit rectangle.)

When rational numbers are represented as fractions, the multiplication algorithm is identical with the definition. As with addition, the algorithm for multiplying decimals is a simple extension of the multiplication algorithm for natural numbers. In this case, however, care must be taken with the placement of the decimal point. To multiply a pair of decimals, we first proceed according to the usual multiplication algorithm and then place the decimal point in the product so that the number of digits to the right of the decimal point is the sum of the number of digits to the right of the respective decimal points of the two numbers being multiplied. For example, consider

$$
\begin{array}{r}
3.00\,2 \\
\times\ 2.5 \\
\hline
1\,5\,01\,0 \\
6\,0\,04 \\
\hline
7.5\,05\,0
\end{array}
$$

Here there are *three* and *one* digits, respectively, to the right of the decimal point in "3.002" and "2.5" and *four* to the right of the decimal point in the product.

This rule may be justified by expressing the decimals to be multiplied as fractions and then multiplying the fractions. When the fractional product is expressed as a decimal, the decimal point will be positioned according to the rule. Although it is not difficult to prove this in general, we show that it is true only in the case of the above example. Exactly the same argument will work with any other pair of decimals. First, we represent "2.5" and "3.002" as $\frac{25}{10}$ and $\frac{3002}{1000}$, respectively. Multiplying, we obtain

$$\frac{25 \times 3002}{10,000} = \frac{75,050}{10,000} = 7.5050, \text{ which is what we expected.}$$

The relationship of this procedure to the rule can be seen more directly by expressing each fraction with denominator in *exponential* form (e.g., $10^n =$

$$\overbrace{10 \times 10 \times \ldots \times 10}^{n \text{ times}}).$$ Here, 25/10 and 3002/1000 become $25/10^1$ and $3002/10^3$, respectively, so

$$\frac{25}{10^1} \times \frac{3002}{10^3} = \frac{25 \times 3002}{10^1 \times 10^3} = \frac{75,050}{10^4}$$

(When we multiply $10^n \times 10^m$, the result is 10^{n+m}, because

$$\overbrace{10 \times 10 \times \ldots \times 10}^{n \text{ times}} \times \overbrace{10 \times 10 \times \ldots \times 10}^{m \text{ times}} = \overbrace{10 \times 10 \times \ldots \times 10}^{(n+m) \text{ times}}$$

Thus, $10^1 \times 10^3 = 10^{1+3} = 10^4$.) Notice here that the exponent of 10 in the denominator of each fraction corresponds to the number of digits to the right of the decimal point in the corresponding decimal. 2.5, 3.002, and 7.5050 correspond, respectively, to $25/10^1$, $3002/10^3$, and $75,050/10^4$.

Without multiplying, tell how many digits must be to the right of the decimal point in the product .000364 × 136.842. Justify the application of the decimal rule by converting the decimals to fractions and multiplying.

EXERCISES 7–6

[S–1.] Show that the positive rational numbers are closed under multiplication (i.e., show that if a/b and c/d represent rational numbers, then $a/b \times c/d$ must represent a rational number).

S–2. a. Use the definition of multiplication to compute $\frac{7}{8} \times \frac{3}{11}$.
   b. Compute the same product using two other representatives of $\frac{7}{8}$ and $\frac{3}{11}$.
   c. Show that the results of parts a and b are equivalent fractions.

S–3. Beginning with a rectangle representing 1 unit, construct both a state-by-state and a state-by-operator interpretation of each of the following products.

   a. $\frac{2}{5} \times \frac{3}{4}$

   b. $\frac{5}{3} \times \frac{1}{2}$

S–4. How many digits will be to the right of the decimal point in the decimal representation of .00036 × 136.842? Justify your answer by converting to fractions and multiplying.

M–5. Write four examination questions designed to test whether your students understand the algorithms for multiplication of positive rational numbers.

E–6. Devise an operator-by-operator interpretation of $\frac{2}{3} \times \frac{4}{5}$, beginning with a unit rectangle. (Draw pictures to illustrate.) Show that your procedure gives the same result as the definition of multiplication.

E–7.   Find fractions x/y such that

a. $\dfrac{x}{y} \times \dfrac{3}{4} \cong \dfrac{3}{4}$

b. $\dfrac{7}{9} \times \dfrac{x}{y} \cong \dfrac{4}{4}$

c. $\dfrac{2}{3} \times \dfrac{x}{y} \cong \dfrac{7}{4}$

For further thought and study: Consult several elementary textbook series to see what algorithms are given for multiplication of rational numbers.

7. DIVISION

As with subtraction, division over the (positive) rational numbers may either be defined independently or in terms of multiplication. Each definition has its advantages so we shall give both and then capitalize on the advantages to simplify our discussion.

The *quotient*, $r_1 \div r_2 = r_3$, of two rational numbers r_1 and r_2 where r_1 is named by x/y and r_2 by z/w, *is defined* to be the rational number named by

$$\left(\frac{x}{y} \div \frac{z}{w}\right) = \frac{xw}{yz}$$

(Compare the quotient here with the *product* of two rationals.) For example

$$\frac{3}{5} \div \frac{2}{4} = \frac{3 \times 4}{5 \times 2} = \frac{12}{10}\left(= 1\frac{2}{10}\right)$$

This definition of course, is just another way of expressing the familiar division algorithm: "Invert the divisor and multiply."

Alternatively, the quotient $r_1 \div r_2 = r_3$ may be defined as that rational number (r_3) such that $r_2 \times r_3 = r_1$. Thus, for example, the quotient named by $\frac{3}{5} \div \frac{2}{4} (= \frac{12}{10})$ satisfies the equation, $\frac{2}{4} \times \square = \frac{3}{5}$. That is, $\frac{2}{4} \times \frac{12}{10} = \frac{24}{40}$ which is equivalent to $\frac{3}{5}$ and, hence, names the same rational number. (To see this, divide both the numerator and denominator of $\frac{24}{40}$ by 8, or, equivalently, "cross-multiply" and show that $\frac{24}{40}$ and $\frac{3}{5}$ satisfy the definition of equivalence.)

Prove to yourself that the former definition always yields a rational number and is well defined. The proofs follow directly the arguments given in the previous sections.

Fortunately, the definition again conforms to what our intuition tells us the quotient of two (rational) numbers should be. In general, we suspect that the quotient should express the number of (perhaps fractional) times the amount corresponding to one rational (the divisor) is contained in the amount represented by the other (the dividend). Consider $\frac{3}{5} \div \frac{2}{4} (= \frac{12}{10})$.

To see how many times the second amount ($\frac{2}{4}$) "goes into" the first, represent both amounts in terms of the same-sized "chunks." We can do this by finding the least common multiple (or any common multiple for that matter) of the denominators, which in this case is $5 \times 4 = 20$. This is represented in Figure 7–16. Now, there are 12 such chunks in $\frac{3}{5}$ and 10, in $\frac{2}{4}$. Hence, the problem

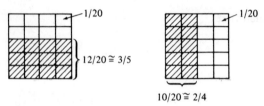

$$10/20 \cong 2/4$$

Figure 7–16

is reduced to finding how many (fractional) times 10 will go into 12. Clearly, it goes $\frac{12}{10} = 1\frac{2}{10}$ times, as our definition says it must (i.e., $\frac{12}{1} \div \frac{10}{1} = \frac{12}{1} \times \frac{1}{10} = \frac{12}{10}$).

In general, given the quotient $a/b \div c/d$, we can use bd as the common multiple. The number of chunks will be ad and bc, respectively. Check these remarks with respect to the quotient named by $\frac{3}{4} \div \frac{1}{2} (= \frac{6}{4})$. Use pictures to go through the argument step by step.

Notice that the rational number (quotient) named by $a/b \div c/d$ is the same as the quotient named by ad/bc. This provides intuitive justification for the division algorithm for rational numbers, because ad/bc can be obtained from a/b and c/d (in $a/b \div c/d$) by inverting the divisor and multiplying (i.e., $a/b \times d/c$).

The division algorithm for decimals parallels that for multiplication. We use the usual division algorithm for natural numbers (see Chapter 6) and properly position the decimal point. For example, in dividing 7.3 into 149.26 we obtain

```
                  2 0.44 . . .
       7.3. )149.2.60
            146 0 00
              3 2 60
              2 9 20
                3 40
                2 92
                  48
```

Perhaps the simplest way to justify the positioning of the decimal point in the divisor and dividend is to observe that the quotient

$$q = 149.26 \div 7.3$$

must satisfy the equation

$$7.3 \times q = 149.26$$

Multiplying both sides by 10, we find that

$$10 \times 7.3 \times q = 10 \times 149.26$$

or

$$73 \times q = 1492.6$$

Hence, $q = 1492.6 \div 73$ as well as $149.26 \div 7.3$. When we move the decimal points over (as above), we in effect transform the given problem into a new one with the same quotient.

In this way, a given division problem can always be reduced to another problem that involves an integral divisor. Satisfy yourself that this is true. Try $25.026 \div .004$. What is an equivalent problem with integral divisor?

The division algorithm itself (including the placement of the decimal point in the quotient) is more difficult to justify; going into all the details would become as much a problem of exposition as one of mathematics. Instead, we shall outline some key ideas and ask some leading questions to help you develop a more complete understanding of the division algorithm for yourself.

Think in terms of the problem $149.26 \div 7.3$. An equivalent problem involving only natural numbers is $14926 \div 730$. (Why?) The division algorithm for natural numbers gives a whole number (20) and a remainder (326) such that $14926 = (730 \times 20) + 326$. The total quotient therefore may be expressed in the form $20 + (326 \div 730)$.

The main problem is to justify finding the decimal part of the quotient (i.e., .44) by the ordinary division algorithm (for natural numbers). The algorithm does not work when the divisor (730) is larger than the dividend (326). Nonetheless, we can always find an equivalent division problem to which the algorithm does apply by reasoning as follows. The decimal $d = 326 \div 730$ must satisfy $730 \times d = 326$ (if our arithmetic is to be consistent). Multiplying both sides by 10^2 (because we want two digits in the decimal answer) we see that d must also satisfy $730 \times d \times 10^2 = 326 \times 10^2$. Equivalently, $d \times 10^2 = (326 \times 10^2) \div 730$.

Dividing 73 into 3260 in the example above, then, is equivalent to dividing 730 into 32600 (i.e., 326×10^2). In both cases we must remember that the result is $d \times 10^2$ ($d \times 100$) and not d.

Dividing 730 into $326 \times 10^2 = 32600$ gives 44 with a remainder of 480, or $44 + \frac{480}{730}$. But 44 is $10^2 = 100$ times greater than $326 \div 730$ so the quotient $326 \div 730 = .44 + \frac{480}{73000}$.

Putting this all together we get

$$149.26 \div 7.3 = 14926 \div 730$$
$$= (14600 + 326) \div 730$$
$$= (14600* \div 730) + (326 \div 730)$$
$$= (20) + \left(\frac{32600* \div 730}{100}\right)$$
$$= 20 + \left(\frac{(32120* + 480) \div 730}{100}\right)$$
$$= 20 + \left(\frac{32120 \div 730}{100}\right) + \left(\frac{480 \div 730}{100}\right)$$
$$= 20 + \frac{44}{100} + \frac{480}{73000}$$
$$= 20.44 + \frac{48}{7300}$$

(Note for the numbers marked by $*$: $14600 = 730 \times 20$; $32600 = 326 \times 10^2$; $32120 = 730 \times 44$.)

Each of these steps can be justified by properties of the system of rational numbers (discussed below). For present purposes, your intuition will suffice.

EXERCISES 7–7

S–1. Construct a concrete (e.g., pictorial) interpretation for $\frac{3}{4} \div \frac{1}{2}$ of the type described in the text.

S–2. Transform $25.026 \div .004$ into an equivalent division problem (i.e., one with the same quotient) with integral dividend *and* divisor.

[S–3.] Show that the quotient of the positive rational numbers $r_1 = x/y$ and $r_2 = z/w$ is also a rational number. What is this property called?

[S–4.] Show that division of positive rational numbers is well defined. That is, if $a/b \cong a'/b'$ and $c/d \cong c'/d'$, show that $a/b \div c/d \cong a'/b' \div c'/d'$.

S–5. Carry out the steps described in the discussion of the division algorithm for decimals, using the quotient $3.255 \div 1.4$ and computing the quotient to the nearest thousandth.

M–6. "Invert the divisor and multiply" has been a magic prescription for division for generations of elementary school children. If a child knows the relation between multiplication and division, it is not hard to explain why this rule works. Complete the following explanation.

If $\frac{2}{3} \div \frac{4}{7} = \frac{a}{b}$, then we know that $\frac{a}{b} \times \frac{4}{7} = \frac{2}{3}$.

So, we can find $\frac{a}{b}$ by multiplying both sides of the latter equation by _____.

Simplifying, we obtain $\frac{a}{b} =$ _____.

[M–7.] One requirement of any good mathematical notation is that it have only one meaning or interpretation. A student computing the quotient $8.65 \div 7.0$ gives the result as $1.23\frac{4}{7}$. Do you know what he means? Can you suggest a better form for his answer?

M–8. How would you have your students demonstrate that they understand the idea of division of rational numbers? Would you expect them to be able to express all of the rules verbally?

8. PROPERTIES OF THE SYSTEM OF POSITIVE RATIONALS

In this section we consider certain important properties of the positive rational numbers under the binary operations of addition, multiplication, subtraction, and division. These properties are important in the sense that other properties of interest can be deduced from them. For example, we have made heavy use of the associative and commutative properties of the system of natural numbers in the proofs involving the arithmetic algorithms. In mathematics, the basic properties are called axioms and the deduced properties are called theorems. Where we want to talk about a property in the abstract, apart from a particular binary operation (i.e., $+$, $-$, $\times$, $\div$), we use the symbol $\odot$ to represent an arbitrary operation.

8.1 Closure

A number system is said to be closed under a binary operation $\odot$ if for every pair of numbers a and b, there is a third number c (in the system) such that $a \odot b = c$.

Under which operations is the system of positive rationals closed? Is addition always defined? What about multiplication? Division? Check the definitions to make sure that no matter what two positive rationals you start with the definition guarantees that there will always be a third, called the sum, product, and quotient, respectively. Check several examples of each to get a feeling for what is going on. Try $\frac{2}{3}$ and $\frac{4}{5}$ for a start.

The system is *not* closed under subtraction. Check this both by verifying the definition and by simply thinking about what subtraction involves. Thus, in the former case, $r_1 - r_2$ is only defined when $r_1 > r_2$ (check this). Clearly, this restriction makes sense, because we cannot take away more than we start with. For example, if we only have a fourth of a pie, we cannot take away a third (of the pie). There is no *positive* rational number equal to $\frac{1}{4} - \frac{1}{3}$.

EXERCISES 7–8.1

S–1. Tell whether or not each of the following is a positive rational number.

a. $\dfrac{2}{3} + \dfrac{4}{5}$ d. $\dfrac{4}{5} - \dfrac{2}{3}$

b. $\dfrac{2}{3} - \dfrac{4}{5}$ e. $\dfrac{2}{3} \div \dfrac{4}{5}$

c. $\dfrac{2}{3} \times \dfrac{4}{5}$ f. $\dfrac{4}{5} \div \dfrac{2}{3}$

S–2. Is the system of *positive* rational numbers closed under each of the operations listed below? If it is not closed, justify each answer by giving an example.

a. addition
b. subtraction
c. multiplication
d. division.

E–3. Repeat Exercise S–2 for the system of *non-negative* rational numbers (i.e., the system of positive rationals together with *zero*).

8.2 Associativity and Commutativity

A binary operation is associative in a number system if for every triple of numbers a, b, and c, $(a \odot b) \odot c = a \odot (b \odot c)$. The operation is *commutative* if for every pair of numbers a, b, $a \odot b = b \odot a$.

The commutative and associative properties of addition and multiplication over the positive rationals follow directly from the properties of the system of natural numbers. We can show commutativity of addition as follows. (For simplicity of exposition, we equate the fractions a/b, c/d, e/f, and so on with the positive rationals they denote.)

$$\frac{a}{b} + \frac{c}{d} = \frac{ad + bc}{bd} \qquad \text{by definition}$$

$$= \frac{bc + ad}{bd} \qquad \text{by the commutative property of addition of natural numbers}$$

$$= \frac{cb + da}{db} \qquad \text{by the commutative property of multiplication of natural numbers}$$

$$= \frac{c}{d} + \frac{a}{b} \qquad \text{by definition of addition of rationals}$$

Check this result, by showing that $\frac{3}{4} + \frac{5}{2} = \frac{5}{2} + \frac{3}{4}$.

Can you prove the commutative property of multiplication? The same type of argument will work.

We could also prove the associative properties of addition, $(r_1 + r_2) + r_3 = r_1 + (r_2 + r_3)$, and multiplication, $(r_1 \times r_2) \times r_3 = r_1 \times (r_2 \times r_3)$, in a similar fashion, but the proofs are not particularly enlightening and so are not included here. The proof for multiplication is less cumbersome than the one for addition, however, and you may wish to try it. In any case, satisfy yourself that these operations are associative. For example, is $(\frac{1}{3} + \frac{2}{5}) + \frac{1}{4}$ equal to $\frac{1}{3} + (\frac{2}{5} + \frac{1}{4})$? What about $(\frac{1}{3} \times \frac{2}{5}) \times \frac{1}{4}$ and $\frac{1}{3} \times (\frac{2}{5} \times \frac{1}{4})$?

Subtraction and division over the positive rational numbers are neither commutative nor associative. Demonstrate this for yourself by checking to see that each of the following statements is false.

1. $\dfrac{1}{3} - \dfrac{1}{4} = \dfrac{1}{4} - \dfrac{1}{3}$

2. $\dfrac{1}{3} \div \dfrac{1}{4} = \dfrac{1}{4} \div \dfrac{1}{3}$

3. $\left(\dfrac{2}{3} - \dfrac{2}{5}\right) - \dfrac{1}{6} = \dfrac{2}{3} - \left(\dfrac{2}{5} - \dfrac{1}{6}\right)$

4. $\left(\dfrac{2}{3} \div \dfrac{2}{5}\right) \div \dfrac{1}{6} = \dfrac{2}{3} \div \left(\dfrac{2}{5} \div \dfrac{1}{6}\right)$

Can you find other examples where these properties are false? Consider commutativity of subtraction, of division. Also consider associativity.[18]

EXERCISES 7–8.2

S–1. Using the definition of addition of positive rational numbers, show that $\frac{3}{4} + \frac{5}{2} = \frac{5}{2} + \frac{3}{4}$. What property does this illustrate?

S–2. Choose specific positive rational numbers and use the definitions of addition and multiplication to illustrate the following properties:

    a.   the commutative property of multiplication

[18] See J. M. Scandura, "Concrete Examples of Commutative Non-associative Systems," *The Mathematics Teacher*, 1966, *59*, 735–736.

b. the associative property of addition
c. the associative property of multiplication

S–3.   Imitate the proof of commutativity of addition of positive rationals to prove that multiplication of positive rationals is also commutative.

S–4.   Show that statements 1 to 4 in the text are false. What properties are shown *not* to hold by each of these examples?

8.3 Identity

A number system has an *identity*, e, with respect to an operation $\odot$, if for all numbers a, $a \odot e = e \odot a = a$. Surprisingly, perhaps, the system of positive rationals does not have an additive identity. The set of positive rationals does not include a rational which corresponds to the whole number 0. We could add an identity to our system, however, by simply defining a new rational number $r_o = \{\frac{0}{1}, \frac{0}{2}, \frac{0}{3}, \ldots, 0/n, \ldots\}$. In this case, addition would still be defined as above, because the definition of addition is given in terms of operations on the natural (or whole) numbers and they work with 0 as well.[19] The system obtained by adding r_o to the positive rationals is called the system of *non-negative rationals*. (We will introduce negative rationals after discussing the integers.) Thus, given any non-negative rational,

$$r = \left\{\frac{a}{b}, \frac{2a}{2b}, \ldots, \frac{ma}{mb}, \ldots\right\}$$

then

$$r_o + r = r$$

because $r_o + r$ is defined to be

$$\left(\frac{0}{1} + \frac{a}{b}\right) = \frac{0 \times b + a \times 1}{1 \times b} = \frac{a}{b}$$

(which, of course, names r).

Parenthetically, we note that r_o is only what is called a *right* identity for subtraction. This means simply that we can subtract r_o from any given positive rational, r and still get r. But we cannot surbtract r from r_o and get a positive rational. For example, $\frac{3}{4} - r_o = \frac{3}{4}$ but $r_o - \frac{3}{4}$ is undefined (over the non-negative rationals).

The identity for multiplication is the *rational number*

$$r_I = \left\{\frac{1}{1}, \frac{2}{2}, \frac{3}{3}, \ldots, \frac{n}{n}, \ldots\right\}$$

Thus, if $r = \{a/b, 2a/2b, \ldots, ma/mb, \ldots\}$, then

$$r \times r_I = r_I \times r = r$$

since

$$\frac{a}{b} \times \frac{1}{1} = \frac{a \times 1}{b \times 1} = \frac{a}{b}$$

[19] Division by 0 and subtraction from 0, however, are *not* allowed.

which is a name for r. Use the definition of multiplication to show that $\frac{7}{4} \times r_I = \frac{7}{4}$. Notice that in choosing the canonical representatives a/b and $\frac{1}{1}$ to compute $r \times r_I$, we use the fact that multiplication is well defined. We have also implicitly assumed the commutative property of multiplication of rationals. Where?

The rational r_I is also a *right* identity for division because

$$\left(\frac{a}{b} \div r_I = \frac{a}{b} \times \frac{1}{1} = \right) \frac{a \times 1}{b \times 1} = \frac{a}{b}.$$

It is not, however, a *left* identity, as you can check by dividing $\frac{1}{1}$ by $\frac{3}{4}$ (i.e., does $\frac{1}{1} \div \frac{3}{4} = \frac{3}{4}$?).

EXERCISES 7–8.3

S–1.   Show that r_I is *not* a left identity for division.

S–2.   Explain why, in the proof in the text, it is permissible to choose the canonical representatives a/b and $\frac{1}{1}$ in computing the product and quotient of the rational numbers

$$r = \left\{\frac{a}{b}, \frac{2a}{2b}, \frac{3a}{3b}, \ldots, \frac{ma}{mb}, \ldots\right\}$$

and

$$r_I = \left\{\frac{1}{1}, \frac{2}{2}, \frac{3}{3}, \ldots, \frac{n}{n}, \ldots\right\}$$

E–3.   If we are to add the new number r_o to the system of positive rational numbers, we should specify its behavior under multiplication as well as addition. Assume the previous definition of multiplication applies to r_o and compute the product of r_o and $r = \{a/b, 2a/2b, \ldots, ma/mb, \ldots\}$.

E–4.   a.   What difficulties are encountered if we try to apply the definition of division of rational numbers to the quotient $r \div r_o$?

b.   Compute $r_o \div r$ using the same definition.

8.4 Inverse

A number system has the *inverse* property with respect to $\odot$, if for every number a, there is a unique number, a', such that $a \odot a' = a' \odot a = e$ (where e is the identity for the operation $\odot$).

Addition over the system of non-negative rationals does not have the inverse property. As with addition of whole numbers, we cannot add two rational numbers which are greater than 0 and obtain 0.

The multiplicative inverse of any *positive* rational[20]

$$r = \left\{\frac{a}{b}, \frac{2a}{2b}, \ldots, \frac{na}{nb}, \ldots\right\}$$

is the rational

$$r' = \left\{\frac{b}{a}, \frac{2b}{2a}, \ldots, \frac{nb}{na}, \ldots\right\}$$

---

[20] r_o is specifically not allowed because it does *not* have a multiplicative inverse even in the system of (all) rationals (see Chapter 9).

Thus, for example, the inverse of $\frac{3}{4}$ is $\frac{4}{3}$, because $\frac{3}{4} \times \frac{4}{3} = \frac{3 \times 4}{4 \times 3} = \frac{12}{12} \cong \frac{1}{1}$ (the multiplicative identity).[21] Can you prove this statement in general? (*Hint:* Apply the definition of multiplication to the arbitrary rationals r and r'.)

The inverses under subtraction and division are simply the numbers themselves. That is, for every positive rational, r, $r - r = 0$ and $r \div r = 1$. For example, $\frac{3}{4} - \frac{3}{4} = 0$ and $\frac{3}{4} \div \frac{3}{4} = 1$.

EXERCISES 7–8.4

S–1.   Show that a/b and b/a are multiplicative inverses in the system of positive rational numbers.

E–2.   a.   Recall the definition of division of positive rationals: $a/b \div c/d = (a \times d)/(b \times c)$. Restate this definition using the concept of multiplicative inverse of c/d.

b.   Could the definition of subtraction for positive rationals be similarly stated? Why?

8.5 Distributivity

One operation, $\odot$, in a number system is said to distribute over a second, $\oplus$, if given any triple of numbers a, b, and c, $a \odot (b \oplus c) = (a \odot b) \oplus (a \odot c)$.

Multiplication and division distribute over both addition and subtraction, but in certain cases this is restricted. In discussing these properties, we limit ourselves to giving examples. Attempt to prove these properties for yourself. The main prerequisites are familiarity with what can be done in the system of *natural* numbers, knowledge of the definitions of the binary operations in the system of positive rationals, and clear thinking.

First, observe that multiplication distributes over addition:

$$r_1 \times (r_2 + r_3) = (r_1 \times r_2) + (r_1 \times r_3)$$

For example

$$\frac{1}{3} \times \left(\frac{3}{4} + \frac{2}{3}\right) = \frac{1}{3} \times \left(\frac{9+8}{12}\right) = \frac{1}{3} \times \frac{17}{12} = \frac{17}{36}$$

and

$$\left(\frac{1}{3} \times \frac{3}{4}\right) + \left(\frac{1}{3} \times \frac{2}{3}\right) = \frac{3}{12} + \frac{2}{9} = \frac{27+24}{108} = \frac{51}{108} = \frac{3 \times 17}{3 \times 36}$$

Because

$$\frac{17}{36} \cong \frac{3 \times 17}{3 \times 36}$$

it follows that

$$\frac{1}{3} \times \left(\frac{3}{4} + \frac{2}{3}\right) = \left(\frac{1}{3} \times \frac{3}{4}\right) + \left(\frac{1}{3} \times \frac{2}{3}\right)$$

which is what we wished to show.

[21] Each fraction corresponding to r' is, as we say, the *reciprocal* of some fraction, representing r.

Second, note that where subtraction is defined, multiplication also distributes over subtraction; that is, if $r_2 > r_3$, then

$$r_1 \times (r_2 - r_3) = (r_1 \times r_2) - (r_1 \times r_3)$$

Show, for example, that

$$\frac{1}{3} \times \left(\frac{3}{4} - \frac{1}{5}\right) = \left(\frac{1}{3} \times \frac{3}{4}\right) - \left(\frac{1}{3} \times \frac{1}{5}\right)$$

What about $\frac{1}{3} \times (\frac{1}{5} - \frac{3}{4})$? (*Hint:* Is it defined?)

Finally, we see that division distributes over both addition and subtraction in the system of positive rationals but in only one direction.[22] For example

$$\left(\frac{3}{4} + \frac{1}{5}\right) \div \frac{1}{3} = \left(\frac{3}{4} \div \frac{1}{3}\right) + \left(\frac{1}{5} \div \frac{1}{3}\right)$$

but

$$\frac{1}{3} \div \left(\frac{3}{4} + \frac{1}{5}\right) \neq \left(\frac{1}{3} \div \frac{3}{4}\right) + \left(\frac{1}{3} \div \frac{1}{5}\right)$$

Can you show this? The equations

$$(r_1 + r_2) \div r_3 = (r_1 \div r_3) + (r_2 \div r_3)$$

and

$$(r_1 - r_2) \div r_3 = (r_1 \div r_3) - (r_2 \div r_3)$$

whenever $r_1 > r_2$, are called the *right* distributive properties of division over addition and over subtraction, respectively.

EXERCISES 7–8.5

S–1. Show that $\frac{1}{3} \times (\frac{3}{4} + \frac{1}{5}) = (\frac{1}{3} \times \frac{3}{4}) + (\frac{1}{3} \times \frac{1}{5})$

S–2. Show that $(\frac{3}{4} + \frac{1}{5}) \div \frac{1}{3} = (\frac{3}{4} \div \frac{1}{3}) + (\frac{1}{5} \div \frac{1}{3})$ but that $\frac{1}{3} \div (\frac{3}{4} + \frac{1}{5}) \neq (\frac{1}{3} \div \frac{3}{4}) + (\frac{1}{3} \div \frac{1}{5})$.

S–3. Use the distributive property to perform each of the following computations in the simplest possible way.

a. $\left(\frac{2}{3} \times \frac{1}{2}\right) + \left(\frac{2}{3} \times \frac{3}{2}\right)$

b. $\left(\frac{11}{17} \times \frac{3}{7}\right) + \left(\frac{11}{17} \times \frac{4}{7}\right)$

c. $\left(\frac{14}{15} \times \frac{3}{2}\right) - \left(\frac{14}{15} \times \frac{1}{2}\right)$

M–4. Without using the terms commutative, associative, and so on, write five questions designed to determine whether your students understand the various properties of the positive rational number system.

---

[22] Where the operation $\odot$ is commutative, the distributive law (p. 269) is equivalent to $(b \oplus c) \odot a = (b \odot a) \oplus (c \odot a)$. Remember that division, however, is not a commutative operation.

8.6 Other Properties: Relationships Between the Natural Numbers and the Positive Rationals

The system of positive rational numbers is closely related to the system of natural numbers. We saw this when we defined fractions in terms of natural numbers and in our discussion of computational algorithms.

Throughout our discussion we have been careful to distinguish between fractions and rational numbers (e.g., $\frac{3}{4}$ and $\{\frac{3}{4}, \frac{6}{8}, \ldots, 3m/4m, \ldots\}$) on the one hand, and natural numbers (e.g., 3) on the other. Clearly, however, each natural number corresponds to a unique rational number in a very "natural" way. Thus, for example, 1 corresponds to $\{\frac{1}{1}, \frac{2}{2}, \ldots, m/m, \ldots\}$; and 5 to $\{\frac{5}{1}, \frac{10}{2}, \ldots, 5m/m, \ldots\}$. In general, n corresponds to $\{n/1, 2n/2, \ldots, mn/m, \ldots\}$.

Furthermore, this correspondence is an embedding (as defined in Chapter 4) and the system of natural numbers is said to be *embedded* in the system of positive rational numbers. Thus, the corresponding binary operations over the naturals and rationals are "preserved" by the correspondence. For example, $5 \times 3 = 15$ corresponds to

$$\left\{\frac{5}{1}, \ldots, \frac{5m}{m}, \ldots\right\} \otimes \left\{\frac{3}{1}, \ldots, \frac{3m}{m}, \ldots\right\} = \left\{\frac{5 \times 3}{1}, \ldots, \frac{(5 \times 3)m}{m}, \ldots\right\}.$$

That is, the rational that corresponds to the result of multiplying 5 by 3 (in the system of natural numbers) is identical with the result (i.e., rational) obtained by multiplying the rational corresponding to 5 by the rational corresponding to 3 (in the system of rational numbers). (We used $\otimes$ to indicate multiplication in the latter system to emphasize that the two multiplication operations are not identical. For most purposes, this distinction may safely be ignored.)

This embedding, however, is not the only correspondence between the natural and positive rational numbers. Oddly, there is a one-to-one correspondence between the set of positive rational numbers and the set of natural numbers. Let us be clear on this. We do not mean that *some* of the rationals correspond to natural numbers (as in the embedding above) but that *every* positive rational corresponds to some natural number (and vice versa). To see this we represent the positive rationals as fractions in the infinite array below.

$\dfrac{1}{1}$	$\dfrac{1}{2}$	$\dfrac{1}{3}$	$\dfrac{1}{4}$	$\cdots$	$\dfrac{1}{n}$	$\cdots$
$\dfrac{2}{1}$	$\dfrac{2}{2}$	$\dfrac{2}{3}$	$\dfrac{2}{4}$	$\cdots$	$\dfrac{2}{n}$	$\cdots$
$\dfrac{3}{1}$	$\dfrac{3}{2}$	$\cdots$			$\dfrac{3}{n}$	$\cdots$
$\cdot$	$\cdot$				$\cdot$	
$\cdot$	$\cdot$				$\cdot$	
$\cdot$	$\cdot$				$\cdot$	
$\dfrac{m}{1}$	$\dfrac{m}{2}$	$\cdots$			$\dfrac{m}{n}$	$\cdots$
$\cdot$					$\cdot$	
$\cdot$					$\cdot$	
$\cdot$					$\cdot$	

We first convince ourselves that this array contains at least one fraction representative of every positive rational number. We need only observe that every positive rational can be represented by some fraction of the form m/n and that we have constructed the array so that it includes every possible fraction. (All fractions are of the form m/n.)

Observe next that certain fractions are equivalent (i.e., name the same rational). We can take care of this by eliminating from the array all fractions except those in which the numerator and denominator are relatively prime (including $\frac{1}{1}$). For example, we would keep $\frac{1}{2}$ but eliminate from the array $\frac{2}{4}, \frac{3}{6}, \frac{4}{8}$, and so on. (We avoid the philosophical problem as to how one might actually go about the task of eliminating the unneeded fractions. Just think of them as being "crossed off.")

Finally, we devise a scheme for matching the remaining fractions (each denotes a different rational number) with the natural numbers as indicated by the path in Figure 7–17. The first *uncrossed* fraction ($\frac{1}{1}$) corresponds to 1, the second ($\frac{1}{2}$) to 2, the third ($\frac{2}{1}$) to 3, the fourth ($\frac{3}{1}$) to 4, the fifth ($\frac{1}{3}$ since $\frac{2}{2}$ is crossed out) to 5, and so on.

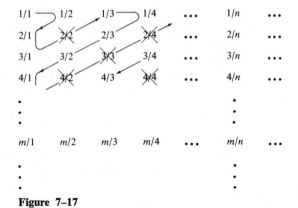

Figure 7–17

Although the idea has deep philosophical roots, many young children can appreciate the argument involved if they are allowed to have a hand in formulating the question. It is perhaps one of the simplest examples of a situation where one's intuition is at variance with the facts. You may want to keep it in mind as a counterexample for the bright youngster who says, "I know they aren't equal because they don't look equal." Many a mathematician got his start at a young age by being challenged with a lesser problem.

In some sense, the positive rationals form a richer number system than the natural numbers. For one thing, every pair of positive rational numbers has a unique quotient which is a positive rational. This is not true of the system of natural numbers. For example, there is no natural number equal to $7 \div 3$—the quotient, as we say, is undefined.

Another property of the rationals is that they form what we call a *dense* set. Given any pair of rational numbers, no matter how close, we can always find a rational number which is between them. Suppose, for example, we are given $\frac{2}{3} \cong \frac{8}{12} \cong \frac{16}{24}$ and $\frac{3}{4} \cong \frac{9}{12} \cong \frac{18}{24}$. Then $\frac{17}{24}$ is between them. In turn, if we consider $\frac{2}{3} \cong \frac{16}{24} \cong \frac{32}{48}$ and $\frac{17}{24} \cong \frac{34}{48}$ we see that $\frac{33}{48}$ is between them. This process may be repeated indefinitely; each time the difference between the two numbers will get smaller. (This process is reminiscent of the rabbit who can jump only half of the way to where he is going. Although he can get as close as he wants, he never gets all the way there. The jumps get smaller and smaller the closer he gets to his goal.)

Clearly, the natural numbers do *not* form a dense set. Given any two consecutive natural numbers, there is no natural number between them. For example, there is no natural number between 3 and 4. In higher mathematics, the properties of being dense or not dense play an important role in proving many important theorems.

Fair is fair, however, and although we gain some things in going from the natural numbers to the rationals, we lose in other ways. For example, the set of natural numbers has a smallest (or minimum) element, denoted 1, from which all of the other elements may be generated by repeated addition. Thus $2 = 1 + 1$,

$$3 = 1 + 1 + 1, \ldots, n = \overbrace{1 + 1 + \ldots + 1}^{n \text{ times}}.$$

In the set of *non-negative* rationals, on the other hand, the corresponding rational, $\frac{1}{1}$ (which is also the multiplicative identity), does not generate all of the rationals. Thus, no matter how many times we add $\frac{1}{1}$ to itself, we always get a fraction of the form $n/1$. More important, there is no other rational number that will work. Given any rational, we can only generate other rationals which may be expressed in a particular form. For example, repeated addition of $\frac{3}{5}$ always gives a rational of the type, $3n/5$.

The set of *positive* rationals does *not* even have a minimum element since it does not contain 0. To see this, we need only to consider what would happen to our unfortunate rabbit if he tried to reach 0. He could get closer and closer, but he would never make it. There is no smallest positive rational. (As we shall see in Chapter 9, there is no smallest *signed rational* either.)

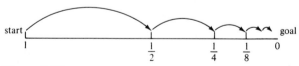

Figure 7–18

EXERCISES 7–8.6

S–1. Explain why every possible fraction is guaranteed an appearance in the array of Figure 7–17.

S–2. a. List two rational numbers between $\frac{1}{3}$ and $\frac{1}{2}$.
   b. List three rational numbers between $\frac{2}{3}$ and $\frac{3}{4}$.
   c. List four rational numbers between $\frac{1}{8}$ and $\frac{1}{9}$.

S–3.   a.   What is the smallest positive rational number?

 b.   What is the smallest non-negative rational number?

∗E–4.   The symbol a/b now has two meanings for us. In Chapter 5, a/b was given as an alternative way of writing $a \div b$, but in Chapter 7, a/b has been taken as a fraction. Show that in view of the embedding of the system of natural numbers into the system of positive rational numbers, these two meanings are consistent. That is, show that if a and b are natural numbers and $a \div b$ is defined, then the natural number $a \div b$ corresponds to the positive rational number named by the fraction a/b.

8

The System of Integers

Although the introduction of integers (signed numbers) has traditionally been postponed as late as ninth-grade algebra, there are several reasons why this should not be done. First, children are already familiar with many situations that call for the use of integers. Debts and losses, temperatures above and below zero, and, in fact, any situation in which both direction and magnitude are involved, provide prime examples. Second, not having integers available to them, children often become accustomed to using natural numbers, together with such descriptors as "debt," "above," and "more," in such situations, thereby making it more difficult for them later to grasp the more general nature and significance of integers. Third, there is no logical or psychological justification for introducing the (positive) rational numbers before the integers. As we shall argue below, the concept of integer appears to be less difficult for young children than the concept of rational number.

In Chapter 7, we saw that fractions may be thought of as properties of pairs of sets. We also showed that they may be represented as ordered pairs of numbers with the understanding that these pairs should be interpreted to mean "relative size" (of two sets). Next, we pointed out that any number of different pairs of sets may have the same fractional property (i.e., correspond to the same pair of natural numbers), and, in particular, that each fraction defines an equivalence class of pairs of sets with that (fractional) property. We defined rational numbers to be equivalence classes of fractions and showed that all equivalent fractions can be thought of as representing the same amount. After that, we defined, justified, and illustrated the four binary operations of addition, subtraction, multiplication, and division. Finally, we identified certain properties of the system of positive rationals.

Our discussion of the system of integers in this chapter proceeds along very much the same lines. This approach helps to clarify the nature of both systems by identifying parallels that are frequently passed over lightly or not at all.

1. MORE-LESS PROPERTY

In introducing fractions in Chapter 7, we suggested that number pairs might be interpreted quite differently but did not elaborate. In this chapter, the corresponding idea may be called the *more-less property*, a property that allows us to talk about the number of elements in one set being so many more or less than the number in a second set. For example, the ordered pair of sets, $(\{a, b\}, \{c, d, e, f\})$, has the property, "2 (elements in the first set) is *2 less* than 4 (elements in the second)."[1]

This same ordered pair of sets, of course, also has the fractional property, "the *ratio* 2 to 4." Hence, when we represent both the more-less property and the fractional property by the same ordered pair (i.e., (2, 4)), we tend to camouflage the difference between them.[2] The special notation, a/b, for fractions helps make this distinction explicit.

In view of the close relationship between the more-less property and fractions, it is not surprising that each corresponding pair of properties defines the *same* equivalence class of pairs of sets. For example, the more-less property, "2 is *2 less* than 4," defines an equivalence class consisting of all pairs of sets in which the first has two elements and the second has four. Because each such pair of sets also has the corresponding fractional property, "the *ratio* of 2 to 4," we see this latter property defines exactly the same equivalence class. As before, we shall call any pair of sets which belongs to the equivalence class defined by an ordered pair an *interpretation* of the ordered pair.

Can you construct five pairs of sets, each of which has both of these properties? (One such pair is given above.) Recall from our discussion of fractions that two pairs of sets have the same fractional property, and, hence, the same more-less property, if the first and second sets of one pair can be put into one-to-one correspondence with the first and second sets, respectively, of the other. (If you feel uncertain about the meaning of this statement, review the corresponding discussion in Chapter 7.)

In Chapter 7, we included a fairly detailed discussion of different kinds of interpretations of fractions. Because this discussion carries over with no essential change (except that here we are concerned with the more-less property), we shall not repeat what was said there. Instead, we shall simply mention certain similarities and differences that are of psychological importance.

In discussing fractions, recall that most interpretations involve set inclusion. That is, the elements of the first set (of proper fractions) are typically *in-*

---

[1] According to convention, we say "*a* less than *b*" rather than "*b* greater than *a*."

[2] As we shall see later on, (two-dimensional) vectors are also concerned with ordered pairs of numbers.

cluded in the second set. This is reasonable, of course, because most people think of fractions as referring to parts of some whole. In fact, fractions were undoubtedly invented for that reason in the first place. For example, the display

corresponds to the interpretation, ({ $\ominus$, $\langle\!|\!\rangle$ }, { $\ominus$, $\langle\!|\!\rangle$, $\Diamond$ }), of the fraction $\frac{2}{3}$. (This pair, of course, also has the property "2 is *1 less* than 3.")

Generally speaking, the sets in interpretations of the *more-less property* are *not* related by inclusion. Instead, the sets are typically *disjoint*. This difference is psychologically important and no doubt is one of the major reasons why the more-less property is intrinsically easier for young children (ages 5–8) to learn than fractions.

Piaget,[3] for example, found that young children (ages 5–8) are generally unable to determine which of two sets is larger when one is contained in the other. For example, consider a situation where a child is given a set of blue objects, most of which are sticks and the rest beads, and is asked whether there are more blue objects or more sticks (see Figure 8–1). When posed with this question, many young children (most 5- and 6-year-olds) will answer that there are more sticks. (The "correct" answer, of course, is blue, because all of the objects are blue.) They will also answer "more sticks," of course, when asked whether there are more sticks or more beads. In effect, many children seem able to tell which of two sets is larger when the sets are disjoint but not when one set is included in the other.

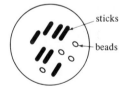

Figure 8–1

An even more important reason why fractions are more difficult to learn than the more-less relation is that *relative amount* is a far more sophisticated notion than is *number more (less)*. The child who has learned to conserve number (i.e., pair equivalent sets in one-to-one fashion) is usually ready to learn about differences but not relative amount. Probably the main reason for this is that in finding differences the basic measuring units are set *elements*. Thus, for example, the set {*a, b, c*} has two *less elements* than {*d, e, f, g, h*}. We can determine this by pairing the (elements in the) two sets and seeing how many are left over. In

[3] For more details see J. Piaget, *The Child's Conception of Number*, New York, Norton, 1965.

order to determine relative amount, on the other hand, the basic measuring units are the sets themselves. Thus, the fraction $\frac{3}{5}$, which corresponds to the two sets above, expresses the relative size of the two sets. Because sets are at a higher level of abstraction than elements, it is not surprising that fractions are harder to learn.

Finally, we indicate briefly how states and/or operators may serve as interpretations of the more-less idea. For example, consider the property, "2 is *1 less* than 3." Clearly, a pair of sets, like ($\{a, b\}$, $\{c, d, e\}$), is a *state* interpretation. We can just as well, however, conceive of the following interpretations which involve one or more operators: (1) the (two) *acts* of forming the two sets above (e.g., by selecting first two and then three English letters); (2) the block

tower $\begin{array}{|c|}\hline B \\\hline A \\\hline\end{array}$, together with the act of building a tower with three such blocks

$\left(\text{yielding } \begin{array}{|c|}\hline E \\\hline D \\\hline C \\\hline\end{array}\right)$.

These examples are suggestive of activities in which young children engage spontaneously and which might be used to illustrate that the more-less relation can be interpreted in a wide variety of ways. The teacher should not overdo this, of course. As with all good things, a little goes a long way.

EXERCISES 8–1

S–1. What are some reasons for *not* postponing the introduction of integers until the ninth grade?

S–2. The ordered pair of sets ($\{a, b\}$, $\{c, d, e, f\}$) is an interpretation of the more-less property of the ordered pair (2, 4). Give five more interpretations of the ordered pair (2, 4).

S–3. Give an interpretation of the more-less property of each of the following ordered pairs.

a.	(3, 2)	e.	(3, 1)
b.	(1, 4)	f.	(0, 2)
c.	(2, 2)	g.	(3, 0)
d.	(1, 1)		

S–4. Give at least two reasons why young children might find integers easier to learn than fractions.

[E–5.] The notation $\frac{2}{4}$ may be used in place of the ordered pair (2, 4) to show that we mean the fractional property of the pair (2, 4), rather than the more-less property. What notation might we use to show that the pair (2, 4) is to represent the more-less property "2 is 2 less than 4"?

E–6. The ordered pair (2, 4) represents the more-less property, "2 is *2 less* than 4," and the ordered pair (5, 1) represents the more-less property "5 is *4 more* than

1." State the more-less property represented by each of the following ordered pairs.

a. (3, 2)   e. (3, 1)
b. (1, 4)   f. (0, 2)
c. (2, 2)   g. (3, 0)
d. (1, 1)

M–7. Consider two towers of blocks consisting of three and five blocks, respectively. What question(s) could you ask children about this interpretation of the ordered pair (3, 5) to lead them to discover the associated more-less property?

M–8. What is the distinction between the more-less property of the display

```
B     E
A     D
      C
```

and the more-less property of the similar display

```
C     E
B     D
A
```

How would you make the difference clear to children?

M–9. How would you make clear to children the distinction between the more-less properties of the ordered pairs (2, 4) and (4, 2)?

2. INTEGERS

In Chapter 7, we saw that rational numbers are concerned with the relative size of pairs of sets, independent of the number of elements in each set. For example, the fractions $\frac{3}{4}$ and $\frac{6}{8}$ can be used to represent the same amount, or proportion of a given whole, even though the first corresponds to a situation where the whole in question is divided into four equal parts and we have three of these parts, and the second, to a situation where the whole is divided into eight parts and we have six. This led us to define rational numbers as equivalence classes of fractions, where two fractions, a/b and c/d, were said to be equivalent if $ad = bc$. Finally, this definition was shown to correspond to our intuition that equivalent fractions should represent the same proportion.

In parallel fashion, *integers* are concerned with the *number of elements more (less)* one set of a pair has as compared with the other, *independent of the (absolute) number of elements in each set.*

The more-less properties of the pairs, $(\{1, 2, 3, 4, 5\}, \{a, b\})$ and $(\{1, 2, 3, 4, 5, 6, 7, 8, 9\}, \{a, b, c, d, e, f\})$, for example, clearly have much in common. The first numbers of the ordered pairs, (5, 2) and (9, 6), which represent these properties, are *3 more* than the corresponding second numbers. The ordered pairs

(3, 0), (4, 1), (5, 2), (6, 3), (7, 4), (8, 5), (9, 6) have this same property of *3-more-ness*, as, in fact, does any ordered pair of the form $(n + 3, n)$, where n is any whole number.

The same general idea applies to all ordered pairs (when used to represent more-less properties). The pairs (1, 0), (2, 1), (3, 2), (4, 3), and so on correspond to the property of *1-moreness*, and (0, 4), (1, 5), (2, 6), (3, 7), . . ., to the property of *4-lessness*.

To see how children might come to recognize this sort of equivalence, consider the game of "giant steps." In this game, the players start at a given line and the leader stands, say, 20 yards in front of the players by a "goal line." Additional lines may be used to mark off yards both in front of and behind the starting line. The leader calls out instructions, each of which consists of a pair of numbers. The first number indicates the number of yards the player is to move toward the goal line and the second number indicates the number of yards he (or she) is to move away from the goal line. The leader requires that the players get to the proper yard line but does not care how they get there. If they end up on the wrong yard line, they may be removed from the game, or, better, be penalized, say, 3 yards. The objective, of course, is to reach the goal line.

After playing the game for awhile, some of the children will begin to recognize that they can take "short-cuts." Thus, for example, instead of going five yards forward (toward the goal line) and six yards backward, a player may simply move one yard backward. Similarly, "(11, 6)" may be converted into a move 5 yards forward.

This sort of behavior can be encouraged with young children by asking such questions as, "Can you find the shortest way to get where you are going?" When children recognize such equivalences more or less spontaneously (perhaps with some encouragement), they can profitably be introduced to more-less properties in the abstract, as divorced from particular pairs of sets. [4]

This idea of equivalence may be formalized as an equivalence relation: Two ordered pairs (a, b) and (c, d) are said to be *more-less equivalent* if and only if $a + d = b + c$. We denote this, $(a, b) \doteq (c, d)$. For example, the ordered pairs (2, 10) and (16, 24) are more-less equivalent by this definition because $2 + 24 = 10 + 16$. Notice, however, that (10, 2) and (16, 24) are *not* more-less equivalent (since $10 + 24 \neq 2 + 16$), even though the difference between the first and second element in each case is 8. Intuitively, then, two pairs are equivalent if and only if both the *magnitude and direction* of the difference between the elements of each pair is the same. (More-less equivalence cannot be defined in terms of the equation $a - b = c - d$, however, since subtraction is not always defined over the whole numbers—e.g., consider the pair (16, 24).)

---

[4] More-less properties of this type may be described by "open" sentences of the form, "*x* is *n* more (less) than *y*," where *n* is fixed and *x* and *y* are allowed to vary so long as *x* is *n* more (less) than *y*. (Recall that by open sentence we mean a statement that is neither true nor false until the variables are assigned values.) For example, the property *2-lessness* may be defined in terms of the set (actually an equivalence class), {(0, 2), (1, 3), (2, 4), . . ., (n, n + 2), . . .}, because every pair in this set satisfies the corresponding open sentence, "*x* is 2 less than *y*" (e.g., "0 is *2* less than 2"). Note, however, that there is no pair outside this class which satisfies this open sentence (e.g., "1 is *not* 2 less than 4").

We are finally ready to define precisely what we mean by an *integer*. Again paralleling our development of the positive rational numbers, we define the integers as the equivalence classes of ordered pairs formed by applying the above definition of more-less equivalence (a binary relation) to all pairs of ordered pairs (of whole numbers). The equivalence classes $\{(0, 1), (1, 2), (2, 3), \ldots,$ $(n, n + 1), \ldots\}$ and $\{(3, 0), (4, 1), (5, 2), \ldots, (n + 3, n), \ldots\}$ provide two examples corresponding to the integers. *1-lessness* and *3-moreness*, respectively. (In effect, we were talking about integers above but we did not call them that.)

Can you verify that the pairs in each equivalence class are equivalent by the definition? Are the differences of the same magnitude and direction? Consider the pairs (0, 1) and (2, 3).

EXERCISES 8–2

S–1. For each of the following, list three ordered pairs of whole numbers that represent the stated more-less property.

    a. 2-moreness
    b. 3-lessness
    c. no-moreness
    d. 1-moreness
    e. no-lessness

S–2. Give the smallest possible whole number that can be used as the *first* element of an ordered pair that represents

    a. 4-lessness
    b. 3-moreness
    c. 6-lessness
    d. 7-moreness
    e. n-moreness, where n is a whole number
    f. m-lessness, where m is a whole number

S–3. What is the more-less property of each of the following?

    a. $(\{*, \Box, \triangle\}, \{2, 4, 6, 8, 10\})$
    b. $(\{0\}, \{3\})$
    c. (0, 3)
    d. $(n + 7, n)$ where n is any whole number

S–4. For each of the following pairs of ordered pairs, show whether or not they are more-less equivalent.

    a. (2, 7) and (13, 18)
    b. (8, 3) and (14, 10)
    c. (0, 0) and (17, 17)
    d. (0, 12) and (17, 33)
    e. (37, 64) and (49, 22)
    f. (7, 7) and (n, n) for any whole number n
    g. $(n + 1, n + 7)$ and $(n - 6, n)$ for any whole number $n \geq 6$

S-5. For each of the following ordered pairs, find three other ordered pairs which are more-less equivalent to it.

a.	(7, 1)	e.	(0, 3)
b.	(6, 2)	f.	(0, 0)
c.	(1, 4)	g.	(4, 3)
d.	(5, 5)		

M-6. Give four test items you would use to determine if a child knows how to tell if two pairs of whole numbers are more-less equivalent.

[M-7.] It is possible to use the game "giant steps" to lead children to detect the more-less properties of ordered pairs of commands (e.g., 3 forward, 2 backward) as a regularity. What question(s) could you ask the child to focus his attention on this regularity?

[M-8.] a. What property is common to all ordered pairs that represent no-moreness?

b. How would you help children to detect this property?

For further thought and study: Examine several textbook series to find the grade level at which integers are introduced. How does each series define "integer"?

3. REPRESENTING INTEGERS

As with the positive rationals, we could represent integers as sets (i.e., equivalence classes) of ordered pairs, as above, or we might use the set-builder notation. In the latter case, the integer, *1-lessness*, might be represented $\{(n, n + 1) \mid n$ is a *whole* number$\}$. Alternatively we could denote integers by certain *canonical* (special) pairs, where one of the elements in the pair is 0. Thus, 1-lessness might be represented by (0, 1). Similarly, 3-moreness could be represented by (3, 0). How would 5-lessness be represented? 9-moreness? No (zero)-moreness?[5]

In point of fact, we use none of these in practice (although we could). Rather, we represent integers by ordinary base 10 numerals, prefaced by a "+" or "−" sign to indicate direction. *The sign "+" corresponds to "moreness" and "−" to "lessness."* Hence, the integer, 1-lessness, will henceforth be denoted, "−1." Similarly, 5-moreness is denoted, "+5." In general, *n*-moreness is denoted "+*n*" and *m*-lessness, by "−*m*," where *n* and *m* are arbitrary *numerals* (i.e., names for natural numbers). To complete matters, we denote the integer, *no-moreness* (= no-lessness) = $\{(0, 0), (1, 1), (2, 2), \ldots, (n, n), \ldots\}$ by "+0" or just "0" for short (unless we wish to distinguish it from the whole number 0).[6]

Notice, that although the non-negative rationals and the integers both refer to equivalence classes of pairs of whole numbers,[7] the rationals are routinely

---

[5] We have used whole numbers, rather than natural numbers, in our definition of the integers for convenience. This allows us to denote integers by canonical pairs in the particularly simple way shown above.

[6] Remember, the whole number 0 is a property of the empty set, whereas "+0" denotes a certain equivalence class of ordered pairs of whole numbers.

[7] Where the corresponding pairs have different connotations.

denoted in a variety of different ways (e.g., $\frac{4}{8}$, $\frac{2}{4}$, $\frac{1}{2}$, .5, all denote the rational one-half), while each integer has a unique name (e.g., $^+2$). The name, "$^+5$," for example, adequately represents both magnitude and direction; this is all that a signed number (i.e., integer) is designed to do.

Each rational number, of course, refers to a unique ratio, and it would certainly be possible to agree on one way to represent them. One obvious possibility would be to restrict oneself to decimal notation. This is not done for two reasons. First, some rationals with simple fractional representations (e.g., $\frac{1}{3}$ and $\frac{2}{7}$) cannot be represented by terminating (finite) decimals. Second, we frequently wish to distinguish between interpretations of equivalent fractions (e.g., $\frac{1}{2}$ and $\frac{2}{4}$ both become .5 in decimal notation). Common usage does not seem to dictate this sort of distinction with respect to the more-less property. When one asks, "How many more?" he is not normally concerned with the absolute amounts in each set but the amount of *gain* or *loss*, say, in some financial transaction. Where one is interested in absolute amounts as well, they are usually indicated by statements such as "50 is 10 more than 40"—no special name like "fractions" is used.

It is frequently convenient (and instructive) to represent integers on the so-called "number line." The basic idea is simple and can easily be seen by studying the number line in Figure 8–2. $^+2$ corresponds to a move of 2 units to the right of 0, and $^-5$ to a move of 5 units to the left. How would you interpret $^+6$? $^-1$? Notice that in each case the arrows indicate both magnitude and direction.

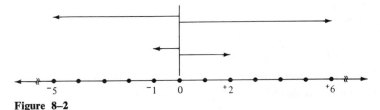

Figure 8–2

The whole numbers and the non-negative rationals can also be represented on a number line. Study the two number lines in Figure 8–3. Notice that in both cases there are no numbers to the left of 0. In addition, one might suspect from the second number line that the rationals "fill up" that part of the number line to the right of 0 but this is not true. (See the discussion of the real numbers.) Are the integers dense in the same way as the non-negative rationals? Elaborate.

Figure 8–3

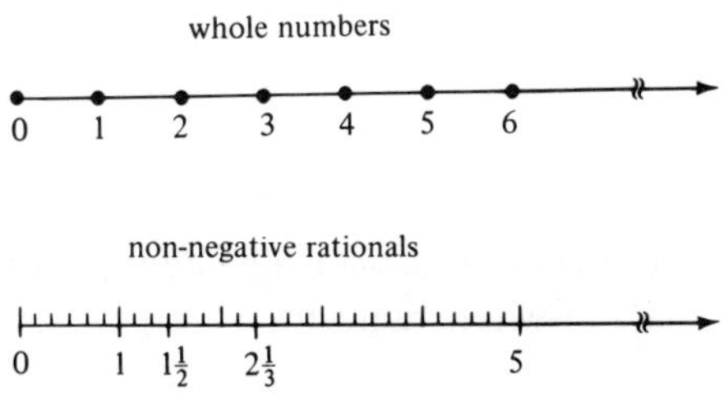

whole numbers

0 1 2 3 4 5 6

non-negative rationals

0   1 1½ 2⅓   5

What would happen if we considered ordered pairs of *integers*, such as (⁻1, ⁺2) and (⁺3, ⁻4), and viewed them like fractions?

As we shall see in the sections that follow, the number line is a very useful devise for representing the arithmetic operations on the set of integers.[8]

EXERCISES 8–3

[S–1.] Draw a number line and use it to represent the following integers.

a.	⁺6	d.	⁺3
b.	⁻1	e.	⁻4
c.	⁻3	f.	0

[S–2.] Represent each of the following integers in set-builder notation (e.g., 1-lessness is represented by $\{(n, n + 1) \mid n$ is a whole number$\}$).

a. 1-moreness
b. 5-moreness
c. 2-lessness
d. no-moreness

[S–3.] Represent each of the following integers (integer representatives) by a *canonical* ordered pair (i.e., an ordered pair containing at least one zero).

a.	(7, 9)	e.	⁺5
b.	(16, 4)	f.	⁻7
c.	(7, 7)	g.	⁻2
d.	(13, 3)	h.	⁺0

S–4. Give three different real-life interpretations of the integer represented by (79, 37). Consider temperature, test grades, game scores, and so on.

E–5. Recall from Chapter 7 that the non-negative rational numbers are *dense* on the number line because between any two rationals we can always find another rational. Are the integers dense on the number line "in the same way" as the non-negative rationals? Why or why not?

∗E–6. It was shown in Chapter 7 that the positive rationals may be placed in one-to-one correspondence with the natural numbers. Set up a one-to-one correspondence between the integers and the natural numbers. (*Hint:* Begin by pairing ⁺0 ↔ 1, ⁺1 ↔ 2, ⁻1 ↔ 3.)

THE SYSTEM OF INTEGERS

4. ADDITION

Continuing our parallel with the positive rationals, we define addition over the integers in terms of representative pairs of whole numbers.

---

[8] It can also be used with other systems.

In particular, to *define* the *sum* $x + y$ of two integers, we let the pair (a, b) represent x and the pair (c, d) represent y.[9] Then the sum, $x + y$ is represented by $(a, b) + (c, d) = (a + c, b + d)$. Stated verbally, this definition says that to add any given pair of integers, select arbitrary ordered pair representatives of each integer, and then add the corresponding first and second numbers in each pair. For example, to add $^+3$ and $^-4$ we select two representatives, say $(3, 0)$ and $(2, 6)$, respectively, and add the pairs as indicated above. This gives $(3 + 2, 0 + 6) = (5, 6)$, which represents the integer $^-1$. (Note that $(5, 6)$ is more-less equivalent to $(0, 1)$.) Thus, $^+3 + {^-4} = {^-1}$.

It would appear that this is a good definition, but to make sure, we must go through the familiar procedure of convincing ourselves that the operation: (1) always yields an integer as a sum (i.e., that the set of integers is closed under the operation of addition), (2) is well defined, and (3) conforms to our intuition of what the sum of two integers should be.

Clearly, (1) holds; no matter what pairs (a, b) and (c, d) we are given, $(a + c, b + d)$ is always an ordered pair of whole numbers[10] and hence defines an integer.

To show that the operation of addition of integers is well defined, we must prove that no matter what pairs of whole numbers we choose to represent the integers x and y, we always obtain equivalent pairs as the sum. Let (a, b) and (a', b') be two arbitrary representatives of x, and let (c, d) and (c', d') be two arbitrary representatives of y. Then, we must prove that the resulting sums, $(a + c, b + d)$ and $(a' + c', b' + d')$ are more-less equivalent.

The gist of the proof is to apply the definition to (a, b) and (c, d) and to (a', b') and (c', d') and then to show that the resulting pairs $(a + c, b + d)$ and $(a' + c', b' + d')$ represent the same sum—that is, that the pairs are more-less equivalent (i.e., $(a + c) + (b' + d') = (b + d) + (a' + c')$). This statement follows because (a, b) and (c, d) represent the same integers (x and y) as (a', b') and (c', d'), respectively—that is, (1) $a + b' = b + a'$, and (2) $c + d' = d + c'$.

The formal proof is as follows:

1. $a + b' = b + a'$ because (a, b) and (a', b') are more-less equivalent (they both represent x). $c + d' = d + c'$ because (c, d) and (c', d') are more-less equivalent (they both represent y).
2. $(a + b') + (c + d') = (b + a') + (d + c')$ by adding equals to equals.
3. $(a + c) + (b' + d') = (b + d) + (a' + c')$ by the commutative and associative properties of addition of whole numbers.
4. $(a + c, b + d) \doteq (a' + c', b' + d')$ by definition of more-less equivalence.

---

[9] We use x and y to represent integers because we do not know which signs ($+$ or $-$) to attach. If we *knew*, for instance, that x was positive and y was negative, we could write $x = {^+m}$ and $y = {^-n}$, where m and n refer to whole numbers. None of this changes the intended meaning, of course.

[10] $a + c$ and $b + d$ are whole numbers, because $a, b, c,$ and d are whole numbers and the sum of any two whole numbers is also a whole number (by the closure property of addition over the whole numbers).

Because step 4 says that $(a + c, b + d)$ and $(a' + c', b' + d')$ are more-less equivalent, we see that they represent the same integer; hence the operation of addition is well defined.

We illustrate with the following example. Let $(3, 0) \doteq (5, 2)$ represent $^+3$ and $(1, 5) \doteq (4, 8)$ represent $^-4$. By our definition of addition, $^+3 + {}^-4$ is represented by $(3 + 1, 0 + 5) = (4, 5)$ and by $(5 + 4, 2 + 8) = (9, 10)$. The two pairs $(4, 5)$ and $(9, 10)$ represent the same sum (integer) because $4 + 10 = 5 + 9$.

For further practice, go through the above argument with another particular pair of integers, say, $^-5$ and $^+2$. *Hint:* Let $(0, 5)$ and $(2, 0)$ be one pair of representatives and show that the sum $(0 + 2, 5 + 0) = (2, 5)$ is more-less equivalent to the sum determined by every other pair of representatives, $(m, m + 5)$ and $(n + 2, n)$, respectively. If that is too difficult, choose another *particular* pair of representatives, say, $(1, 6)$ and $(5, 3)$.

Although it is rarely used in practice, this definition provides what is perhaps the simplest algorithm for adding integers: *Pick a representative for each integer, add the corresponding whole numbers as usual, and find the integer corresponding to the sum.* To simplify matters, we can always pick representatives in which one of the entries is 0. (Why?) For example, to add $^+7$ and $^-13$, we choose $(7, 0)$ and $(0, 13)$ as representatives. This gives us $(7, 13)$ as the sum, which corresponds to $^-6$. Hence, $^+7 + {}^-13 = {}^-6$.

How does our definition, and hence the algorithm, fit our intuition? We would like our definition to correspond to combining corresponding pairs of sets and, happily, it does. Suppose, for example, that Joey and Janie each get a certain number of cookies each day, according to how well they have behaved. On the first day, Joey gets four cookies and Janie gets three. Here, we can represent the relative amounts by the pair $(4, 3)$ which is interpreted to mean that Joey has one more cookie than Janie. This would make Joey very happy. On the next day, however, Joey only gets three cookies and Janie gets four. We represent the amount by the pair $(3, 4)$. This sort of result does not always lead to peace around the campfire, but by drawing on their (limited) knowledge of signed numbers and counting on good luck, we might be able to convince Joey and Janie that they both made out all right. In effect, Joey received seven cookies over the two-day period and so did Janie—and we use the pair $(7, 7)$, which represents 0. According to our definition, $(4, 3) + (3, 4) = (4 + 3, 3 + 4)$, or equivalently, $^+1 + {}^-1 = 0$.

For one reason or another, most people learn to add integers in the following manner:

1. If both signs are +, add the magnitudes (as if they were whole numbers) and affix "+" to the result.

2. If both signs are −, add the magnitudes and affix "−" to the result.

3. If the sign of the integer with the larger magnitude[11] is + and that of the smaller magnitude is −, then subtract the smaller magnitude and affix "+" to the result.

[11] If both integers are of equal magnitude, either may be picked as the smaller (larger) in order to apply the algorithm.

4. If the sign of the integer with the larger magnitude[11] is − and that of the smaller magnitude is +, then subtract the smaller magnitude and affix "−" to the result.

When written in this way, we see that the algorithm is composed of what essentially amounts to four distinct algorithms, one corresponding to each possible combination of signs: ++, −−, +−, and −+. For this reason, we call this algorithm the *four-case algorithm*. Use these algorithms to find each of the following: $^+2 + {}^+3$, $^-2 + {}^-3$, $^+3 + {}^-2$, $^-3 + {}^+2$.

The meaning of these four cases is perhaps easiest to see on the number line. Thus, case 1 corresponds to

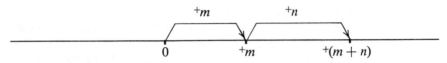

Case 2 corresponds to

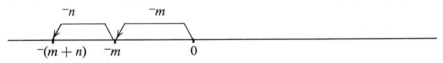

Case 3 corresponds to

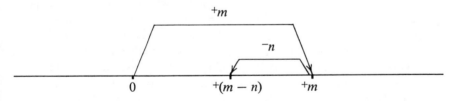

Case 4 corresponds to

Use the number line to determine $^+2 + {}^+3$, $^-2 + {}^-3$, $^+3 + {}^-2$, and $^-3 + {}^+2$.

As we would hope, our definition and the four-case algorithm give exactly the same sums. Thus, each case can be justified by a simple application of our definition.

1. Let ^+m be represented by the pair $(m, 0)$ and ^+n be represented by the pair $(n, 0)$. Then by our definition, $^+m + {}^+n$ is represented by $(m + n, 0)$, which also represents $^+(m + n)$. But $^+(m + n)$ is exactly what application of case 1 gives as the sum. For example, if $(3, 0)$ and $(4, 0)$ represent $^+3$ and $^+4$, respectively, then the sum is represented by $(7, 0)$. But $(7, 0)$

represents $^+7$, which is the same answer as given by case 1 of the four-case algorithm.

2. Let ^-m be represented by the pair $(0, m)$ and ^-n be represented by the pair $(0, n)$. Then by our definition, $^-m + {}^-n$ is represented by $(0, m + n)$, which also represents $^-(m + n)$. But $^-(m + n)$ is exactly what application of case 2 gives. For example, if $(0, 3)$ and $(0, 4)$ represent $^-3$ and $^-4$, respectively, then the sum is represented by $(0, 7)$. But $(0, 7)$ represents $^-7$, which is the same answer as given by case 2.

3. Let ^+m be represented by the pair $(m, 0)$ and ^-n be represented by the pair $(0, n)$, where $m \geq n$. Then, by our definition, $^+m + {}^-n$ is represented by (m, n). But, since $m \geq n$, (m, n) is more-less equivalent to $(m - n, 0)$, and both pairs represent the integer $^+(m - n)$. Case 3 also gives $^+(m - n)$. For example, if $^+7$ and $^-3$ are represented by $(7, 0)$ and $(0, 3)$, respectively, then $^+7 + {}^-3$ is represented by $(7, 3)$. But $(7, 3)$ is more-less equivalent to $(4, 0)$ and hence represents $^+4$. Check the result against case 3.

4. We leave this case as an exercise. *Hint:* Consider ^-m and ^+n where $m \geq n$.

The four-case algorithm is important in that it is suggestive of another type of application of integers—namely, applications where both magnitude and direction are involved. For example, one might want to represent changes in altitude (up or down), movements (forward or backward), or changes in temperature (up or down) by integers. The integer $^+3000$ might be used to indicate an *increase* in altitude of 3000 feet, while $^-2500$ might indicate a *decrease* of 2500 feet. Addition of two such integers would correspond to determining that change in altitude which corresponds to first making the change associated with the first integer and then the change associated with the second. The sum $(^+3000 + {}^-2500 = {}^+500)$ of $^+3000$ and $^-2500$, therefore, would correspond to an increase in altitude of 500 feet, which is equivalent to the change resulting from first going up 3000 feet and then, down 2500.

In order to represent this addition in terms of ordered pairs, it would be necessary to know the altitudes after, as well as before, each change. Thus, $^+3000$ might correspond to the pair $(8000, 5000)$ and $^-2500$, to $(5500, 8000)$. In this case, the sum would be represented by the pair $(8000 + 5500, 5000 + 8000) = (13,500, 13,000)$ by our definition of addition. This pair corresponds to the integer $^+500$, which agrees with what we obtained above.

The directed magnitudes approach, however, appears more natural and hence is preferred in applications where the quantities involved have both direction and magnitude. This approach supplements the definition of addition proposed above, and both approaches should be introduced into the elementary school curriculum wherever possible.

Which approach to addition do you prefer, the ordered pair approach or the four-case algorithm? Why? What are their relative advantages and disadvantages of each? Consider intrinsic simplicity, computational efficiency, and type of interpretation.

EXERCISES 8–4

S–1. Use the definition of addition to find the indicated sums. Write your answers in the form ^+n or ^-m.

  a. $(7, 2) + (3, 6)$
  b. $(3, 0) + (16, 10)$
  c. $(9, 12) + (4, 6)$
  d. $(7, 1) + (3, 3)$

S–2. Add each of the following pairs of integers by selecting ordered pair representatives of each and applying the definition of addition.

  a. $^+6 + {}^+7$ c. $^-5 + {}^+5$
  b. $^-3 + {}^-6$ d. $^+9 + {}^-4$

S–3. Show that the definition of addition of integers is *well defined* in the following special case. Represent $^-5$ by $(3, 8)$ and $(1, 6)$, and $^+6$ by $(10, 4)$ and $(8, 2)$. Show that $(3, 8) + (10, 4)$ is more-less equivalent to $(1, 6) + (8, 2)$.

∗S–4. Fill in the details for step 3 of the proof that addition of integers is well defined (see the text). Begin by showing why $(a + b') + (c + d') = (a + c) + (b' + d')$ by repeated use of the associative and commutative properties for addition of whole numbers. Similarly, show why $(b + a') + (d + c') = (b + d) + (a' + c')$.

S–5. Let $(5, 0)$ and $(0, 2)$ represent the integers $^+5$ and $^-2$, respectively. Use the definition of addition to obtain the sum of $^+5$ and $^-2$, which is represented by $(5, 0) + (0, 2) = (5, 2)$. Now let $(m + 5, m)$ and $(n, n + 2)$ represent $^+5$ and $^-2$, respectively, where m and n are any whole numbers. Show that $(m + 5, m) + (n, n + 2)$ is more-less equivalent to the pair $(5, 2)$ obtained above.

S–6. Use the four-case algorithm to find each of the following, and give a number line interpretation of each.

  a. $^+2 + {}^+3$ c. $^-2 + {}^+3$
  b. $^-2 + {}^-3$ d. $^+2 + {}^-3$

[S–7.] Verify four-case algorithm 4 by using the definition for addition of integers. *Hint:* Consider ^-m and ^+n where $m \geq n$ and m, n are whole numbers.

S–8. Formulate the following problems in terms of integers and solve by using the definition of addition of integers.

  a. If the temperature at 8:00 AM is $62°$ and by 11:00 AM the temperature rises $15°$, what is the temperature at 11:00 AM?
  b. If a submarine is cruising at 150 feet below sea level and ascends 73 feet, how far is it below the surface of the water?
  c. Suppose you overdraw on your checking account by $3. If you then deposit $6, what will be the resulting balance?

E-9.   Consider the set P of all integers which have an ordered pair representation of the form (a, b) where $a \geq b$. Try to set up an isomorphism under addition between this set and the set of whole numbers. (See Chapter 4 for the definition of isomorphism.) *Hint:* The members of P are simply $^{+}0$, $^{+}1$, $^{+}2$, $^{+}3$, ...

M-10.   Give four test questions you would use to determine if a student can, using the four-case algorithm, add any two integers whose magnitude is less than 10.

[M-11.]   What are some concrete examples or situations which you might use in class to illustrate addition of integers?

For further thought and study: Examine several textbook series to see how each defines addition of integers. (You may have to look at secondary-level texts.)

5. SUBTRACTION

To *define* the *difference*, $x - y$, of two integers, let the pair (a, b) represent x and the pair (c, d) represent y, with the important restriction that $a \geq c$ and $b \geq d$. Then the difference, $x - y$, is represented by $(a, b) - (c, d) = (a - c, b - d)$. In effect, the definition exactly parallels that for addition except that the choice of representative of the minuend (x) is partly dependent on the choice of representative of the subtrahend (y). For example, to subtract (take away) $^{+}7$ from $^{+}3$, we first choose a representative of $^{+}7$, say, $(7, 0)$. Then, we choose a representative of $^{+}3$; but here we notice that such representatives as $(3, 0)$ and $(4, 1)$ will not do, because it is meaningless to try to subtract the whole number 7 from 3 or 4. The definition requires that the whole numbers in the representative we choose be at least as large as 7 and 0 (in $(7, 0)$), respectively. In this case, we could choose any of the following: $(7, 4)$, $(8, 5)$, $(9, 6)$, ... Suppose we pick $(8, 5)$. Then $^{+}3 - {}^{+}7$ is represented by $(8, 5) - (7, 0) = (8 - 7, 5 - 0) = (1, 5)$, which represents $^{-}4$. As a second example, consider $^{-}1 - {}^{+}4$. If $^{+}4$ is represented by $(4, 0)$, then $(4, 5)$ is the "smallest" allowable representative of $^{-}1$. Then $^{-}1 - {}^{+}4$ is represented by $(4 - 4, 5 - 0) = (0, 5)$. That is $^{-}1 - {}^{+}4 = {}^{-}5$.

Of course, subtraction may also be defined as the "inverse" of addition. Thus, $z - x = y$ if and only if $x + y = z$. For example, if $^{+}3 + {}^{-}4 = {}^{-}1$ (by the definition of addition), we know immediately that $^{-}1 - {}^{+}3 = {}^{-}4$. In effect, knowing an addition fact and the relationship between addition and subtraction is tantamount to knowing the corresponding subtraction fact as well.[12]

We leave the task of showing that the former definition is a good one largely to the reader. First, does the definition always yield an integer as the difference? That is, can $(a - c, b - d)$ ever fail to be a pair of whole numbers when $a \geq c$ and $b \geq d$? Why not? (Remember that a, b, c, and d are whole

---

[12] This simple observation has deep roots and is illustrative of the type of problem involved in giving behavioral reality to "mathematical knowledge." For a detailed discussion of this question, see J. M. Scandura, *Mathematics and Structural Learning*, Englewood Cliffs, N.J., Prentice-Hall, to be published.

numbers and the "$-$" in $(a - c, b - d)$ refers to subtraction over the whole numbers.)

Although we shall refer to the property again where we discuss the system of integers, it is worth noting here that, unlike the natural numbers, the *difference* between two integers is always defined. For example, whereas $3 - 7$ is undefined, $^+3 - {}^+7 = {}^-4$. Try additional examples to convince yourself that this is always true.

Second, is the definition well defined? Show that no matter what pair of (allowable) representatives of $^+3$ and $^-2$ you pick, the definition of $^+3 - {}^-2$ always yields a representative of $^+5$. In particular, consider the representatives (a) (5, 2) and (0, 2) and (b) (10, 7) and (4, 6), respectively. Apply the definition to (a) and (b) and show that the (two) defined differences are equivalent. Why can't you use the representative (3, 0) for $^+3$ and (0, 2) for $^-2$?

If you can prove in general that the definition is well defined, so much the better. *Hint:* The proof parallels that given for addition. In adapting the proof, keep in mind that we are "taking away" instead of adding and that the integer representatives (a, b) and (c, d) must be such that $a \geq c$ and $b \geq d$.

Third, as with addition, the definition above provides a highly efficient algorithm for computing differences. *Pick a representative of the integer to be subtracted. Then pick a representative of the other, so that each whole number in the second pair is greater than or equal to the corresponding number in the first pair. Subtract the corresponding numbers of the pairs and, finally, find the integer represented by the difference.* For example, in order to subtract $^-6$ from $^-2$ we first select (0, 6) to represent $^-6$. Then, we pick (4, 6) for $^-2$. (*Note:* (4, 6) is the "smallest" allowable representative.) Subtracting, we get $(4 - 0, 6 - 6) = (4, 0)$, which is a representative of $^+4$. Hence, $^-2 - {}^-6 = {}^+4$. Can you find $^-3 - {}^-2$? $^-7 - {}^+6$?

Does this algorithm (and, hence, the definition) conform to your intuition? To convince yourself, consider the following situation: Joey and Julie are given 14 and 10 presents, respectively, on Christmas. Joey has four more presents and we represent this $^+4$. Julie is not very happy about this, of course, and a series of skirmishes develop (including ambushes, commando raids, and about everything else that a good military commander might think of). Taking our cue from the United Nations, we try to ignore this on Christmas Day, but when the hostilities continue into the following week, we finally decide that something must be done. So we penalize both children by taking away all of the toys that are causing difficulty. This amounts to taking six toys from Joey and one from Julie. Because we take away five more from Joey than from Julie, we represent this by $^+5$. Now, this leaves Joey with $14 - 6 = 8$ toys and Julie with $10 - 1 = 9$ toys. We represent this difference by $^-1$. Happily, the definition of subtraction also gives $^+4 - {}^+5 = {}^-1$ and with two exceptions we are all happy. If you are still not convinced, set up another confrontation and see what happens.

Turning more serious, we now consider the four-case algorithm for subtraction. It can be stated very simply in terms of the four-case algorithm for addition, as follows: *Change the sign of the number being subtracted and add, using the appropriate four-case algorithm for addition.* Thus, the four possible cases are:

1. $^+m - {}^+n = {}^+m + {}^-n$
2. $^+m - {}^-n = {}^+m + {}^+n$
3. $^-m - {}^+n = {}^-m + {}^-n$
4. $^-m - {}^-n = {}^-m + {}^+n$

The number line interpretations for these are simply the related number line interpretations for the addition algorithms (see Section 4).

Apply the appropriate algorithm and the corresponding number line argument to determine each of the following: $^+3 - {}^+2$, $^-3 - {}^-2$, $^+3 - {}^-2$, $^-3 - {}^+2$. Try $^+5 - {}^-5$ for good measure.

As expected, each of the four-case algorithms can be justified directly in terms of the definition. We show this only for case 4 and leave the others as exercises.

The proof of 4 goes as follows: Let ^-m and ^-n be represented by $(n, n + m)$ and $(0, n)$, respectively. (Why can't we let ^-m be represented by $(0, m)$? *Hint:* Compare n in $(0, n)$ and m in $(0, m)$. Is m necessarily larger than n?) Applying the definitions, we get $(n, n + m) - (0, n) = (n - 0, n + m - n) = (n, m)$. But, $(n, m) = (0, m) + (n, 0)$, which represents $^-m + {}^+n$. Case 4 also gives $^-m + {}^+n$. Try applying both the definition and case 4 to $^-3 - {}^-5$. If you do not get identical results, try again. The difference is $^+2$.

Before attempting to justify cases 1, 2, and 3, you may find it helpful to try some particular subtraction problems, such as those listed above. In this case, you need only check to see that the same difference is obtained by applying the definition and the appropriate four-case algorithm.

The four-case algorithm for subtraction can also be justified by using the fact that subtraction is the inverse of addition. For example, consider the following special case of case 2. To subtract $^-3$ from $^+5$, we change $^-3$ to $^+3$ and add $^+3$ to $^+5$ to obtain $^+8$ (by four-case algorithm 1 for addition).

Now, because subtraction is the inverse of addition, $^+5 - {}^-3 = {}^+8$ if and only if $^+5 = {}^-3 + {}^+8$. Adding $^+3$ to both sides of the latter equation, we get $^+5 + {}^+3 = ({}^-3 + {}^+8) + {}^+3 = ({}^+8 + {}^-3) + {}^+3 = {}^+8 + ({}^-3 + {}^+3) = {}^+8 + 0 = {}^+8$. So, $^+8 = {}^+5 + {}^+3$, which is what we got by changing the sign of $^-3$ and adding (to $^+5$). Using the same technique, attempt to justify case 2 for the general case $^+m - {}^-n$.

EXERCISES 8–5

S–1. Use the definition of subtraction to find the indicated differences. Write your answers in the form ^+m or ^-n.

  a. $(7, 6) - (1, 4)$
  b. $(7, 3) - (3, 2)$
  c. $(3, 5) - (0, 1)$
  d. $(5, 7) - (1, 3)$
 *e. $(7, 3) - (6, 5)$

S–2.   Subtract each of the following by selecting (allowable) ordered pair representatives of each and applying the definition of subtraction.

a. $^{+}13 - {^{+}3}$ c. $^{+}7 - {^{-}7}$
b. $^{-}6 - {^{+}5}$ d. $^{-}1 - {^{-}10}$

S–3.   Why can't we use the representative (3, 0) for $^{+}3$ and (0, 2) for $^{-}2$ to compute $^{+}3 - {^{-}2}$ using the definition of subtraction?

S–4.   Justify four-case algorithm 3 by using the definition of subtraction of integers. (*Hint:* Let ^{-}m be represented by $(n, n + m)$ and ^{+}n by $(n, 0)$. Apply the definitions of subtraction and addition to show that $^{-}m - {^{+}n} = {^{-}m} + {^{-}n}$.)

E–5.   Consider any allowable representatives of $^{+}3$ and $^{-}2$. In particular consider

a.   (5, 2) and (0, 2)
b.   (10, 7) and (4, 6)

Subtract $^{-}2$ from $^{+}3$ by applying the definition of subtraction to *a* and *b*. Show that the resulting differences are more-less equivalent.

S–6.   Use the four-case algorithm to compute each of the following:

a. $^{+}2 - {^{+}6}$ c. $^{-}5 - {^{-}5}$
b. $^{-}6 - {^{+}3}$ d. $^{+}4 - {^{-}4}$

S–7.   Interpret each of the problems in Exercise S–6 on a separate number line.

E–8.   Let x, y, and z be integers. Because subtraction is the inverse of addition, $z - x = y$ if and only if $z = x + y$. Thus, $^{+}12 - {^{+}16} = {^{-}4}$ if and only if $^{+}12 = {^{+}16} + {^{-}4}$. This gives a method for verifying that a subtraction problem was done correctly. Verify your answers for Exercise S–6 above by using this method.

S–9.   Formulate the following problems in terms of integers and solve by using the definition of subtraction of integers.

a.   If water freezes at $^{+}32°F$ and boils at $^{+}212°F$, what is the difference between the boiling and freezing points?
b.   The highest point in town is 313 feet above sea level and the lowest point is 13 feet below sea level. What is the difference between the highest and lowest points?
c.   The temperature extremes last January 5 were $^{-}9°F$ and $^{+}21°F$. By how much did the temperature increase as it rose from its lowest point to its highest?

M–10.   Give four questions you would use to determine whether or not a child is able to subtract with integers whose magnitudes are less than 10, using the four-case algorithm.

[M–11.]   How could you pair addition and subtraction problems in order to lead your students to detect as a regularity the inverse relationship between addition and subtraction?

For further thought and study: Examine several textbook series to see how subtraction of integers is defined. Are number lines used? Is subtraction related to addition in the textbook series? (You may need to look at secondary-level texts.)

6. MULTIPLICATION

The situation with multiplication and division of integers is somewhat more complicated than with addition and subtraction and, in this sense, reverses the situation with rationals where addition and subtraction were more complicated. Nevertheless, continuing our parallel, to *define* the *product*, $x \cdot y$, of two integers, we let the pairs (a, b) and (c, d) represent x and y, respectively. Then the product, $x \cdot y$, is represented by $(a, b) \cdot (c, d) = (a \cdot c + b \cdot d, a \cdot d + b \cdot c)$. For example, let $^+3$ and $^+4$ be represented by $(3, 0)$ and $(4, 0)$, respectively. Then the product, $^+3 \times {}^+4$ is represented by $(3, 0) \cdot (4, 0) = (3 \cdot 4 + 0 \cdot 0, 3 \cdot 0 + 0 \cdot 4) = (12, 0)$, which represents $^+12$. Similarly, the product: (a) $^+3 \times {}^-4$ is represented by $(4, 1) \cdot (1, 5) = (4 \cdot 1 + 1 \cdot 5, 4 \cdot 5 + 1 \cdot 1) = (9, 21)$, which represents $^-12$, (b) $^-3 \times {}^+4$ is represented by $(2, 5) \cdot (6, 2) = (2 \cdot 6 + 5 \cdot 2, 2 \cdot 2 + 5 \cdot 6) = (22, 34)$, which represents $^-12$, (c) $^-3 \times {}^-4$ is represented by $(3, 6) \cdot (3, 7) = (3 \cdot 3 + 6 \cdot 7, 3 \cdot 7 + 6 \cdot 3) = (51, 39)$, which represents $^+12$.

As these examples tend to suggest, the definition always gives an integer as product (no matter what the signs of the given integers) and is well defined. To prove the former, we need only observe that a, b, c, and d are *whole numbers* so that $a \cdot c + b \cdot d$ and $a \cdot d + b \cdot c$ are also whole numbers. Hence, $(a \cdot c + b \cdot d, a \cdot d + b \cdot c)$ represents an *integer* (because it is an ordered pair of whole numbers).

Showing that the definition is well defined involves cumbersome algebra. If you wish, you may omit the proof since it is not used in the subsequent development.

Let (a, b) and (a', b') be arbitrary representatives of x, and (c, d) and (c', d') arbitrary representatives of y. Then $(a, b) \doteq (a', b')$ and $(c, d) \doteq (c', d')$ or $a + b' = b + a'$ and $c + d' = d + c'$, respectively. What we want to show is that the resulting representatives of $x \cdot y$, namely $(ac + bd, ad + bc)$ and $(a'c' + b'd', a'd' + b'c')$ are more-less equivalent—or, equivalently, that $(ac + bd) + (a'd' + b'c') = (ad + bc) + (a'c' + b'd')$.

To see this, we multiply the equations $a + b' = b + a'$ and $c + d' = d + c'$ as follows.

1. Multiply $a + b' = b + a'$ by c: $ac + b'c = bc + a'c$
2. Multiply $b + a' = a + b'$ by d: $bd + a'd = ad + b'd$
3. Multiply $d' + c = c' + d$ by a': $a'd' + a'c = a'c' + a'd$
4. Multiply $c' + d = d' + c$ by b': $b'c' + b'd = b'd' + b'c$.

Adding and regrouping we get

$$ac + bd + a'd' + b'c' + (b'c + a'd + a'c + b'd)$$
$$= ad + bc + a'c' + b'd' + (b'c + a'd + a'c + b'd)$$

Subtracting $(b'c + a'd + a'c + b'd)$ from both sides, we get

$$ac + bd + a'd' + b'c' = ad + bc + a'c' + b'd'$$

which is what we wanted to show.

The above definition provides a perfectly good algorithm for finding products of integers. Most people, however, have relatively little intuition concerning what the algorithm means. To get some feeling for what is involved, recall first that the product of two *natural* numbers is the number of ordered pairs that can be formed by pairing elements of corresponding sets. For example, $2 \times 3 = 6$ is the number of pairs (elements) in the Cartesian product set $\{a, b\} \times \{c, d, e\} = \{(a, c), (a, d), (a, e), (b, c), (b, d), (b, e)\}$ (where $\{a, b\}$ has 2 elements and $\{c, d, e\}$ has 3).

Similarly, given two pairs of sets corresponding to the signed numbers, x and y, the product, $x \cdot y$, answers the question: How many more (or less) like pairs of elements are there than unlike pairs? For example, given the pairs $(\{Jimmy, Joey\}, \{Jeanne, Janie, Julie\})$ and $(\{John, Don\}, \{Ann, Mary, Sue\})$, both of which represent $^-1$, there are $(2 \times 2) + (3 \times 3) = 13$ *like* pairs (both names in each pair refer either to boys or to girls), and $(2 \times 3) + (3 \times 2) = 12$ *unlike* pairs (one name in each pair refers to a boy and one name refers to a girl), as shown in Figure 8–4. In this case, there is *one* more *like* pair than *unlike* pair. Representing this difference by $^+1$, we get $^-1 \times {}^-1 = {}^+1$.

		Boys		Girls		
		John	Don	Ann	Mary	Sue
Boys	Jimmy Joey	(Jimmy, John) (Joey, John)	(Jimmy, Don) (Joey, Don)	(Jimmy, Ann) (Joey, Ann)	(Jimmy, Mary) (Joey, Mary)	(Jimmy, Sue) (Joey, Sue)
		[like pairs]		[unlike pairs]		
Girls	Jeanne Janie Julie	(Jeanne, John) (Janie, John) (Julie, John)	(Jeanne, Don) (Janie, Don) (Julie, Don)	(Jeanne, Ann) (Janie, Ann) (Julie, Ann)	(Jeanne, Mary) (Janie, Mary) (Julie, Mary)	(Jeanne, Sue) (Janie, Sue) (Julie, Sue)
		[unlike pairs]		[like pairs]		

Figure 8–4

Show that $^-1 \times {}^-1 = {}^+1$ by representing $^-1$ by the pair $(2, 3)$ and applying the definition. Notice that $(2, 3)$ denotes the common more-less property of the two pairs (of sets), $(\{Jimmy, Joey\}, \{Jeanne, Janie, Julie\})$ and $(\{John, Don\}, \{Ann, Mary, Sue\})$.

To introduce multiplication of integers to young children (say, in the fifth or sixth grade), one might convert a situation such as the one above into a problem. Instead of talking about "like" and "unlike" pairs, however, we might describe the situation as follows. Two families, each having two boys and three girls, are planning to have a picnic. To make sure that things go smoothly, the mothers want to plan a series of activities. For a variety of reasons, they decide on games which pair children from *different* families. During their conversation, the question arises as to how many more (or less) ways there are of pairing children from different families according to sex than there are by crossing sex.

(Presumably this question came up because the mothers wanted to know whether to concentrate on sex-related games, like wrestling and playing dolls, or coed activities, like dancing.) Happily, one of the mothers recalls the meaning of integer multiplication and determines that the children can be paired according to sex in one more way ($^-1 \times {}^-1 = {}^+1$) than by crossing sex. More serious activities of this sort may arise in planning similar activities between different schools where large numbers of children are involved and facilities are limited.

Can you think of similar problems which can be solved by multiplying integers? Consider two camping groups, one consisting of 40 cars and 30 trailers and the other of 25 cars and 30 trailers. What sort of question would multiplying $^+10$ and $^-5$ answer? (Do not worry if the problem appears to be artificial. The important thing is that you understand what is happening. In any case, "immediate relevance" is rarely of importance to elementary school children. They tend to find such problems interesting.)

Now let us consider the four-case algorithm for multiplication of integers m and n.

1. If *both* signs are $+$, multiply the magnitudes (as if they were whole numbers) and affix "$+$" to the result.

repeat ^+n, m times

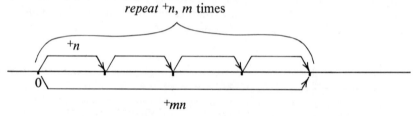

2. If the sign of m is $-$ and n, $+$, multiply the magnitudes and affix "$-$" to the result.

undo ^+n, m times[13]

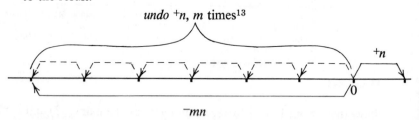

3. If the sign of m is $+$ and n, $-$, multiply the magnitudes and affix "$-$" to the result.

repeat ^-n, m times

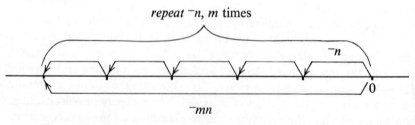

[13] "Undoing" a jump means "doing" a jump of the same magnitude in the opposite direction. For example, undoing a jump of $^-3$ means the same as doing a jump of $^+3$.

4. If *both* signs are −, multiply the magnitudes and affix "+" to the result.

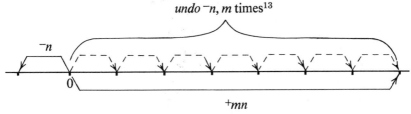

(If either integer is $^+0$, any one of the four-case algorithms will give the desired result—that is, $^+0$. Check this with $^+5 \times {}^+0$.)

Apply the appropriate algorithm and the corresponding number line argument to determine $^+4 \times {}^+2$, $^-5 \times {}^-3$, $^+3 \times {}^-2$, and $^-5 \times {}^+2$. (The answers are $^+8$, $^+15$, $^-6$, and $^-10$, respectively.) What does $^+6 \times 0$ equal? $0 \times {}^-2$?

Again, it is a simple matter to justify the four-case algorithm in terms of our definition. We demonstrate this for case 4 and leave the others as exercises.

Let ^-m and ^-n be represented by $(0, m)$ and $(0, n)$, respectively. Then, according to our definition, $^-m \times {}^-n$ is represented by $(0, m) \times (0, n) = (0 \times 0 + m \times n, 0 \times n + m \times 0) = (mn, 0)$ which represents ^+mn. This is exactly what four-case algorithm 4 gives.

Prove that case 3 works. In doing this, it will be convenient to let $(m, 0)$ represent ^+m and $(0, n)$ represent ^-n. The other proofs are equally easy. Can you construct them?

As might be suspected from the number line interpretations above, the four-case algorithm corresponds to viewing multiplication of the natural numbers as repeated addition. In this case, however, we need to consider *undoing* a directed magnitude, as well as *doing* one.

Any number of situations (interpretations) of this type may be considered.[14] For example, consider a situation in which the directed action involves filling or emptying a tank of liquid at a given rate. We agree to indicate filling the tank at a rate of a gallons per minute by "^+a," and emptying the tank at the same rate by "^-a." We represent *doing* this for b minutes by "^+b" (which also represents "b minutes from now"). We represent *undoing* this for b minutes by "^-b" (which also represents "b minutes ago"). In this case, the product $x \times y$ (where x is an integer representing the *rate of flow* and y is an integer representing *time from the present*) corresponds to the state of the tank relative to what it is "now."

For example, if we fill the tank at $^+7$ gallons per minute, the number of gallons in the tank 6 minutes from now ($^+6$) will be 42 more than now. Hence the product $^+7 \times {}^+6 = {}^+42$. If, on the other hand, we empty the tank at $^-3$ gallons per minute, then the number of gallons in the tank 4 minutes *ago* ($^-4$) was 12 more than now. Hence, $^-3 \times {}^-4 = {}^+12$. Can you construct similar interpretations of the other two cases? Suppose the rate is $^-4$ and the time, 3 minutes from now ($^+3$); suppose the rate is $^+6$ and the time, 2 minutes ago ($^-2$).

---

[14] Unfortunately, it would be impractical for us to distinguish systematically between interpretations which involve states and/or operators in different ways. None of the distinctions is crucial to the mathematics, however. The only difference is in how one looks at the various interpretations.

Other situations in which the same ideas apply involve financial gains or losses (e.g., debts) and numbers of repeats or cancellations. For example, if a person is presented with two (+2) debts of $3 (−3), he is $6 worse off (−6) than he was before. On the other hand, if presented with cancellations of two (−2) debts of $3 (−3), he is $6 better off (+6) than he was before. Postman stories, in which the postman delivers or reclaims checks and IOUs, are based on this same idea.

Other widely used illustrations have to do with machines which can run either forward or backward, and transmissions (i.e., sequences of gears) which "multiply" the speed of the machine either directly or by putting everything into reverse. In this case, we might ask such questions as: How rapidly will a wheel revolve if it is driven by a (backward-forward) machine going at +160 revolutions per minute hooked up to a sequence of gears (transmission) which puts every-thing into reverse and multiplies the effect tenfold (−10)? (The answer, of course, is +160 × −10 = −1600.)

Make up postman stories for the following problems: +6 × −2 = −12, −7 × +3 = −21, −16 × −3 = +48, +2 × +24 = +48. Can you also devise machine-transmission combinations that reflect these facts? Try to think of some other situations that reflect multiplication of integers.

In general, the four-case algorithm applies in the following type of situa-tion. We are given some directed state or operator (e.g., debts and credits, flow in and out of a tank) which is acted on by a higher-order directed operator (e.g., number of times an operator is repeated, time in the future and time "ago") and want to know the resulting directed state or operator (e.g., debt or credit, amount of liquid in a tank).

EXERCISES 8–6

S–1. Use the definition of multiplication to find each of the indicated products. Write your answers in the form +m or −n.

a. (5, 3) × (3, 2)
b. (6, 8) × (4, 1)
c. (3, 6) × (1, 5)
d. (7, 7) × (6, 2)

S–2. Multiply each of the following pairs of integers by selecting ordered-pair representatives of each and applying the definition of multiplication.

a. +4 × +7
b. −9 × +3
c. −52 × −4
d. +16 × −3

S–3. Multiply each of the following by using the appropriate four-case algo-rithm.

a. +3 × +2
b. −4 × −1
c. +5 × −2
d. −3 × +3
e. −16 × +13
f. −18 × +0
g. +63 × −11
h. −4 × −132

S–4. Give a number line interpretation for each of the following.

a. +3 × +2
b. −3 × +1
c. +2 × −4
d. −2 × −3

S–5. Represent $^+3$ by (6, 3) and (4, 1), and $^-4$ by (2, 6) and (0, 4). Show that the result of multiplying (6, 3) × (2, 6) is more-less equivalent to the result of multiplying (4, 1) × (0, 4).

S–6. Justify case 3 of the four-case algorithm by using the definition of multiplication of integers. *Hint:* Let ^+m and ^-n be represented by $(m, 0)$ and $(0, n)$, respectively.

S–7. Formulate the following problems in terms of integers and solve by using the definition of multiplication of integers.

  a. The temperature decreases at the rate of 5° each hour for 4 hours. What is the total temperature change?

  b. During the day a lake loses 3 gallons of water per hour due to evaporation. How much more water was in the lake at 11:00 AM than at 3:00 PM on the same day?

  c. The gears linking a motor to a drive shaft not only reverse the direction but also double the speed of the rotations of the motor. If the motor turns clockwise at three revolutions per second, at what speed and in what direction does the drive shaft turn? (Let clockwise be the "positive" direction.)

[M–8.] a. What regularity might children detect in the array below that would enable them to fill in the blank spaces?

$^+3$ × $^+2$ = $^+6$ $^-1$ × $^+2$ = _____

$^+2$ × $^+2$ = $^+4$ $^-2$ × $^+2$ = _____

$^+1$ × $^+2$ = $^+2$ $^-3$ × $^+2$ = _____

$^+0$ × $^+2$ = $^+0$

  b. What similar array would you construct to aid children in multiplying the following?

$^-1$ × $^+3$ = _____

$^-2$ × $^+3$ = _____

$^-3$ × $^+3$ = _____

  c. Construct arrays similar to that given in part *a* that would aid children in multiplying the following?

$^-1$ × $^-2$ = _____ $^-1$ × $^-3$ = _____

$^-2$ × $^-2$ = _____ $^-2$ × $^-3$ = _____

$^-3$ × $^-2$ = _____ $^-3$ × $^-3$ = _____

M–9. Devise a real-life situation that is represented by:

  a. $^+6$ × $^-2$ = $^-12$
  b. $^-16$ × $^-3$ = $^+48$
  c. $^+2$ × $^+24$ = $^+48$

Hint: Consider postman stories, machine-transmission combinations, water tanks which can be filled and/or emptied.

M-10.   Give a set of questions which you could use to evaluate a child's ability to multiply integers whose magnitudes are less than 10, using the four-case algorithm.

For further thought and study: How does R. B. Davis introduce multiplication of signed numbers? (See *Discovery in Mathematics: A Text for Teachers*, Reading, Mass.: Addison-Wesley, 1964.) What about Z. P. Dienes? (See *Mathematics in the Primary School*, Melbourne: Macmillan, 1966.) P. Suppes? (See *Sets and Numbers*, New York: Singer/Random House, 1965.) Select an introductory algebra text and ask the same question.

7. DIVISION

In the previous section, we saw that there is a certain artificiality about the kinds of situations in which multiplication of integers (as ordered pairs) applies. Not only did the problems themselves seem somewhat far-fetched, but their solutions could have been obtained as easily by working exclusively with natural numbers and using a little common sense.

As one might expect, the situation with division is like this, only more so. It is of considerable theoretical interest, nonetheless, to know how division fits in and we include a brief discussion of it. The theoretically oriented reader will have no difficulty in filling in the details, many of which under proper conditions can even be introduced in the elementary school classroom. It should be emphasized in this regard that most fifth- and sixth-graders are far less concerned with immediate practicality than are adults. The esthetic appeal of an integrated body of knowledge is frequently far more attractive.

One problem in defining division in terms of representative ordered pairs is that division of integers is not always defined—in fact, division is defined only in certain very special cases. Furthermore, it is not possible to state simply the conditions under which division is defined. In order to keep the definition from being unwieldy, we need to introduce the notion of the *magnitude* of the difference between two whole numbers, a and b—more commonly known as the *absolute value* and denoted $|a - b|$. The absolute value $|a - b|$ means $a - b$ when $a \geq b$, and it means $b - a$ when $b > a$. For example, $|7 - 5| = 7 - 5 = 2$, and $|6 - 10| = 10 - 6 = 4$.

With this as background, let the integers x and y be represented by (a, b) and (c, d), respectively. Then the *quotient* $x \div y$ is *defined* if and only if $|a - b| \div |c - d|$ is a whole number. If this is so, let $|a - b| \div |c - d| = z$. Let $p = 1$ if $a \geq b$ and $c > d$, or $a \leq b$ and $c < d$; let $p = 0$ if $a \geq b$ and $c < d$, or $a \leq b$ and $c > d$. (Why can't $c = d$?) Then $x \div y$ is the integer represented by $(p \cdot z, (1 - p) \cdot z)$. (Notice that this definition could be separated into four parts corresponding to the four conditions.)

For example, to divide $^+4$ by $^-2$, we first check the conditions. Choosing representatives, $(4, 0)$ and $(0, 2)$, respectively, we see that division is possible, that is, $|4 - 0| = 4 - 0 = 4$ and $|0 - 2| = 2 - 0 = 2$ so $z = 4 \div 2 = 2$ is defined as the definition requires. Next, we see that $p = 0$ because $a = 4 \geq 0 = b$ and $c = 0 < 2 = d$. Applying the definition, we see that $^+4 \div {}^-2$ is represented by

$$(0 \cdot 2, 1 \cdot 2) = (0, 2)$$

which represents $^-2$. Hence, $^+4 \div {}^-2 = {}^-2$.

Find $^+9 \div {}^-3$. Can you apply the definition in the other cases? Try $^+6 \div {}^+2$, $^-8 \div {}^+2$, $^-10 \div {}^-5$. What about $^+12 \div {}^-5$? Does the definition apply?

The reader who wishes to do so may skip the next couple paragraphs, picking up again where we define division as the inverse of multiplication.

Although division is somewhat cumbersome as defined, it always yields an integer as quotient and is well defined. Thus $p \cdot z$ and $(1 - p) \cdot z$ are always whole numbers because p and z are. (Why are p and z always whole numbers?) Hence, $(p \cdot z, (1 - p) \cdot z)$ always represents an integer.

To see that division is well defined, let (a, b) and (a', b') be any two representatives of x and let (c, d) and (c', d') be any two representatives of y. Then by definition of more-less equivalence $a + b' = b + a'$ and $c + d' = d + c'$. From this it follows that

$$|a - b| = |a' - b'|$$

and

$$|c - d| = |c' - d'|$$

We want to show that $(a, b) \div (c, d) = (p \cdot z, (1 - p) \cdot z)$ is more-less equivalent to

$$(a', b') \div (c', d') = (p' \cdot z', (1 - p') \cdot z')$$

(where z' and p' have meanings analogous to z and p). Now since

$$|a - b| = |a' - b'|$$

and

$$|c - d| = |c' - d'|$$

it follows that

$$z = |a - b| \div |c - d| = |a' - b'| \div |c' - d'| = z'$$

p can be shown equal to p' by examination of the various cases but we shall not do so here. (Notice, for example, that $a \geq b$ if and only if $a' \geq b'$.) Hence,

$$(p \cdot z, (1 - p) \cdot z) = (p' \cdot z', (1 - p') \cdot z')$$

by substitution of equals. (Since $(p \cdot z, (1 - p) \cdot z)$ and $(p' \cdot z', (1 - p') \cdot z')$ are equal, both they and the quotients they represent are certainly more-less equivalent.)

As one might expect, division can also be defined as the *inverse* of multiplication. Thus, $x \div y = w$ is the quotient of two integers x and y if and only if $y \cdot w = x$. For example, $^+10 \div {}^-5 = {}^-2$ since $^-5 \cdot {}^-2 = {}^+10$. Similarly, $^-18 \div {}^+6 = {}^-3$ since $^+6 \cdot {}^-3 = {}^-18$.

As suggested above, the problem situations to which the definition of division applies are of no great practical importance. Nonetheless, the close relationship between division and multiplication, together with the way multiplication was interpreted earlier, is suggestive of one type of question that can be answered by dividing integers. Suppose we are given the number of boys and girls in one family and also the difference between the number of like and unlike pairs that can be formed by pairing the children from this family with children from another family. (See the corresponding example in the section on multiplication.) By representing both quantities as integers, the quotient represents how many more (or less) boys than girls there are in the second family. This does not sound like a particularly likely sort of question to ask. In fact, if it were asked we would doubtlessly go about solving it in a roundabout way. It is rather interesting, nonetheless, that the answer could be obtained by a straightforward application of the definition of division of integers.

The four-case algorithm for division corresponds rather directly to the four conditions of the above definition. It is, however, much simpler to apply. In order to apply the four-case algorithm, one must check first to see that the magnitude of the divisor divides the magnitude of the dividend (e.g., given ^+m and ^-n, that $m \div n$ is defined). Then

1. If both signs are $+$, divide the magnitudes as if they were natural numbers and affix "$+$" to the result.

take away ^+n from ^+m, $m \div n$ times

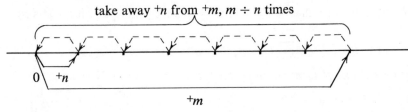

2. If the sign of m is $-$ and n, $+$, divide the magnitudes and affix "$-$" to the result.

undo take away (add) ^+n to ^-m, $m \div n$ times

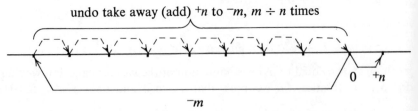

3. If the sign of m is $+$ and n, $-$, divide the magnitudes and affix "$-$" to the result.

undo take away (add) ^-n to ^+m, $m \div n$ times

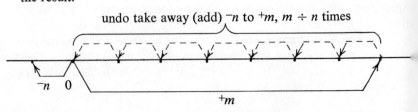

4. If both signs are −, divide the magnitudes and affix "+" to the result.

take away ^-n from ^-m, $m \div n$ times

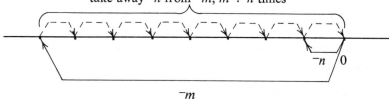

^-m

Compare these number line interpretations with those given in the section on multiplication. Do you notice any parallels? Differences? Which cases correspond? (Compare 1 above with 1, 2 with 2, 3 with 4, and 4 with 3.)

As before, the four-case algorithm can be justified in terms of the definition. We sketch how this can be done because the exercise of filling in the details may help you improve your deductive skills, not because the proofs themselves are important to remember. We only consider case 4. Application of the definition where both signs are negative gives:

$$^-m \div {}^-n \text{ is represented by } (0, m) \div (0, n) = (1 \cdot (m \div n), (1 - 1) \cdot (m \div n))$$

$$= (m \div n, 0), \text{ which represents } {}^+(m \div n).$$

Four-case algorithm 4 gives exactly the same result (check this). Justify each step in this proof. See if you can justify one or more of the other four-case algorithms.

The four-case algorithm for division is somewhat more useful in concrete situations than the definition given in terms of ordered pairs, but not much more. Recall that the four-case algorithm for multiplication corresponds to a directed operator acting on a directed magnitude giving a directed magnitude as a result. The four-case algorithm for division is useful in the same sort of situations.[15]

We can illustrate this by considering, once again, the problem of filling and emptying a tank, as described in the previous section. Suppose 10 minutes ago ($^-10$) we had 30 more gallons ($^+30$) than we do now. Then, the tank must have been changing at a rate of $^+30 \div {}^-10 = {}^-3$ gallons per minute.[16] Similarly, we could ask at what time there will be (were) 30 more gallons ($^+30$) than now if the tank is being emptied at 3 gallons per minute ($^-3$). In this case, our answer would be $^+30 \div {}^-3 = {}^-10$ (10 minutes ago), as emptying the tank implies having more liquid in the past than now. The reader should formulate problems which illustrate the other possibilities.

EXERCISES 8–7

S–1. Compute each of the following by using the appropriate four-case algorithm.

a. $^+10 \div {}^-5$

b. $^-12 \div {}^+4$

c. $^-6 \div {}^-3$

d. $^+8 \div {}^+4$

e. $^+26 \div {}^-13$

f. $^+48 \div {}^+4$

g. $^-63 \div {}^-21$

h. $^-16 \div {}^+1$

---

[15] Hence, it corresponds to a generalized version of repeated subtraction.

[16] One may also think of this as extracting 3 more gallons per minute than are being added.

S–2.   Interpret each of the following problems on a number line.

   a. $^+6 \div ^+2$ c. $^+4 \div ^-2$
   b. $^-8 \div ^+4$ d. $^-6 \div ^-2$

∗S–3.   Use the definition of division to divide each of the following integer representatives. Write your answers in the form ^+n or ^-m.

   a. $(3, 9) \div (6, 4)$
   b. $(8, 2) \div (3, 2)$
   c. $(6, 0) \div (1, 4)$
   d. $(6, 6) \div (7, 2)$
   e. $(7, 3) \div (8, 5)$ (Be careful.)

∗S–4.   Divide each of the following by selecting ordered pair representatives of each integer and applying the definition of division.

   a. $^+9 \div ^-3$ d. $^-10 \div ^-5$
   b. $^+6 \div ^+2$ e. $^+12 \div ^-5$ (Be careful.)
   c. $^-8 \div ^+2$

E–5.   Let x, y, and w be integers. Then, because division is the inverse of multiplication, $x \div y = w$ if and only if $x = y \cdot w$. Thus $^+9 \div ^-3 = ^-3$ if and only if $^+9 = ^-3 \times ^-3$. This gives a method for verifying that a division was done correctly. Verify your answers for Exercise S–1 by using this method.

[M–6.]   Give three real-life problems which can be solved by using division of integers. Consider a water tank being emptied or filled, debts and credits, pairing children by sex, and so on.

M–7.   List four questions which you could use to assess a child's ability to divide integers of magnitude less than 10, using the four-case algorithm.

[M–8.]   How could you pair certain multiplication and division problems to lead your students to detect as a regularity the inverse relationship between multiplication and division?

[M–9.]   Find real-life problems whose solutions could be represented by the following equations.

   a. $^+12 + ^-7 = $ _____   e. $^-4 \times ^+9 = $ _____
   b. $^-17 + ^-8 = $ _____   f. $^+18 \div ^-3 = $ _____
   c. $^-5 - ^-9 = $ _____   g. $^-30 \div ^-5 = $ _____
   d. $^+8 \times ^+14 = $ _____

For further thought and study: Examine several textbook series to see how division of integers is defined. Are number lines used? Is division related to multiplication?

8. PROPERTIES OF THE SYSTEM OF INTEGERS

Like the system of positive rationals, the system of integers "enriches" the system of natural numbers in the sense that it makes possible certain things not possible in the latter system. For example, quadratic equations, like

1. $x^2 + 5x + 6 = 0$ (It is understood that $5 \leftrightarrow {}^+5$, $6 \leftrightarrow {}^+6$, etc.)

and

2. $x^2 + 3x + 2 = 0$

have *no* solution in the system of natural numbers (or in the system of positive rationals) but they do in the system of integers. Thus, the solution set of equation (1) is $\{^-3, {}^-2\}$. Substituting $^-3$ in (1), we get $(^-3)^2 + (5 \cdot {}^-3) + 6 = {}^+9 + {}^-15 + 6 = {}^-6 + 6 = 0$. Similarly, substituting $^-2$ in (1), we get

$$(^-2)^2 + (5 \cdot {}^-2) + 6 = {}^+4 + {}^-10 + 6 = {}^-6 + 6 = 0$$

Can you find the solution set of equation 2? Try each of the following: $^+2$, $^-2$, 0, $^+1$, $^-1$. Two of them satisfy the equation.

Quadratic equations of this type arise frequently in ninth-grade algebra and beyond, and have application to a large variety of practical and scientific problems (see any ninth-grade algebra text for examples). They have also been used successfully in the elementary school to help introduce the topic of signed integers, and children generally like solving them.[17]

One way to introduce signed numbers is first to have children learn to solve quadratic equations with whole-number roots (solutions).[18] After they have come to expect whole number solutions to such equations, we then present equations that have two negative roots or one positive and one negative root. (As it turns out, every quadratic equation has two roots over some suitable system of numbers—in particular, all quadratic equations have two roots over the system of complex numbers. See Chapter 9 for a definition of complex numbers.) This alerts the child to the fact that something has gone awry. Children tend to gain satisfaction when they eventually find out that such equations have negative roots.

In this section, we consider certain basic properties of the system of integers in a systematic way, and discuss some of the main relationships between this system and those discussed previously.

[17] For example, see R. B. Davis, *Discovery in Mathematics*, Reading, Mass., Addison Wesley, 1963.

[18] It is easy to construct quadratic equations of this type by selecting any two natural numbers a and b, and forming the product, $(x - a)(x - b) = x^2 - (a + b)x + ab$. Briefly, the rationale is as follows:

1. $x = a$ and $x = b$ are to satisfy the equation.
2. Hence, $x + {}^-a = 0$ and $x + {}^-b = 0$.
3. So $(x + {}^-a) \cdot (x + {}^-b) = 0 \times 0 = 0$.
4. Hence, $x^2 + {}^-bx + {}^-ax + ({}^-a \cdot {}^-b) = x^2 + ({}^-b + {}^-a)x + {}^-a \cdot {}^-b$
   $= x^2 - (a + b)x + a \cdot b = 0$.

(This argument also shows how the roots of a quadratic equation are related to the coefficient of x, $-(a + b)$, and the constant, $a \cdot b$.)

EXERCISES 8–8

S–1.   Find the solution set for the quadratic equation $x^2 + 3x + 2 = 0$. (*Hint:* Try $^+2$, $^-2$, 0, $^+1$, and $^-1$. Two of them satisfy the equation.)

E–2.   Find the quadratic equation which has the roots $^+1$ and $^+3$. (Recall that if a and b are two natural numbers, then $(x - a)(x - b) = x^2 - (a + b)x + ab$ is a quadratic equation whose truth set is $\{a, b\}$.)

[S–3.]   Form several quadratic equations such that

    a.   both roots are positive integers (try $\{2, 3\}$, $\{1, 2\}$).
    b.   both roots are negative integers (try $\{^-1, ^-2\}$, $\{^-3, ^-4\}$).
    c.   one root is a positive integer, and the other a negative integer (try $\{^-5, ^+2\}$, $\{^+4, ^-3\}$).

All such equations can be written in the form $x^2 + px + q = 0$, where p and q are integers. Try to detect a regularity that will enable you to predict whether p and q will be positive or negative, depending upon the signs of the roots.

E–4.   Why is it *not* possible for the solution set of $x^2 + 5x + 6 = 0$ to contain a positive integer?

For further thought and study: Consult a book on the history of mathematics to find the *chronological* order of man's development of natural numbers, zero, integers, non-negative rationals. Do you think these topics should be taught in the same order as they were developed? Why or why not?

8.1 Closure

A number system is said to be closed under a binary operation $\odot$, if for every pair of numbers a and b there is a unique third number c such that $a \odot b = c$. (Recall that the symbol $\odot$ represents an arbitrary binary operation.)

    Under which operations is the system of integers closed (i.e., defined for every pair of integers)? Is the sum of two integers always defined? The difference? Product? Quotient? (Try to answer without looking back. If you cannot remember the definitions, consider some examples (e.g., $^-4 + ^+6$, $^-3 - ^-21$, $^-2 \times ^-16$, $^+8 \div ^-2$, $^-8 \div ^+3$). Then check the definitions given in the preceding sections.)

    Is division over the *whole* numbers closed? Consider $8 \div 4$, $15 \div 6$, $4 \div 3$. Does division over the *integers* depend on division over the *whole* numbers (recall the four-case algorithm)? Which of the following *integer* quotients are defined: $^+8 \div ^-4$, $^-15 \div ^+6$, $^-4 \div ^-3$? How about $^+4 \div ^-3$? $^-4 \div ^+3$? Compare with the whole number quotients $8 \div 4$, $15 \div 6$, $4 \div 3$. Is it surprising that division over the integers is *not* closed?

EXERCISES 8–8.1

S–1.   Is the system of integers closed under

    a.   addition?
    b.   subtraction?

c. multiplication?
d. division?

Verify your answers. Refer to Sections 4, 5, 6, and 7 if you have difficulty.

S–2. Consider the set, ^-N, of all integers that have an ordered pair representation of the form (a, b), where $a \leq b$. Is this set closed under

a. addition?
b. subtraction?
c. multiplication?
d. division?

Support your answers by selecting specific members of ^-N and performing the indicated operations.

8.2 Associativity and Commutativity

A number system is *associative* under the binary operation $\odot$, if for every triple of numbers a, b, and c, $(a \odot b) \odot c = a \odot (b \odot c)$. It is *commutative* if for every pair a and b, $a \odot b = b \odot a$.

As with the positive rationals, the commutative and associative properties of addition and multiplication over the integers follow directly from the commutative and associative properties, respectively, of the whole-number system. We illustrate by proving that addition of integers is *associative*.

Let x, y and z be arbitrary integers represented by (a, b), (c, d), and (e, f), respectively, where a, b, c, d, e, f are whole numbers. Then

1. $(x + y) + z$ is represented by $[(a, b) + (c, d)] + (e, f)$ by definition of x, y, z
2. $= (a + c, b + d) + (e, f)$
   by the definition of addition of integers
3. $= ((a + c) + e, (b + d) + f)$
   by the definition of addition of integers
4. $= (a + (c + e), b + (d + f))$
   by the associative property of whole numbers
5. $= (a, b) + (c + e, d + f)$
   by the definition of addition of integers
6. $= (a, b) + ((c, d) + (e, f))$
   by the definition of addition of integers
7. which represents $x + (y + z)$.
8. Hence, $(x + y) + z = x + (y + z)$.

The commutative property of addition follows in the same way. Can you prove it?

What about the associative and commutative properties of multiplication? The proofs are similar.

Whether or not you decide to prove them, satisfy yourself that these properties hold by checking a variety of instances—for example, does $^-3 + (^+4 + {}^-6)$

equal $(^-3 + ^+4) + ^-6$? What about $^-3 + ^+4$ and $^+4 + ^-3$? $^-3 \times (^+4 \times ^-6)$ and $(^-3 \times ^+4) \times ^-6$? $^-3 \times ^+4$ and $^+4 \times ^-3$?

Is subtraction over the integers associative? Was subtraction over the whole numbers associative? Subtraction over the positive rationals? Is there any reason to suspect that subtraction over the integers would be associative? If there is any doubt in your mind check to see if $^+3 - (^-4 - ^+5)$ is equal to $(^+3 - ^-4) - ^+5$.

Ask the same question of division. Is division over the integers associative? Does $(^+12 \div ^-6) \div ^-2$ equal $^+12 \div (^-6 \div ^-2)$?

Now consider commutativity. Is subtraction over the integers commutative? Division? Show that the following statements are false.

1. $^+3 - ^-4 = ^-4 - ^+3$
2. $^-13 - ^+4 = ^+4 - ^-13$
3. $^+12 \div ^-6 = ^-6 \div ^+12$
4. $^-4 \div ^+2 = ^+2 \div ^-4$

Can you find other examples where these properties are false?

EXERCISES 8–8.2

S–1.   Show that the commutative property of subtraction of integers does not hold by showing that the following sentence is false. $^+3 - ^-4 = ^-4 - ^+3$. Find another counterexample.

S–2.   Show that the associative property of subtraction of integers does *not* hold by showing that the following sentence is false. $^-13 - (^+4 - ^-2) = (^-13 - ^+4) - ^-2$. Find two other counterexamples.

[S–3.]   Find specific examples (as in Problems S–1 and S–2) to prove that the associative and commutative properties of division of integers do *not* hold.

S–4.   Verify that the associative property of addition holds for each of the following:

    a. $^+3 + (^-7 + ^+2) = (^+3 + ^-7) + ^+2$
    b. $^-3 + (^+4 + ^-6) = (^-3 + ^+4) + ^-6$
    c. $^-5 + (^-9 + ^-1) = (^-5 + ^-9) + ^-1$

S–5.   Verify that the commutative property for addition of integers holds for each of the following:

    a. $^+7 + ^-5 = ^-5 + ^+7$
    b. $^-9 + ^-2 = ^-2 + ^-9$
    c. $^-6 + ^+4 = ^+4 + ^-6$

S–6.   Verify that the associative property of multiplication of integers holds for each of the following:

    a. $^-3 \cdot (^+5 \cdot ^-6) = (^-3 \cdot ^+5) \cdot ^-6$
    b. $^+7 \cdot (^-1 \cdot 0) = (^+7 \cdot ^-1) \cdot 0$
    c. $^+3 \cdot (^-3 \cdot ^+6) = (^+3 \cdot ^-3) \cdot ^+6$

S–7. Verify that the commutative property for multiplication of integers holds for each of the following:

    a. $-7 \cdot +5 = +5 \cdot -7$
    b. $+8 \cdot +9 = +9 \cdot +8$
    c. $-4 \cdot +12 = +12 \cdot -4$

S–8. Find three integers a, b, and c such that $(a \div b) \div c = a \div (b \div c)$. Does this prove that division of integers is associative (see Problem S–3)?

8.3 Identity

A number system has an identity, e, with respect to the operation $\odot$, if for each number a, $a \odot e = e \odot a = a$.

The additive identity in the system of integers is

$$+0 = \{(0, 0), (1, 1), (2, 2), (3, 3), \ldots, (n, n), \ldots\}$$

(Note that $+0$ is not the same as the whole number 0.)

To see that $x + +0 = x$ we let x be represented by (a, b) and $+0$ by $(1, 1)$. Then, $x + +0$ is represented by $(a, b) + (1, 1) = (a + 1, b + 1)$. But $(a + 1, b + 1)$ is more-less equivalent to (a, b), and hence represents x.

Check this result by proving the special case, $-3 + +0 = -3$. (*Hint:* Let -3 and $+0$ be represented by $(0, 3)$ and $(1, 1)$, respectively.) Can you prove that $+0 + x = x$? (The proof closely parallels that given above.)

We also note that $+0$ is a *right* identity for subtraction. That is, we can subtract $+0$ from any integer x and still get x. But, we cannot get x by subtracting x from $+0$. Thus, for example, $-3 - +0 = -3$ but $+0 - -3 = +3$. What is $+0 - +3$?

The identity for multiplication is the integer $+1$. Thus, $+1 \times x = x \times +1 = x$ for every integer x. For example, $-7 \times +1 = +1 \times -7 = -7$. Make up and test some additional examples.

The integer $+1$ is also a *right* identity for division. Thus, for example, $-6 \div +1 = -6$. On the other hand, $+1$ is *not* a left identity. To see this, consider $+1 \div -2$. $+1 \div +4$. (The quotients are undefined.)

EXERCISES 8–8.3

S–1. Illustrate that $+0$ is both a left and right additive identity by doing the following problems:

    a. $-12 + +0 =$ _____   c. $+0 + -6 =$ _____
    b. $+3 + +0 =$ _____   d. $+0 + +0 =$ _____

S–2. a. Show that $+0$ is not a left identity for subtraction by computing $+0 - +5$.

    b. Is there any integer a such that $+0 - a = a$?

S–3. Verify that $+1$ is both a left and right multiplicative identity by doing the following problems:

    a. $+1 \cdot -7 =$ _____   c. $+8 \cdot +1 =$ _____
    b. $+1 \cdot +4 =$ _____   d. $-6 \cdot +1 =$ _____

S–4.   Let ^-N be the set of all integers which have an ordered pair representation of the form (a, b) ,where $a \leq b$.

    a.   Find an element in ^-N that will serve as an additive identity.

    b.   Try to find an element in ^-N that will serve as a multiplicative identity. If you are unsuccessful, show that there is no such element in ^-N.

[M–5.]   Give a set of addition exercises that could lead a child to detect as a regularity the addition properties of the integer $^+0$.

[M–6.]   Give a set of multiplication exercises that could lead a child to detect as a regularity the multiplication property of $^+1$.

8.4 Inverse

A number system has the *inverse* property with respect to $\odot$, if for every number a, there is a unique number a', such that $a \odot a' = a' \odot a = e$ (where e is the identity for $\odot$).

    Let x be represented by (a, b). Then x', the additive inverse of x, is represented by (b, a), because

$$(a, b) + (b, a) = (a + b, b + a)$$

which represents $x + x' = {}^+0$. For example, the additive inverse for $^-3$ is $^+3$, because we can represent $^-3$ and $^+3$ by $(0, 3)$ and $(3, 0)$, respectively, and $(0, 3) + (3, 0) = (0 + 3, 3 + 0) = (3, 3)$, which represents $^-3 + {}^+3 = 0$. Can you find the additive inverses of $^+4$, $^+7$, and $^-2$ if these are represented by $(4, 0)$, $(7, 0)$, and $(0, 2)$, respectively? What about the additive inverses of $^+6$, $^-10$, $^+12$, $^-13$, and $^+9$? Have you discovered an easy way to find the additive inverse? If not, go back and compare the signs of the given integers and their inverses. If the sign of the given integer is $+$ (e.g., $^+5$), what is the sign of its inverse? What if the integer is negative (e.g., $^-5$)?

    Multiplication in the system of integers does not have the inverse property. To convince yourself of this, try to find an integer, x, such that $^+5 \cdot x = {}^+1$; or, $^-6 \cdot x = {}^+1$; or, $^+4 \cdot x = {}^+1$. Of course, some integers do have multiplicative inverses (e.g., $^+1 \cdot {}^+1 = {}^+1$, $^-1 \cdot {}^-1 = {}^+1$); the point is that this is not true in general.

    The inverses under subtraction and division are particularly easy to find. Each integer is its own inverse. That is, for every integer x, $x - x = {}^+0$. Further, if $x \neq 0$, then $x \div x = {}^+1$. What is the inverse of $^-4$ under subtraction? Under division?

EXERCISES 8–8.4

S–1.   In the system of integers what is the inverse (if any) of $^-4$ under

    a.   addition?   c.   multiplication?

    b.   subtraction?   d.   division?

S–2.   What integer(s) (if any) has (have) *no* inverse under

    a.   addition?   c.   multiplication?

    b.   subtraction?   d.   division?

[S–3.] a. List the inverses under addition for ⁻4, ⁺7, ⁺0, ⁻5, ⁺3.

  b. List the inverses under subtraction for ⁻4, ⁺7, ⁺0, ⁻5, ⁺3.

  c. If x is an integer how are the inverses of x under addition and subtraction related? (Try to detect a regularity in your answers to a and b above that will enable you to answer this question.)

[M–4.] Give a set of problems that might lead a child to detect the regularity that every integer is its own inverse under subtraction.

8.5 Distributivity

One operation $\odot$ in a number system is said to *distribute* over a second operation $\oplus$ if given any triple of numbers a, b, and c, $a \odot (b \oplus c) = (a \odot b) \oplus (a \odot c)$.

Multiplication distributes over both addition and subtraction. For example, consider $^-6 \times (^+4 + ^-2) = (^-6 \times ^+4) + (^-6 \times ^-2)$ and $^+7 \times (^-3 + ^+6) = (^+7 \times ^-3) + (^+7 \times ^+6)$. Similarly, for subtraction, consider $^-6 \times (^+4 - ^-2) = (^-6 \times ^+4) - (^-6 \times ^-2)$ and $^+7 \times (^-3 - ^+6) = (^+7 \times ^-3) - (^+7 \times ^+6)$. (Check each equation by simplifying the expressions on both sides of the equal sign.)

Each of these distributive properties follows directly from the corresponding distributive property of the whole numbers. For example, consider multiplication over addition.

Let x, y, and z be represented by (a, b), (c, d), and (e, f), respectively, where a, b, c, d, e, f are *whole* numbers. Then

1. $x \times (y + z)$ is represented by $(a, b) \times [(c, d) + (e, f)]$
2. $= (a, b) \times [(c + e, d + f)]$
   by the definition of addition
3. $= (a(c + e) + b(d + f), a(d + f) + b(c + e))$
   by the definition of multiplication
4. $= (ac + ae + bd + bf, ad + af + bc + be)$
   by the distributive property of *whole numbers*
5. $= ((ac + bd) + (ae + bf), (ad + bc) + (af + be))$
   by the associative and commutative properties of whole numbers
6. $= (ac + bd, ad + bc) + (ae + bf, af + be)$
   by the definition of addition
7. $= [(a, b) \times (c, d)] + [(a, b) \times (e, f)]$
   by the definition of multiplication, which represents
8. $(x \times y) + (x \times z)$.
9. Hence, $x \times (y + z) = (x \times y) + (x \times z)$.

Can you show that multiplication distributes over subtraction? (The argument closely parallels that given above.)

Finally, we simply mention that, in the system of integers, division does *not* distribute over addition or subtraction. For example, consider $(^+5 + ^-3) \div ^+2$ and $(^+5 \div ^+2) + (^-3 \div ^+2)$. The former expression is defined and equals $(^+2) \div ^+2 = ^+1$, but the latter expression is not defined. Hence, the expressions certainly cannot be equal (over the integers).

EXERCISES 8–8.5

S–1.   Show that the distributive property holds for each of the following by computing each side of the equality and showing that the answers are the same.

    a. $+3 \cdot (-6 + +7) = (+3 \cdot -6) + (+3 \cdot +7)$
    b. $-5 \cdot (+1 + -4) = (-5 \cdot +1) + (-5 \cdot -4)$
    c. $-3 \cdot (+6 + -7) = (-3 \cdot +6) + (-3 \cdot -7)$

E–2.   Show that multiplication does *not* distribute over division. (*Hint:* Consider $+3 \cdot (-12 \div +4)$. Does this equal $(+3 \cdot -12) \div (+3 \cdot +4)$?)

M–3.   Suppose one of your students proposes the following argument: "I think multiplication *does* distribute over division, because $+1 \cdot (-12 \div +4) = (+1 \cdot -12) \div (+1 \cdot +4)$," This is an example of fallacious reasoning, but how would you convince the student of this?

8.6 Multiplication by +0

Here we introduce another property and show how it can be derived from the *above properties* of the system of *integers*. All other proofs in this section have been based on the system of *whole* numbers. It should be emphasized, however, that we prove the property in this way for illustrative purposes only. It can also be proved by referral to properties of the whole numbers.

The property is simply this: For every integer, x, $x \cdot +0 = +0$. The proof goes as follows.

1. $+0 = +0 + +0$
   by the additive identity property of *integers* applied to $+0$
2. $x \cdot +0 = x \cdot (+0 + +0)$
   by multiplying both sides of step 1 by x
3. $x \cdot +0 = (x \cdot +0) + (x \cdot +0)$
   by the distributive property of multiplication over addition in the system of *integers*
4. $(x \cdot +0) + -(x \cdot +0) = [(x \cdot +0) + (x \cdot +0)] + -(x \cdot +0)$
   by adding the same number (i.e., $-(x \cdot +0)$) to both sides of the equation
5. $(x \cdot +0) + -(x \cdot +0) = (x \cdot +0) + [(x \cdot +0) + -(x \cdot +0)]$
   by the associative property of addition of *integers*
6. $+0 = (x \cdot +0) + [+0]$
   by the additive inverse property of *integers* applied to
   $(x \cdot +0) + -(x \cdot +0) = +0$
7. $+0 = x \cdot +0$
   by the additive identity property of *integers*

To further understand this argument, substitute a particular integer for x and prove the theorem for that special case. For example, can you show that $-3 \times +0 = +0$? What about $+4 \times +0 = +0$?

Can you prove that $x \cdot +0 = +0$ by referring to properties of the *whole numbers*? (*Hint:* Let x and $+0$ be represented by (a, b) and $(0, 0)$, respectively; apply the definition of multiplication to find $x \cdot +0$.)

Can you prove that $^+0 \cdot x = {}^+0$? (*Hint:* Apply the commutative property of multiplication of *integers* to the theorem we proved above.)

EXERCISES 8–8.6

S–1. Show that $^-3 \cdot {}^+0 = {}^+0$ by referring only to properties of *integers*. (*Hint:* Model your proof after the one given for $x \cdot {}^+0 = {}^+0$.)

S–2. Show that $^-3 \cdot {}^+0 = {}^+0$ by referring only to properties of *whole numbers* and to the definition of multiplication of integers. (*Hint:* Let $^-3$ and $^+0$ be represented by (0, 3) and (0, 0), respectively.)

[E–3.] Consider the proof of $x \cdot {}^+0 = {}^+0$ in the text.

    a. What are the premises (assumptions) that are used?
    b. What is the conclusion?

8.7 Other Properties: Relationships Between the Natural Numbers, Positive Rationals, and Integers

The system of integers is closely related to the system of natural numbers. We have seen this in everything from the way we defined integers in the first place to our discussion of the various properties of the system of integers.

As in the case of the positive rationals, there is an even more obvious type of relationship between the integers and the natural numbers. Each natural number corresponds to a *unique* integer. For example, the natural number 1 corresponds to the integer $^+1$, and 5 corresponds to $^+5$. In general, each natural number n corresponds to the integer ^+n.

Furthermore, this correspondence is an *embedding* (see Chapter 4) of the system of natural numbers into the system of integers. The corresponding binary operations over the naturals and the integers are preserved by this correspondence. For example, $5 \times 3 = 15$ in the system of naturals corresponds to $^+5 \times {}^+3 = {}^+15$ in the system of integers and $6 + 7 = 13$ corresponds to $^+6 + {}^+7 = {}^+13$. Stated more precisely, the integer, which corresponds to the product of 5 and 3 in the system of natural numbers, is identical with the product (in the system of integers) of the integers $^+5$ and $^+3$, which correspond to the natural numbers 5 and 3. Similarly, the integer corresponding to the sum of 6 and 7 is identical with the sum of $^+6$ and $^+7$, which correspond to the natural numbers 6 and 7.

Although the system of natural numbers can be embedded in both the system of positive rationals and the system of integers, we must not mistakenly think that the latter two systems are also *homomorphic* (see Chapter 4) to one another. In particular, there are no positive rationals corresponding to the negative integers (in a way which preserves the operations) and no integers corresponding to positive rationals such as $\frac{1}{3}, \frac{5}{6}, \frac{101}{3}$, and so on.

Nonetheless, there are one-to-one correspondences between these systems that do *not* preserve the operations. To see this, we show first that the set of natural numbers can be paired in one-to-one fashion with the set of integers. (We have already seen that the natural numbers can be paired in a one-to-one fashion with the positive rationals.) We can display such a correspondence directly by writing the integers in alternating form (see Figure 8–5).

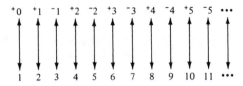

Figure 8–5

Without bothering to extend the corresponding argument in Chapter 7, why do we now know that the sets of integers and positive rationals can be paired in one-to-one fashion? Because both sets can be paired one-to-one with the set of naturals, could we pair together those integers and positive rationals which correspond to the same natural numbers? Refer to Chapter 7 and show explicitly how this might be done.

Although they can be paired one-to-one with the positive rationals, the set of integers is *not* dense. For example, there is no integer between ⁻3 and ⁻4. In this sense, the set of integers is more like the set of natural numbers. Every integer (and every natural number) has a unique successor. Thus, for example, the unique successor of 5 is 6 and that of ⁺5 is ⁺6. What is the successor of ⁺2? ⁺1? ⁺0? ⁻1? ⁻15?

Unlike the set of whole numbers, however, each integer also has a unique *predecessor*. (Recall that 0 in the set of whole numbers does not have a predecessor.) The corresponding integer ⁺0, however, has the predecessor ⁻1. Similarly, the unique predecessors of ⁺3, ⁻7, ⁻6, ⁺10 are ⁺2, ⁻8, ⁻7, ⁺9, respectively.

Having a unique predecessor precludes the possibility of having a minimal or least element. (The least element in the set of naturals is 1, and in the set of wholes is 0.) Suppose there were a least integer, ⁻m. Then the unique predecessor ⁻(m + 1) of ⁻m is less than ⁻m. This is absurd because it contradicts the assumption that ⁻m is the least integer. Hence, the set of integers has no least element. (This is an example of an indirect proof.)

EXERCISES 8–8.7

S–1.   Give the successor of each of the following integers.

   a.   ⁺2   d.   ⁻15
   b.   ⁺1   e.   ⁻136
   c.   ⁺0

∗E–2.   Show that the fraction $\frac{3}{4}$ does *not* have a unique successor. (*Hint:* Suppose a/b is the unique successor of $\frac{3}{4}$. Can you find a fraction that is *between* $\frac{3}{4}$ and a/b. Recall that the rationals are dense. Why does this contradict the assumption that $\frac{3}{4}$ has a unique successor?)

E–3.   Distinguish carefully between the natural number 2, the rational number represented by $\frac{2}{1}$ and the integer ⁺2. (If necessary, refer to Chapters 5 and 7.)

9

The Rationals, Reals, and Further Extensions

We have developed the systems of positive rationals and integers separately, because these systems are of primary interest in the elementary school. To leave things here, however, would give a stilted view of number systems. In particular, the algebraic systems more typically studied in advanced mathematics—indeed, some of the number systems we use every day—have in some sense nicer properties than, for example, the system of positive rationals, where subtraction is not always defined, or the system of integers, where division is not always defined.

In this chapter, we shall try to round off some of the rough edges by introducing further extensions of the number systems we have studied. Any attempt to deal with these extended systems in a complete or systematic way, however, would be both impractical and unnecessary in a book of this sort. We shall try, instead, to consider some of the highlights of these systems, to show what they add or subtract from existing systems and, generally, to give the reader a more comprehensive view of number systems.

In turn, we shall discuss the system of rational numbers, the system of real numbers, and then vectors, including the special case of complex numbers. Also included is a brief treatment of the nature of geometry as it relates to algebra, in general, and the real number system, in particular.

1. THE SYSTEM OF (SIGNED) RATIONALS

In the preceding chapters, we have seen that the system of positive (non-negative) rationals and the system of integers leave major gaps on the number line, although the gaps are of quite different sorts. In the former system, no numbers correspond to points on the number line to the left of 0 (i.e., there are no negative numbers). The positive rationals to the right of 0, however, were shown to be *dense*. In the system of integers, on the other hand, there are integers corresponding to points to the left of 0, but gaps exist between *every* integer and its successor (and predecessor). Stated differently, the integers are *not* dense.

As it turns out, each of these systems can be extended to new number systems which have most of the advantages of both (systems). Thus, just as we extended the system of natural numbers to the positive rationals, we can also extend the *integers* to the system of *signed rationals* (*rationals*, for short). Likewise, we can extend the *non-negative rationals* to the system of *rationals* in much the same way we developed the integers from the whole numbers. Perhaps not too surprisingly, the two extended systems turn out to be essentially equivalent to one another. More precisely, they turn out to be *isomorphic*. The inherent relationships can be represented schematically.

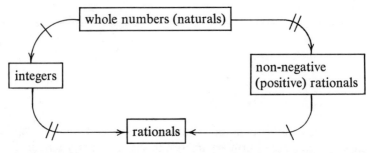

It is beyond the scope of this treatment to systematically develop the system of rational numbers. The interested and energetic reader will find this to be both an interesting and rewarding project, however, and we shall sketch some of the main ideas.

First, we consider the development of the rationals from the integers. In this case, we define *signed fractions* as ordered pairs of *integers*, written (x, y) or x/y where $y \neq {}^+0$. (Each integer, of course, is an equivalence class of pairs of whole numbers.) Examples of signed fractions are ${}^+1/{}^-2$, ${}^+3/{}^-6$, and ${}^-2/{}^-5$. Paralleling the development of the positive rationals, two signed fractions x/y and x'/y' are (defined to be) *equivalent* if and only if $x \cdot y' = y \cdot x'$. For example, ${}^+1/{}^-2$ and ${}^+3/{}^-6$ are equivalent because ${}^+1 \cdot {}^-6 (= {}^-6) = {}^-2 \cdot {}^+3 (= {}^-6)$. The signed fractions ${}^+3/{}^-6$ and ${}^-2/{}^-5$ are not equivalent, however, because ${}^+3 \cdot {}^-5 (= {}^-15) \neq {}^-6 \cdot {}^-2 (= {}^+12)$.

Rational numbers, then, are *defined* to be equivalence classes of signed fractions. For example, the signed rational ${}^+1/5 = \{{}^+1/{}^+5, {}^-1/{}^-5, {}^+2/{}^+10, {}^-2/{}^-10, {}^+3/{}^+15, \dots\}$.

We define addition and multiplication of rationals as before, in terms of representative (signed) fractions. Given that x/y represents the rational number R, and x'/y' represents the rational number R', where x, y, x', and y' are integers, the *sum* $R + R'$ is *defined* to be the rational number, represented by

$$\left(\frac{x}{y} + \frac{x'}{y'} = \right) \frac{(x \cdot y' + y \cdot x')}{y \cdot y'}$$

Similarly, the *product, $R \cdot R'$* is *defined* to be the rational number represented by

$$\left(\frac{x}{y} \cdot \frac{x'}{y'} = \right) \frac{x \cdot x'}{y \cdot y'}$$

(Compare these definitions with those in Chapter 7 for the positive rationals.)

For example, the sum of the *signed rationals* denoted $\frac{+3}{5}$ and $\frac{-2}{3}$ may be found by selecting representative *signed fractions,* $\frac{+3}{+5}$ and $\frac{+2}{-3}$, respectively, and substituting into the definition, giving $\frac{(+3 \cdot {}^-3) + (+5 \cdot {}^+2)}{+5 \cdot {}^-3} = \frac{-9 + {}^+10}{-15} = \frac{+1}{-15}$

which represents the rational, $\frac{-1}{15}$. Similarly, the product of $\frac{+3}{5}$ and $\frac{-2}{3}$ is found

to be $\frac{+3 \cdot {}^+2}{+5 \cdot {}^-3} = \frac{+6}{-15}$, which represents the rational $\frac{-6}{15}$.

Subtraction and division may be defined in parallel fashion, or more simply, as the inverses of addition and multiplication, respectively. We leave it to the reader to define subtraction and division and to show that all of the operations are well defined. (Remember that division by $^+0$ is not allowed.)

Alternatively, we may develop the system of rationals from the non-negative rationals. Here, *rationals* are *defined* as equivalence classes of ordered pairs of *non-negative rationals.* For example, we would equate the signed rationals $^+2/5$ and $^-3/4$ with the equivalence classes $\{(a/b, c/d) \mid a/b - c/d = 2/5\}$ and $\{(a/b, c/d) \mid c/d - a/b = 3/4\}$, respectively, where a/b and c/d are to be thought of as *rationals* (not fractions). (Note the order of subtraction in the defining sets. Because the subtraction is defined on the non-negative rationals, either $a/b - c/d$ or $c/d - a/b$ is undefined unless $a/b = c/d$.)

Addition and multiplication, in this case, are defined in a manner which parallels the corresponding definitions in Chapter 8 on the integers. Consider the two rationals, $R = \{(a/b, c/d) \mid a/b - c/d = r$ or $c/d - a/b = r\}$ and $R' = \{(a'/b', c'/d') \mid a'/b' - c'/d' = r'$ or $c'/d' - a'/b' = r'\}$, where r and r' are non-negative rationals representing the magnitudes of R and R', respectively. Then, the *sum $R + R'$* is *defined* as the rational represented by the pair $(a/b + a'/b', c/d + c'/d')$.[1] For example, the sum of $^+2/5$ and $^-3/4$, where $^+2/5$ is represented by $(2/5, 0)$ and $^-3/4$ by $(0, 3/4)$, is defined to be the rational $^+2/5 + {}^-3/4$ represented by $(2/5 + 0, 0 + 3/4) = (2/5, 3/4)$. This rational is $^-7/20$, because $3/4 - 2/5 = 15/20 - 8/20 = 7/20$ (where the subtraction is performed in the system of non-negative rationals). Hence, $^+2/5 + {}^-3/4 = {}^-7/20$.

The product $R \times R'$ may be defined similarly as the rational represented by $((a/b \cdot a'/b') + (c/d \cdot c'/d'), (a/b \cdot c'/d') + (c/d \cdot a'/b'))$. For example, the product of $^+2/5$ and $^-3/4$ (as above) is defined to be the rational $^+2/5 \times {}^-3/4$ represented by

$$\left(\left(\frac{2}{5} \cdot 0\right) + \left(0 \cdot \frac{3}{4}\right), \left(\frac{2}{5} \cdot \frac{3}{4}\right) + (0 \cdot 0)\right) = \left(0, \frac{2}{5} \cdot \frac{3}{4}\right) = \left(0, \frac{6}{20}\right)$$

---

[1] In this definition, $(a/b, c/d)$ and $(a'/b', c'/d')$ are viewed as *arbitrary* pairs of non-negative rationals, representing R and R', respectively.

Hence, the product $^+2/5 \times {}^-3/4 = {}^-6/20$. Show that these operations are well defined and devise definitions for subtraction and division.

In sketching these alternative developments, we have implicitly assumed that the extensions are essentially identical—that is, that we can talk about *the* system of rationals rather than *a* system of rationals. It is not immediately obvious that the two extensions we have described are, indeed, isomorphic to one another. The "rationals" defined in these systems are quite different entities, as are the operations of addition and multiplication. In the one case, for example, the "rationals" are equivalence classes of pairs of integers and, in the other, they are equivalence classes of pairs of nonnegative rationals.

The task of proving the equivalence of the two systems of "rationals" we have defined is left as a challenging project for the ambitious reader. The basic goal is to show that the two extensions defined as above are isomorphic to one another—that is, that there is a one-to-one correspondence, between the elements of the two extensions, which preserves the respective operations. (Showing that "addition" and "multiplication" are preserved will be sufficient.) Expect no surprises; the correspondence is what we would expect it to be, between elements which correspond to the same points on the number line. For example

$$\frac{^+1}{5} = \left\{ \frac{^+1}{^+5}, \frac{^-1}{^-5}, \frac{^+2}{^+10}, \frac{^-2}{^-10}, \frac{^+3}{^+15}, \cdots \right\}$$

in the first extension, corresponds to

$$\frac{^+1}{5} = \left\{ \left(\frac{a}{b}, \frac{c}{d} \right) \middle| \frac{a}{b} - \frac{c}{d} = \frac{1}{5} \right\}$$

in the second extension. The main task is in showing that the corresponding operations of "addition" and "multiplication" are preserved by this correspondence. (Although it should be clear by now, we emphasize that, in spite of our use of $+$ and $\times$ (or $\cdot$) to denote "addition" and "multiplication" in both extensions, the corresponding operations, themselves, are *not* identical. For one thing, they operate on different kinds of elements.)

As promised, the system of rationals has certain convenient features of both the system of integers and the system of nonnegative rationals. It is like the integers in that subtraction is always defined, and it is like the nonnegative rationals in that division (except by 0) is always defined. These properties are a direct consequence of the fact that the system of rationals satisfies all of the properties (axioms) of what is known in mathematics as an (algebraic) *field*. That is:

1. The system is *closed* under both addition and multiplication.
2. Addition and multiplication are both *associative*.
3. Addition and multiplication are both *commutative*.
4. There is an *identity* for both addition (0) and multiplication (1).
5. Every element has an *additive inverse*.
6. Every element *except* the additive identity (0) has a *multiplicative inverse*.
7. Multiplication *distributes* over addition.

(Remember that strictly speaking 0 denotes a whole number only; $r_0 = \frac{0}{1}$ denotes the corresponding nonnegative rational, and $^+0$ the corresponding integer. We shall overlook this distinction below where the intended meaning is clear.)

In any field it can be shown that subtraction and division (except by 0) are closed. In addition we can prove that $0 \cdot x = 0$ as we did in Chapter 8. There are, of course, any number of other properties which can be proved from the field axioms. Try to state some of them.

As for all of the number systems we have discussed so far, a *strict linear order* relation may be defined for the rational numbers. (This is also true of the reals, which we discuss in the next section, but not of vectors and complex numbers, which will be introduced later.) In particular, the rational number R is said to be *greater than* the rational R' if and only if $R - R'$ is positive. Like the integers, however, there is no smallest element and like the positive rationals there are no *unique* successors or predecessors. The set of rationals is *dense* over the entire number line.

Finally, we mention that the integers and positive rationals, as well as the natural numbers, are *embedded* in (i.e., are isomorphic with a subset of) the system of rationals. The essential relationships may be represented by points on number lines. We leave it to the interested reader to prove that such embeddings do exist. (Also, see the starred exercises.)

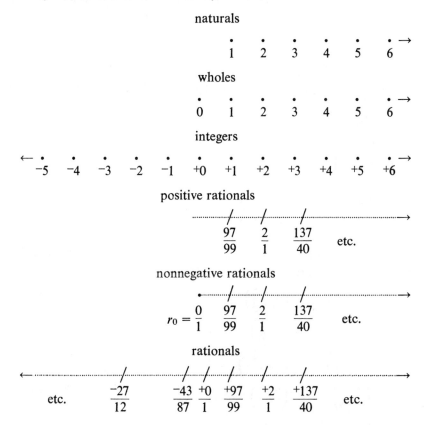

EXERCISES 9–1

S–1. Show that the two fractions in each of the following pairs of signed fractions are equivalent:

a. $\dfrac{+8}{+2}, \dfrac{+4}{+1}$

b. $\dfrac{-3}{-6}, \dfrac{+7}{+14}$

c. $\dfrac{+0}{+2}, \dfrac{+0}{+4}$

d. $\dfrac{-10}{+2}, \dfrac{+20}{-4}$

S–2. Test each of the following pairs of signed fractions to determine whether or not they are equivalent:

a. $\dfrac{+15}{+3}, \dfrac{+5}{+1}$

b. $\dfrac{-5}{+13}, \dfrac{+5}{-13}$

c. $\dfrac{-3}{+7}, \dfrac{+3}{+7}$

d. $\dfrac{-2}{-3}, \dfrac{+4}{+6}$

e. $\dfrac{+7}{+13}, \dfrac{+9}{+15}$

f. $\dfrac{+0}{+6}, \dfrac{+0}{-3}$

g. $\dfrac{+12}{-17}, \dfrac{-36}{+51}$

S–3. Determine the value(s) of the integer n in each of the following so that the signed fractions are representative of the same rational number:

a. $\dfrac{+4}{+5}, \dfrac{n}{+30}$

b. $\dfrac{+3}{-7}, \dfrac{n}{-119}$

c. $\dfrac{-3}{-5}, \dfrac{+21}{n}$

d. $\dfrac{-3}{+2n}, \dfrac{n}{-96}$

E–4. Show that each pair of ordered pairs of non-negative rationals is more-less equivalent:

a. $\left(\dfrac{10}{1}, \dfrac{12}{2}\right)$ $\left(\dfrac{8}{2}, \dfrac{0}{9}\right)$

b. $\left(\dfrac{2}{1}, \dfrac{3}{2}\right)$ $\left(\dfrac{5}{4}, \dfrac{3}{4}\right)$

c. $\left(\dfrac{7}{9}, \dfrac{7}{9}\right)$ $\left(\dfrac{15}{6}, \dfrac{5}{2}\right)$

d. $\left(\dfrac{0}{3}, \dfrac{2}{3}\right)$ $\left(\dfrac{7}{9}, \dfrac{13}{9}\right)$

S–5. Test the following ordered pairs of non-negative rationals to determine whether or not they are more-less equivalent:

a. $\left(\dfrac{14}{2}, \dfrac{9}{3}\right)$ $\left(\dfrac{17}{3}, \dfrac{5}{3}\right)$

b. $\left(\dfrac{5}{6}, \dfrac{1}{3}\right)$ $\left(\dfrac{1}{3}, \dfrac{5}{6}\right)$

c. $\left(\dfrac{0}{10}, \dfrac{3}{5}\right)$ $\left(\dfrac{2}{15}, \dfrac{11}{15}\right)$

d. $\left(\dfrac{2}{5}, \dfrac{3}{8}\right)$ $\left(\dfrac{4}{7}, \dfrac{5}{9}\right)$

S–6. Show that the following pairs of signed fractions are equivalent:

a. $\dfrac{+a}{+b}, \dfrac{-a}{-b}$

b. $\dfrac{+a}{-b}, \dfrac{-a}{+b}$

[S–7.] Give two representations of each of the following rational numbers (one a signed fraction, the other an ordered pair of non-negative rationals):

a. $\dfrac{+1}{4}$ d. $\dfrac{-2}{7}$

b. $\dfrac{-1}{4}$ e. 1

c. $\dfrac{+19}{3}$ f. 0

S–8. Compute each of the following sums using the first definition for addition of rational numbers:

a. $\dfrac{+1}{4} + \dfrac{-2}{7}$ c. $\dfrac{-1}{4} + \dfrac{-2}{7}$

b. $\dfrac{+1}{4} + \dfrac{-1}{4}$ d. $\dfrac{+1}{4} + 0$

S–9. Compute each of the sums in Exercise S–8 using the second definition of addition of rational numbers.

S–10. Compute each of the following products using the first definition for multiplication of rational numbers:

a. $\left(\dfrac{+1}{4}\right) \cdot \left(\dfrac{+19}{3}\right)$ c. $\left(\dfrac{-1}{4}\right) \cdot \left(\dfrac{-2}{7}\right)$

b. $\left(\dfrac{+1}{4}\right) \cdot \left(\dfrac{-2}{7}\right)$ d. $\left(\dfrac{+1}{4}\right) \cdot 1$

S–11. Compute each of the products in Exercise S–10 using the second definition for multiplication of rational numbers.

E–12. Define subtraction and division of rational numbers.

∗E–13. Show that addition and multiplication of rational numbers (with reference to definitions made in the text) are well defined. *Hint:* Show that if $\dfrac{x}{y}$ and $\dfrac{x'}{y'}$ are both representatives of rational number R, and $\dfrac{w}{z}$ and $\dfrac{w'}{z'}$ are both representatives of rational number R', then $\dfrac{x \cdot z + y \cdot w}{y \cdot z}$ and $\dfrac{x' \cdot z' + y' \cdot w'}{y' \cdot z'}$ are both representatives of $R + R'$, and $\dfrac{x \cdot w}{y \cdot z}$ and $\dfrac{x' \cdot w'}{y' \cdot z'}$ are both representatives of $R \cdot R'$.

∗E–14. Show that the two systems of "rationals" defined in this section are isomorphic. *Hint:* Define a one-to-one correspondence between the elements of the two systems which preserves the operations of addition and multiplication. For the rational number $^+a/b$, consider the corresponding distinguished representatives $^+a/^+b$ and $(a/b, 0)$.

S–15. We can define the additive inverse of a rational number R to be that rational number R' which satisfies the equation $R + R' = 0$. In particular, $^-2/3$ is the additive inverse of $^+2/3$ because $^+2/3 + {}^-2/3 = 0$.

a. Give the additive inverses of the following: $^+4/5, {}^-3/7, 1, 0$
b. The system of rationals has all of the properties concerning additive inverses that the system of integers has. List two such properties and provide an example of each to illustrate that the property holds in the system of rationals.
c. Define *multiplicative inverse* for the system of rationals. What other system have you studied which had this concept? Does every rational number have a multiplicative inverse? What is the multiplicative inverse of $^+5/4$? Of $^-5/4$?

S–16. Indicate which field property is being illustrated in each of the following:

a. $\dfrac{^+2}{3} + 0 = \dfrac{^+2}{3}$

b. $\left(\dfrac{^+2}{3}\right) \cdot \left(\dfrac{^-3}{5}\right) = \left(\dfrac{^-3}{5}\right) \cdot \left(\dfrac{^+2}{3}\right)$

c. $\left(\dfrac{^+2}{3}\right) \cdot \left[\dfrac{^+1}{2} + \dfrac{^-3}{4}\right] = \left(\dfrac{^+2}{3}\right) \cdot \left(\dfrac{^+1}{2}\right) + \left(\dfrac{^+2}{3}\right) \cdot \left(\dfrac{^-3}{4}\right)$

d. $\left(\dfrac{^+2}{3}\right) + \left(\dfrac{^-2}{3} + \dfrac{^+3}{4}\right) = \left(\dfrac{^+2}{3} + \dfrac{^-2}{3}\right) + \left(\dfrac{^+3}{4}\right)$

e. $\left(\dfrac{^+2}{3}\right) \cdot \left[\left(\dfrac{^+1}{2}\right) \cdot \left(\dfrac{^+3}{4}\right)\right] = \left[\left(\dfrac{^+1}{2}\right) \cdot \left(\dfrac{^+3}{4}\right)\right] \cdot \left(\dfrac{^+2}{3}\right)$

[S–17.] Prove the commutative properties for addition and multiplication of rational numbers.

E–18. If R and R' are rational numbers with signed fraction representatives a/b and c/b (where b is a positive integer), then $R < R'$ if and only if $a < c$. In particular, $^+1/6 < {}^+1/2$ because by examining their respective signed fraction representatives $^+1/^+6$ and $^+3/^+6$ (each having the same positive integer for a denominator), we see $^+1 < {}^+3$.

a. Arrange the following in order: $^+4/7, 0, {}^-1/5, {}^-1/6, {}^+1/2, {}^+5/9$.
b. Find a rational number R such that $^+1/6 < R < {}^+1/5$.
∗c. Show that the rationals have the density property by showing that the "average" $A = \dfrac{R + R'}{2}$ of two rational numbers R and R' (where $R < R'$) has the property that $R < A < R'$. *Hint:* If x and y are integers and $x < y$ then $2x < x + y < 2y$.

d. Suppose R and R' are rational numbers with signed fraction representatives a/b and c/d. Find several examples which prove that the following is *not* a theorem for the rationals: $R < R'$ if and only if $ad < bc$.

e. Add a restriction to the statement in d so that it becomes a theorem.

S–19. Indicate which of the following systems are fields: If a system is not a field, state the field properties it fails to satisfy.

a. The set of natural numbers with the operations addition and multiplication.

b. The set of integers with the operations addition and multiplication.

c. The set of positive rationals with the operations addition and multiplication.

d. The set $\{0, 1\}$ with the operations addition and multiplication as on a clock with only two positions, 0 and 1 (see Chapter 4).

∗E–20. Show that the integers and positive rationals can be embedded in the rationals.

M–21. How would you represent the rational number line with elementary school children?

M–22. Some elementary school texts use the term *opposite* in place of additive inverse. Give some arguments for and against this practice.

For further thought and study: The interpretation of rational numbers as points on the number line helps us to deal more effectively with this system.

(A) List some concepts you are able to visualize more clearly with the help of the number line.

(B) How could addition of rational numbers be interpreted on the number line? (*Hint:* Consider the corresponding discussion in Chapter 8.)

(C) How is the additive inverse of a rational number found by using the number line?

2. THE SYSTEM OF REAL NUMBERS

Because the rationals are dense over the entire number line, we might suspect that every point on the number line can be represented by a unique rational number. For example, not only can we represent special points on the number line by rational numbers, but we can represent points that are arbitrarily close to one another. We can indicate this by a succession of number lines, each one magnifying a segment of the preceding segment (of the number line).

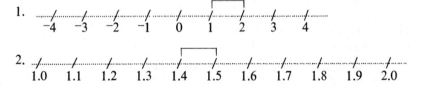

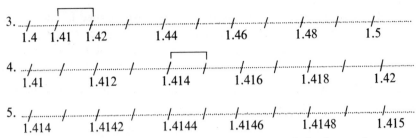

After carrying out this process for a time, we may, indeed, convince ourselves that each point on the number line can be represented by a rational number.

If we think about the matter a little more, however, we may begin to wonder if the statement really is true. In particular, we may wonder what happens if we continue this process indefinitely. As the number lines above suggest, we can conceive of an infinite *sequence* of rationals, each one closer to the preceding rational than the one before—for example, the sequence 1., 1.4, 1.41, 1.414, 1.4142, 1.41421, ... The terms (rationals) in the sequence correspond to points that appear to approach some particular point on the number line. But how do we know for sure that this *limit point* is a rational number?

The fact is, we do not. We can prove that there are points on the number line that do *not* correspond to rational numbers. First, however, we must introduce two new operations, squaring and taking the square root.

The *square* of a number, n, denoted by n^2, is the product of n with itself (i.e., $n^2 = n \times n$). For example, $(^+2)^2 = {}^+2 \times {}^+2 = {}^+4$; $(^-4)^2 = {}^-4 \times {}^-4 = {}^+16$, $(^-1/2)^2 = {}^-1/2 \times {}^-1/2 = {}^+1/4$; $(^+3/5)^2 = {}^+3/5 \times {}^+3/5 = {}^+9/25$. In general, the square of any number, whether positive or negative, is always positive. Convince yourself of this by trying additional examples and/or checking the statement against the definition of multiplication. (Of course, $0^2 = 0$ is not positive since 0 itself is neither positive or negative.) The squaring operation can be generalized in an obvious way to include triple products and beyond, but we need not concern ourselves with this here (e.g., $3^3 = 3 \times 3 \times 3 = 27$ and $2^4 = 2 \times 2 \times 2 \times 2 = 16$).

The inverse operation of squaring a number is taking the *square root*. Thus, a square root of a number, m, is a number which when squared yields m.

The square root of a rational number, however, is not necessarily unique. Thus, if a rational number, ^+m, has a positive square root, $^+\sqrt{m}$, then it also has a negative square root, $^-\sqrt{m}$. For example, the rational $^+4 (= {}^+4/1)$ has square roots $^+\sqrt{4} = {}^+2$ (because $^+2 \times {}^+2 = {}^+4$) and $^-\sqrt{4} = {}^-2$ (because $^-2 \times {}^-2 = {}^+4$). For our purposes, it will suffice to consider only positive square roots. Thus, when we write $\sqrt{2}$, $\sqrt{3}$, or $\sqrt{4}$, we always mean the positive square root.

Not all rational numbers have rational square roots, however. The negative rationals (e.g., $^-1$, $^-4$, $^-1/4$ etc.), for example, do *not* have rational square roots. This follows directly from the fact that the square of every non-zero rational number is positive (see above).[2]

---

[2] As we shall see later on, the only numbers, whether rational or *real*, that have square roots over the *real* number system are *non-negative*. In the system of *complex numbers*, however, square roots always exist.

We are now ready to prove that there is a point on the number line for which no rational number exists. Consider, for example, the right triangle (see Figure 9–1) where the sides CA and CB are both 1 unit long. We indicate this: $\overline{CA} = 1$ and $\overline{CB} = 1$. By the Pythagorean theorem, we know that $\overline{AB}^2 = \overline{CA}^2 + \overline{CB}^2 = 1^2 + 1^2 = 1 + 1 = 2$. (*Note:* $\overline{AB}^2$ means $\overline{AB} \times \overline{AB}$.) Hence, $\overline{AB} = \sqrt{2}$.

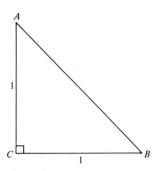

Figure 9–1

Now, if we think of the three sides as segments of the (same) number line (perhaps rotated and moved a bit), it should be clear that each length corresponds to a particular point on the number line. Thus, starting from 0, we may lay off each side as indicated.

The rational number 1 corresponds to the lengths of the two "legs" of the triangle (CA and CB). For lack of a better name, we may let $\sqrt{2}$ correspond to the point representing a distance from 0 equal to the length $\overline{AB}$.

The key question is: How do we know that $\sqrt{2}$ is a rational number? To show that it is *not*, we reason indirectly as follows: Suppose $\sqrt{2}$ is a rational number. Then it can be represented by a fraction (i.e., a pair of natural numbers, because we are only considering positive lengths). Furthermore, it can be represented by a fraction (p/q) in which the numerator and denominator (both natural numbers) are *relatively prime*. That is, p and q have no common factors (except 1). Hence

$$\sqrt{2} = \frac{p}{q}$$

Squaring should not change anything, so

$$2 = \left(\frac{p}{q}\right)^2$$

Rewriting the latter expression as a quotient, we get[3]

$$2 = p^2 \div q^2 \left(= \frac{p^2}{q^2} \right)$$

Because division is the inverse of multiplication, this equation has the same solution set as

$$2 \times q^2 = p^2$$

The last equation says (among other things) that p^2 has 2 as a prime factor. But the only way this can happen is for p to have 2 as a factor. So there must exist a natural number, r, such that

$$p = 2 \times r$$

Hence

$$2 \times q^2 = 2 \times 2 \times r^2 \, (= (2 \times r)^2 = p^2)$$

Dividing both sides by 2, we get

$$q^2 = 2 \times r^2$$

This says that q^2 and, hence, q also has 2 as a prime factor. But we assumed at the beginning that p and q were relatively prime (i.e., they did *not* have any prime factors in common). Hence, our assumption that $\sqrt{2}$ is a rational, and can therefore be represented as a fraction, is *false*. That is, there *are* points on the number line for which no corresponding rational number exists.

Many other points on the number line do not have rational representations, for example, $\sqrt{5}$, $\sqrt{26}$, $\pi = 3.14159265\ldots$. As a matter of fact, there are so many nonrational points on the number line that, contrary to what one might expect, if we selected an arbitrary point, it would be extremely rare that that point would correspond to a rational number. In advanced mathematics, the set of rationa numbers is said to have "measure zero."

Our problem, then, becomes one of extending the system of rationals so that every point on the number line corresponds to a unique number in the extension (and vice versa). As before, of course, we will also want the rationals to be embedded in the extension.

Judging from our past successes, we might suspect that we could generate such an extension by considering equivalence classes of ordered pairs of rationals. Unfortunately, this approach simply does not work. There would still be points on the number line which could not be represented.[4]

As it turns out, there are a number of different ways of constructing the *real* number system. We will consider one briefly.

---

[3] This step follows because every rational number, a/b, can be represented as the quotient $a' \div b'$ (where a' and b' are the rational numbers corresponding to the natural numbers a and b, respectively, and $\div$ is to be interpreted as over the rationals). We have not proved this formally, but this can be done with the "machinery" we have available. In most proofs, this theorem is simply taken for granted and no reason is given for the step taken above.

[4] We shall have to take this statement on faith. It is beyond the scope of this book to give a proof here.

Recall from our earlier discussion that there was no guarantee that the "limit" approached by a sequence of rational numbers is necessarily rational. This gives us the germ of an important idea! Perhaps the numbers in the extension we desire can be represented as *sequences* of rational numbers or equivalence classes of sequences of rationals.

To see what might be involved, let us consider some examples of sequences of rationals that "converge" to a limit point (i.e., that get closer and closer to a given number). (Obviously, a sequence such as $(1, 2, 3, 4, 5, \ldots, n, \ldots)$ does not converge; the terms merely become larger and larger.) We must add, however, that the question of sequence convergence is complex, and we can do no more here than scratch the surface.

(1) The sequence $(1/2, 1/4, 1/8, 1/16, 1/32, \ldots, 1/2^n, \ldots)$ is an infinite sequence that converges to the rational 0. We can also say the sequence approaches 0 in the *limit*. That is, each successive term in the sequence gets closer and closer to 0 and, in fact, we can get as close to 0 as we want. The expression, $1/2^n$, tells us how to get the nth term of the sequence for any natural number n. Thus, for example, the fourth term (substituting $n = 4$) is $1/2^4 = 1/16$; the tenth term is $1/2^{10} = 1/1024$; and the twelfth term is $1/2^{12} = 1/4096$. Notice that as n gets larger $1/2^n$ becomes progressively closer to zero.

We can represent this on the number line.

$$0 \quad \frac{1}{32} \, \frac{1}{16} \, \frac{1}{8} \qquad \frac{1}{4} \qquad\qquad \frac{1}{2} \qquad\qquad\qquad\qquad\qquad\qquad 1$$

(2) $(1, 1/2, 1/3, 1/4, \ldots, 1/n, \ldots)$ is another sequence that converges to 0, only this time more slowly.

$$0 \qquad\qquad \frac{1}{7} \, \frac{1}{6} \, \frac{1}{5} \, \frac{1}{4} \qquad \frac{1}{3} \qquad\qquad \frac{1}{2} \qquad\qquad\qquad\qquad\qquad 1$$

(3) The sequence $(1., .1, .01, .001, .0001, .00001, \ldots)$ also converges to 0. Notice that the terms of this sequence are terminating decimals and that each successive term involves one additional decimal place.

(4) The simplest sequence that converges to 0, however, is the sequence of terminating decimals $(0, .0, .00, .000, .0000, \ldots)$. Not only does each successive term involve one more decimal place than the one which precedes it, but we can form these terms in a particularly simple way *by adding one digit* to the right of what we had before (in this case the digit is always 0). This type of sequence plays a special role in some developments and we shall call such a sequence *distinguished*.

(5) Of course, not all sequences approach 0. Thus, the sequences

(a) $\left(\dfrac{1}{2}, \dfrac{3}{4}, \dfrac{7}{8}, \dfrac{15}{16}, \ldots, \dfrac{2^n - 1}{2^n}, \ldots \right)$

(where the numerator is always one *less than* the denominator)

(b) $\left(\dfrac{3}{2}, \dfrac{5}{4}, \dfrac{9}{8}, \ldots, \dfrac{2^n + 1}{2^n}, \ldots \right)$

(where the numerator is always one *greater than* the denominator)

(c) (.9, .99, .999, .9999, .99999, ...)

(d) (1.1, 1.01, 1.001, 1.0001, 1.00001, ...)

and

(e) (1, 1.0, 1.00, 1.000, ...)

all approach 1. Sequences (a) and (c) approach 1, as we shall say, "from below" and (b) and (d), "from above." Sequence (e) may be said to approach 1 "exactly."

Sequences (c) and (e) are both distinguished in the sense described in Example (4) above. From now on, in those cases where the same limit is approached by two distinguished sequences of decimals, such as (c) and (e), we shall *agree to use* the one which involves 0's. This will make it possible for us to represent each limit point by a unique sequence of decimals. (It can be proved that every limit point can be represented by a distinguished sequence.)

Not all sequences converge to whole numbers, of course, nor are the "distinguished" decimal sequences which represent the limit points necessarily restricted to a single repeating digit.

(6) Thus, (.5, .57, .571, .5714, .57142, $.\overline{571428}$, $.\overline{5714285}$, $.\overline{57142857}$, ..., $.\overline{57142857142857}$, ...) can be viewed as a distinguished decimal sequence in which the string of digits 571428 repeats periodically. This sequence in the limit approaches the rational number, $\frac{4}{7}$, as can be seen by dividing 7 into 4.0. Another convergent sequence which approaches $\frac{4}{7}$ as the limit is $(\frac{3}{7}, \frac{7}{14}, \frac{11}{21}, \ldots, (4n-1)/7n, \ldots)$.

To see that this latter sequence has $\frac{4}{7}$ as a limit, we need only to rewrite the general term $(4n-1)/7n$ in the more suggestive form

$$\frac{4n-1}{7n} = \frac{4n}{7n} - \frac{1}{7n} = \frac{4}{7} - \frac{1}{7n}$$

Because n becomes larger as we continue out in the sequence, the term $1/7n$ approaches 0. Hence, the limit of the sequence may be represented by

$$\frac{4}{7} - 0 = \frac{4}{7}$$

We indicate this by writing

$$\frac{4n-1}{7n} = \frac{4}{7} - \frac{1}{7n} \to \frac{4}{7}$$

As we have argued from the beginning, convergent sequences do *not* necessarily converge to *rational* numbers. The six examples above illustrate that, given the limit of a convergent sequence, we can always find some term far enough out in the sequence which is *as close to that limit as one might want it to be.* If the limit is not a rational number, however, it would not seem reasonable[5] to ask how close a term of a sequence (a rational number) is to the limit.

---

[5] At least until we define what such limits might actually be.

Fortunately, it is not necessary to know the limit of a sequence in order to know that the sequence converges. In particular, a sequence converges if the difference between successive terms can be made as small as we like—that is, if the "difference approaches 0."

More precisely, a sequence of rationals is said to be *Cauchy convergent* if for *any* rational number, $r > 0$, there exists a term of the sequence, t_n, such that the magnitude of the difference between t_n and any term to the right of it in the sequence (i.e., t_{n+m} where $m > 0$) is less than the given rational number r.

For example, consider the "distinguished" sequence (1., 1.4, 1.41, 1.414, 1.4142, 1.41421, ...). If we take $r = .01$ as the given rational, then all the terms after 1.41 differ from 1.41 by less than $r = .01$ (e.g., $1.414 - 1.41 = .004$, $1.4142 - 1.41 = .0042$, and so on). If we take $r = .001$ as the given rational, then all terms after 1.414 differ from 1.414 by less than $r = .001$. If we take $r = .0001$, we can let $t_n = 1.4142$, and so on.

This distinguished sequence is clearly Cauchy convergent. Furthermore, it can be made to converge to a limit which is *not* rational. (In writing successive terms, we need to be sure that we do not get lazy and start repeating ourselves. Remember, all repeating decimals are rational numbers.)[6] In fact, this sequence can be made to converge to a point, which we may denote by $\sqrt{2}$. A quick check indicates that the squares of the indicated terms do get closer and closer to 2. Thus, $(1.4)^2 = 1.96$, $(1.41)^2 = 1.9881$, and so on.

(7) The distinguished sequence

(3, 3.1, 3.14, 3.141, 3.1415, 3.14159, 3.141592, 3.1415926, 3.14159265, ...)

is another convergent sequence which can be continued so as not to approach a rational number. In particular, it can be made to approach π, the constant used so widely in trignometry and geometry (e.g., in computing areas of circles, $A = \pi r^2$).

(8) The distinguished sequence

(1, 1.7, 1.73, 1.732, ...)

is another. It can be made to approach $\sqrt{3}$, which is not rational.

We can now *define* an equivalence class of convergent sequences of rational numbers. *Two convergent sequences belong to the same equivalence class if the difference sequence formed by subtracting corresponding elements in the two sequences approaches 0.* (Where the limit is a rational number, this is the same as saying that two sequences belong to the same equivalence class if they have the same limit.) In this sense, the sequences in Examples (1)–(4) and (5a)–(5e), respectively, belong to common equivalence classes. To see that sequences (5a) and (5b) belong to the same equivalence class, we form the difference sequence $\left(1, \frac{1}{2}, \frac{1}{4}, \frac{1}{8}, \ldots, \frac{1}{2^{n-1}}, \ldots\right)$ by subtracting the terms of (5a) from those of (5b),

---

[6] It is easy to represent a sequence in which the same string of digits repeat themselves. We simply repeat the string a few times to make sure that the reader gets the idea and then add the *ellipsis* "...". With nonrational decimals, there can be no repeating string.

term by term $\left(\text{e.g.}, \dfrac{3}{2} - \dfrac{1}{2} = 1, \dfrac{5}{4} - \dfrac{3}{4} = \dfrac{1}{2}, \dfrac{9}{8} - \dfrac{7}{8} = \dfrac{1}{4}, \text{ and so on}\right)$. Since the general difference term $\dfrac{1}{2^{n-1}}$ may be made as small as we like by taking n large enough, we see that this difference sequence converges to 0.

An important point that we accept without proof is that every equivalence class so defined contains a unique distinguished decimal sequence of the sort we have discussed. Conversely, every distinguished decimal sequence defines a unique equivalence class of convergent sequences. These statements become plausible by noting that every *distinguished decimal sequence converges*. To see this we need only observe that all such sequences satisfy the condition for Cauchy convergence. Thus, given any distinguished decimal sequence

$$(.a_1, .a_1a_2, .a_1a_2a_3, \ldots, .a_1a_2a_3 \ldots a_n, \ldots, .a_1a_2a_3 \ldots a_{n+m}, \ldots)$$

the difference

$$
\begin{array}{l}
 .a_1a_2a_3 \ \ldots \ a_na_{n+1} \ldots a_{n+m} \\
- .a_1a_2a_3 \ \ldots \ a_n \\
\hline
 \underbrace{.0\,0\,0 \ \ldots \ 0}_{n \text{ zeros}} a_{n+1} \ldots a_{n+m}
\end{array}
$$

is less than

$$\underbrace{.0\,0 \ldots 0}_{(n-1)\text{ zeros}} 1$$

Hence, by choosing n large enough we can make the difference as small as we like.[7] Finally, we remark that each distinguished decimal sequence can be unambiguously represented as an *infinite* (repeating or nonrepeating) decimal, and conversely each such decimal denotes a unique distinguished decimal sequence.

Several equivalence classes and their representations as infinite decimals are given in Table 9–1.

With this background, we can *define* the *real numbers* as *equivalence classes of convergent rational sequences*. Alternatively, the *real numbers* may be thought of as *distinguished decimal sequences*, or, still more simply, as *infinite decimals*. Thus, in the latter view, $.000 \ldots, .125000 \ldots, 827.173333 \ldots, 2.\overline{65265265} \ldots,$ $3.14159265 \ldots, 1.414214 \ldots,$ and so on are all real numbers.

Infinite decimals are useful in *representing* real numbers, but neither they nor the distinguished decimal sequences they represent can be used directly to define the operations of addition and multiplication. The sum of two distinguished decimal sequences, for example, is not *always* a distinguished decimal sequence.[8] Fortunately, these operations can be defined on pairs of equivalence

---

[7] An actual proof that the sequence converges would require more detail. This argument is merely a sketch.

[8] To see why this is so, it is enough to observe that adding the distinguished decimal sequences $\pi = (3.1, 3.14, 3.141, 3.1415, 3.14159, \ldots)$ and $\sqrt{3} = (1.7, 1.73, 1.732, 1.7320, 1.73205, \ldots)$ term by term gives $\pi + \sqrt{3} = (4.8, 4.87, 4.873, 4.8735, 4.87364, \ldots)$ which is *not* a distinguished decimal sequence. (Note that 5 changes to 6 in the sum sequence.)

Table 9–1 Equivalence Classes of Convergent Rational Sequences and Their Decimal Representations

Equivalence Class of Convergent Sequences (Distinguished decimal sequences are marked with *.)	Infinite Decimal Representation
$\left(1, \frac{1}{2}, \frac{1}{3}, \frac{1}{4}, \ldots, \frac{1}{n}, \ldots\right)$	
$\left(-1, \frac{-1}{2}, \frac{-1}{4}, \frac{-1}{8}, \frac{-1}{16}, \ldots, \frac{-1}{2^{n-1}}, \ldots\right)$	
$\left(\frac{1}{2}, \frac{1}{4}, \frac{1}{8}, \frac{1}{16}, \ldots, \frac{1}{2^n}, \ldots\right)$	
$\left(-1, 1, \frac{-1}{2}, \frac{1}{2}, \frac{-1}{3}, \frac{1}{3}, \frac{-1}{4}, \frac{1}{4}, \ldots,\right.$	0.00000 ...
$\left. \frac{2(-1)^n}{n} \text{ if } n \text{ is even and } \frac{2(-1)^n}{n+1} \text{ if } n \text{ is odd, } \ldots \right)$	
*(0, 0.0, 0.00, 0.000, ...)	

$\left(\frac{1}{2}, \frac{3}{4}, \frac{7}{8}, \frac{15}{16}, \ldots, \frac{2^n - 1}{2^n}, \ldots\right)$	
(1.1, 1.01, 1.001, 1.0001, ...)	1.0000 ...
$\left(\frac{3}{2}, \frac{5}{4}, \frac{9}{8}, \frac{17}{16}, \ldots, \frac{2^n + 1}{2^n}, \ldots\right)$	
*(1, 1.0, 1.00, 1.000, 1.0000, ...)	

$\left(\frac{6}{4}, \frac{9}{8}, \frac{12}{12}, \frac{15}{16}, \frac{18}{20}, \ldots, \frac{3n + 3}{4n}, \ldots\right)$	
*(.75, .750, .7500, .75000, ...)	.75000 ...
(.751, .7501, .75001, .750001, ...)	

(1.5, 1.42, 1.415, 1.4143, ...)	
(1.45, 1.415, 1.4145, 1.41425, ...)	$\sqrt{2} = 1.414213 \ldots$
*(1.4, 1.41, 1.414, 1.4142, ...)	

*(1, 1.7, 1.73, 1.732, 1.7320, 1.73205, ...)	$\sqrt{3} = 1.732 \ldots$
(2, 1.8, 1.74, 1.733, ...)	

classes (of convergent sequences) in a particularly simple way. Thus, *addition* is *defined* as term-by-term addition of *arbitrary* convergent sequences selected from the given equivalence classes. For example, to add the real .25000 ..., which can be represented by $(\frac{1}{2}, \frac{3}{8}, \frac{1}{3}, \ldots, (n + 1)/4n, \ldots)$, and the real .5000 ..., which can be represented by $(1, \frac{3}{4}, \frac{4}{6}, \ldots, (n + 1)/2n, \ldots)$, we simply add the corresponding terms, giving the sequence, $(\frac{1}{2} + 1, \frac{3}{8} + \frac{3}{4}, \frac{1}{3} + \frac{4}{6}, \ldots, (n + 1)/4n + (n + 1)/2n, \ldots)$ or $(\frac{6}{4}, \frac{9}{8}, \frac{12}{12}, \ldots, (3n + 3)/4n, \ldots)$. As we would hope, this sequence converges and, furthermore, is a representative of the equivalence class denoted by .75000.... Hence, according to this definition, .25000 ... + .50000 ... = .75000....

To see that these limits are what we say they are, we rewrite the general terms of these sequences as indicated.

(a) $\dfrac{(n+1)}{4n} = \dfrac{n}{4n} + \dfrac{1}{4n} = \dfrac{1}{4} + \dfrac{1}{4n} \rightarrow \dfrac{1}{4} = .25000$ ... as n becomes large without bound,

(b) $\dfrac{(n+1)}{2n} = \dfrac{n}{2n} + \dfrac{1}{2n} = \dfrac{1}{2} + \dfrac{1}{2n} \rightarrow \dfrac{1}{2} = .5000$ ... as n becomes large without bound,

(c) $\dfrac{(3n+3)}{4n} = \dfrac{3n}{4n} + \dfrac{3}{4n} = \dfrac{3}{4} + \dfrac{3}{4n} \rightarrow \dfrac{3}{4} = .75000$ ... as n becomes large without bound.[9]

A less arbitrary and sometimes more convenient way to add real numbers is to select as representatives the respective distinguished decimal sequences. In the above example we would add the corresponding terms of (.2, .25, .250, .2500, ...) and (.5, .50, .500, .5000, ...). This gives (.7, .75, .750, .7500, ...) which can be denoted .75000 ... as above.

Can you find the sum of the reals .333 ... (i.e., $\frac{1}{3}$) and $.\overline{571428}\ 571428$ ... (i.e., $\frac{4}{7}$)? (*Hint:* First use the sequences $(0, \frac{1}{6}, \frac{2}{9}, \frac{3}{12}, \ldots, (n-1)/3n, \ldots)$ and $(\frac{3}{7}, \frac{7}{14}, \frac{11}{21}, \ldots, (4n-1)/7n, \ldots)$. Then try (.3, .33, .333, ...) and (.5, .57, .571, 5714, ...). When are distinguished decimal sequences easier to work with?)

Fortunately (for us, because otherwise we should have to begin again), it can be proved that the sequences generated by this operation *always* converge and that the operation is well defined.

Multiplication over the reals may be similarly *defined* as term-by-term multiplication. Thus, with respect to the above example, we define the product, .25000 ... $\times$ .5000 ..., in terms of the sequence

$$\left(\frac{1}{2} \times 1,\ \frac{3}{8} \times \frac{3}{4},\ \frac{1}{3} \times \frac{4}{6},\ \ldots,\ \frac{(n+1)}{4n} \times \frac{(n+1)}{2n},\ \ldots \right)$$

which equals $(\frac{1}{2}, \frac{9}{32}, \frac{4}{18}, \ldots, (n+1)^2/8n^2, \ldots)$. Again, we are fortunate in that this sequence converges and belongs to the equivalence class, denoted .125000 ... = .25000 ... $\times$ .5000 .... To see that the limit of this sequence is .12500 ... (which corresponds to the *real* number denoted $\frac{1}{8}$) we rewrite the expression $(n+1)^2/8n^2$ as

$$\frac{(n+1)^2}{8n^2} = \frac{n^2 + 2n + 1}{8n^2} = \frac{n^2}{8n^2} + \frac{2n}{8n^2} + \frac{1}{8n^2} = \frac{1}{8} + \frac{1}{4n} + \frac{1}{8n^2}$$

As n becomes large, the terms $1/4n$ and $1/8n^2$ approach 0. So the limit of the sequence is $1/8 + 0 + 0 = 1/8$ or .125000 .... (This product can be found more directly using the distinguished decimal sequences. Show this.)

[9] The fractions $\frac{1}{4}$, $\frac{1}{2}$, and $\frac{3}{4}$ are used here to represent *real* rather than rational numbers.

Can you find the product of the reals, .333 ... and $\overline{.571428571428}$ ...?

Using only the machinery that has been developed, why can't we find the *exact* sum or product of, say, π and $\sqrt{2}$? (Do we know infinite sequences which have these numbers as their limits?)[10]

Subtraction and division may be defined similarly in term-by-term fashion, but it is simpler just to define them as the inverses of addition and multiplication, respectively. (Division by 0.000 ..., of course, is not allowed.)

As we might guess, the system of real numbers has most of the properties of the system of rational numbers. In particular, the system of real numbers is a *field* and, thus, has all of the field properties (see the previous section). Further-more, as with the rationals, a strict linear order relation may be defined for the reals. Unlike the rationals, however, the *set* of real numbers cannot be paired in one-to-one fashion with the natural numbers (or the integers, the positive ra-tionals, or the rationals). There are simply too many of them and, although both sets contain an infinite number of elements, the number of reals is qualitatively greater.

Although not critical for the elementary school teacher, the argument is fairly simple and is of particular interest. We outline the major ideas and leave the details as a valuable exercise in reasoning.

We first assume that the natural numbers can be put in one-to-one corre-spondence with the real numbers between 0 and 1. Then we show that this assumption leads to a contradiction and, hence, must be false. Finally, we reason that because the number of reals between 0 and 1 is qualitatively greater than the number of natural numbers, the set of *all* real numbers must also be qualitatively greater than the number of natural numbers.

If the real numbers between 0 and 1 could be put into one-to-one corre-spondence with the natural numbers, then, representing each real as an infinite decimal, we could pair those between 0 and 1 as follows

$$1 \longleftrightarrow R_1^{\#} = .a_{11}a_{12}a_{13}a_{14} \ldots a_{1n} \ldots$$

$$2 \longleftrightarrow R_2^{\#} = .a_{21}a_{22}a_{23}a_{24} \ldots a_{2n} \ldots$$

$$3 \longleftrightarrow R_3^{\#} = .a_{31}a_{32}a_{33}a_{34} \ldots a_{3n} \ldots$$

$$4 \longleftrightarrow R_4^{\#} = .a_{41}a_{42}a_{43}a_{44} \ldots a_{4n} \ldots$$

$$\cdot \qquad \qquad \cdot$$
$$\cdot \qquad \qquad \cdot$$
$$\cdot \qquad \qquad \cdot$$

$$m \longleftrightarrow R_m^{\#} = .a_{m1}a_{m2}a_{m3}a_{m4} \ldots a_{mn} \ldots$$

$$\cdot \qquad \qquad \cdot$$
$$\cdot \qquad \qquad \cdot$$
$$\cdot \qquad \qquad \cdot$$

[10] Fortunately, analytic expressions (i.e., formulas) exist for generating sequences that have these real numbers as limits. Using these sequences, we can generate other convergent sequences that represent the sum $\pi + \sqrt{2}$ and the product $\pi \times \sqrt{2}$.

Clearly, if we continue the list far enough we can find a real number between 0 and 1 to correspond to any particular natural number, m. We have *assumed*, of course, that we can also find a natural number which corresponds to any given real number between 0 and 1.

Consider, however, the real number

$$Z^{\#} = .z_1 z_2 z_3 z_4 \ldots z_n \ldots$$

where

z_1 differs from a_{11} (so that $Z^{\#}$ differs from the real $R_1^{\#}$ and, hence, cannot correspond to 1),

z_2 differs from a_{22} (so $Z^{\#}$ differs from $R_2^{\#}$ and cannot correspond to 2),

z_3 differs from a_{33} (so $Z^{\#}$ differs from $R_3^{\#}$ and cannot correspond to 3),

z_4 differs from a_{44} (so $Z^{\#}$ differs from $R_4^{\#}$ and cannot correspond to 4)

. .

. .

. .

z_m differs from a_{mm} (so $Z^{\#}$ differs from $R_m^{\#}$ and cannot correspond to m)

. .

. .

. .

We can continue this process indefinitely so that $Z^{\#}$ is different from every real number in the list. What natural number does $Z^{\#}$ correspond to? (Remember, we want a *one-to-one* pairing.) Because we have paired all of the natural numbers already, there is none. What does this say about our assumption?[11]

The fact that the set of real numbers can be put into one-to-one correspondence with the set of *all* points on the number line is a very important property of the system of reals, one which is unique relative to the other systems so far discussed.

Another very important property of the system of reals is that the system of rationals (integers, positive rationals, natural numbers) can be embedded in it. We should expect no surprises. The reals correspond to precisely those rationals (integers, positive rationals, natural numbers) we would expect them to correspond to. Thus, for example

---

[11] This type of argument has enabled mathematicians and logicians to prove a number of very impressive theorems.

natural numbers	positive rationals	integers	rationals	reals[12]
1	$\frac{1}{1} = 1.0$	+1	$\frac{+1}{1} = +1.0$	+1.000 ...
2	$\frac{2}{1} = 2.0$	+2	$\frac{+2}{1} = +2.0$	+2.000 ...
	$\frac{1}{2} = .5$		$\frac{+1}{2} = +.5$	+.500 ...
	$\frac{1}{3} = .333\ ...$		$\frac{+1}{3} = +.333\ ...$	+.333 ...
		−3	$\frac{-6}{2} = -3.0$	−3.000 ...
			$\frac{-4}{7} = -.\overline{571428}\ ...$	$-.\overline{571428}\ ...$

$$-\sqrt{2} = -1.4142\ ...$$
$$+\pi = +3.14159\ ...$$

An interesting and worthwhile way to review the basic ideas is to verify that addition and multiplication over each number system have been preserved with each extension introduced. This is true even where *intermediary* extensions exist, as between the natural numbers and the reals. Thus, for example, it is an easy task to show that the product of the real number corresponding to the whole number 0 with the real number corresponding to 1 is the real number corresponding to the (natural number) product $0 \times 1 = 0$.

Show that the sum of the *rationals* +2/1 and +1/3 corresponds to the sum of the corresponding *reals*, +2.000 ... and +.333 .... Do the same for the products. (In dealing with real numbers, we must be careful to choose representative sequences that we know how to deal with.)

Select a pair of systems that you have not already considered (e.g., the naturals and rationals, or integers and reals) and show that one can be embedded in the other. In doing this, first experiment by trying out particular pairs of sums and products (such as $0 \times 1 = 0$ above) in the systems you select. In general, each of the following number systems, (1) naturals, (2) positive rationals and integers, (3) rationals, and (4) reals, in turn can be embedded in any system with a higher designated number (e.g., the integers (2) may be embedded in the reals (4)).

EXERCISES 9–2

S–1. Give the first five terms of a distinguished decimal sequence that converges to each of the following:

a. 5

b. $\frac{3}{7}$

c. .4500 ...

d. .123123 ...

[12] Ordinarily, we omit the + sign in denoting *positive* integers, rationals, and reals.

S–2.   Give the first four terms and the general term, in fractional notation, of a sequence that converges to each of the following:

a. $\dfrac{1}{5}$

b.   2

c. $\dfrac{3}{7}$

d.   .4500 ...

S–3.   Separate the following list into sublists so that the sequences in each sublist belong to the same equivalence class of convergent sequences of rational numbers:

a. $\left(\dfrac{1}{20}, \dfrac{1}{200}, \dfrac{1}{2000}, \cdots\right)$

b. $\left(1, \dfrac{4}{6}, \dfrac{5}{9}, \ldots, \dfrac{(n+2)}{3n}, \ldots\right)$

c. $(.3, .33, .333, \ldots)$

d. $\left(\dfrac{11}{30}, \dfrac{101}{300}, \dfrac{1001}{3000}, \cdots\right)$

e. $(.02, .002, .0002, \ldots)$

f. $(.03, .030, .0303, \ldots)$

g. $(.32, .332, .3332, \ldots)$

h. $\left(\dfrac{1}{3}, \dfrac{1}{9}, \dfrac{1}{27}, \ldots, \dfrac{1}{3^n}, \ldots\right)$

S–4.   Indicate which of the sequences in Exercise S–3 are distinguished sequences.

∗S–5.   Find the general term of the sequences in a and d of Exercise S–3.

S–6.   If the terms of a sequence of positive numbers become smaller and smaller, must the sequence converge to zero? Explain.

∗S–7.   Show that the sequences $(\dfrac{2}{5}, \dfrac{5}{10}, \dfrac{8}{15}, \ldots, (3n-1)/5n, \ldots)$ and $(\dfrac{4}{5}, \dfrac{7}{10}, \dfrac{10}{15}, \ldots, (3n+1)/5n, \ldots)$ belong to the same equivalence class of convergent rational sequences.

S–8.   Which of the following infinite decimals represent rational numbers?

a.   .4444 ...   (same pattern continues)
b.   .004400440044 ...   (continues)
c.   .040040004 ...   (continues)

d. $.\overline{0004}$ ...
e. $.57\overline{4404}$ ...
f.   .484484448 ...

[S–9.]   Give two infinite decimals which repeat the same string of digits but which represent two different numbers.

∗[S–10.]   a.   Use the Pythagorean theorem to show that there is a point on the number line corresponding to $\sqrt{5}$.

b.   Show that $\sqrt{5}$ is not rational. (*Hint:* Consider the proof in this section that $\sqrt{2}$ is not rational.)

S–11.   Find the sum of the real numbers $1.\overline{0}$ ... and $.\overline{571428}$ ... by using the sequences $(2, \dfrac{3}{2}, \dfrac{4}{3}, \ldots, (n+1)/n, \ldots)$ and $(\dfrac{3}{7}, \dfrac{7}{14}, \dfrac{11}{21}, \ldots, (4n-1)/7n, \ldots)$.

S–12.   Find the sum of .333 ... and .666 ... by using the sequences $(.3, .33, .333, \ldots)$ and $(.6, .66, .666, \ldots)$.

S–13.   Prove for any real number with infinite decimal representation $.a_1 a_2 a_3 a_4$ ... that 0.000 ... $+ .a_1 a_2 a_3 a_4$ ... $= .a_1 a_2 a_3 a_4$ ....

S–14.   Use the definition for multiplication of reals in the text to find the following products:

  a. $(1.000 \ldots) \times (.\overline{571428} \ldots)$
  b. $(.5000 \ldots) \times (.\overline{571428} \ldots)$
  c. $(.5000 \ldots) \times (.3333 \ldots)$

[S–15.]   Prove that $1.000 \ldots$ is the multiplicative identity for the reals.

E–16.   Find a real number "between" the two given real numbers in the following:

  a. $.4545\overline{45} \ldots$ and $.4546\overline{45} \ldots$
  b. $.07007000700007 \ldots$ and $.07008000700007 \ldots$
  c. $.4545988 \ldots$ and $.4546088 \ldots$
  *d. $.a_1a_2a_3a_4a_5 \ldots$ and $.a_1a_2b_3a_4a_5 \ldots$ where $a_3 < b_3$ and there is an $i > 3$ such that $a_i > 0$.
  e.   What important property of the reals do your answers suggest?

M–17.   To understand the proof that $\sqrt{2}$ is irrational requires a good understanding of deductive reasoning, and may be too difficult for most elementary school children. By what other method could you help your students to be aware that there are points on the number line which do not correspond to any rational number?

3.   FURTHER EXTENSIONS

For some purposes, we need number systems that are even more general than those discussed so far. In business, for example, one typically needs to simultaneously represent quantities of a variety of different products. To consider one case, suppose a car rental agency handles three different models of Fords: Fairlanes, Thunderbirds, and Mustangs. From experience with car servicing and the like, the owner knows that:

   1.   He cannot rent a total of more than five Fairlanes and Mustangs at the same time or a total of more than six Mustangs and Thunderbirds.

   2.   He always rents two more Mustangs than Thunderbirds.

The profit per car is $50 per Fairlane, $75 per Mustang, and $100 per Thunderbird. The owner wants to know how many cars of each model he should rent at one time in order to maximize his profit. The solution to such a problem (where one exists) involves three numbers rather than one. In this case, the solution is to rent one Fairlane, four Mustangs, and two Thunderbirds at a time. (This solution can be arrived at by the methods of linear programming, but the solution procedure itself is unimportant for our purposes.)

   In representing such quantities, one uses n-tuples of numbers $(a_1, a_2, \ldots, a_n)$, which are called *vectors*. In the example above, we used vectors (with $n = 3$) whose components were natural numbers. Integers (see Chapter 8) are very similar to vectors (with $n = 2$) over the natural numbers.[13] The major difference is that integers involve equivalence classes. (Strictly speaking, natural numbers

---

[13] Addition of integers and addition of vectors ($n = 2$) over the natural numbers, for example, are defined in the same way.

are not allowed; the components of a vector must come from a *field* (e.g., the reals, rationals). This is important to insure closure in *scalar multiplication* (defined below). These components, however, may also come from fields which are not number systems.)

The primary operations on vectors are *addition* of vectors and *scalar multiplication*. Vectors are added component-wise. Thus

$$(a_1, a_2, \ldots, a_n) + (b_1, b_2, \ldots, b_n) = (a_1 + b_1, a_2 + b_2, \ldots, a_n + b_n)$$

For example, $(0, 1, 6) + (2, 3, 4) = (2, 4, 10)$.

Scalar multiplication, on the other hand, is the component-wise product of a vector with an element, called a scalar, from the field over which the vectors are defined. Thus

$$b \cdot (a_1, a_2, \ldots, a_n) = (b \cdot a_1, b \cdot a_2, \ldots, b \cdot a_n)$$

where b is a scalar. For example

$$\frac{1}{2} \cdot \left(\frac{3}{4}, \frac{2}{1}, \frac{7}{15}\right) = \left(\frac{1}{2} \cdot \frac{3}{4}, \frac{1}{2} \cdot \frac{2}{1}, \frac{1}{2} \cdot \frac{7}{15}\right) = \left(\frac{3}{8}, \frac{1}{1}, \frac{7}{30}\right)$$

where the vectors are defined over the rationals.

(Another binary operation is called vector multiplication, mapping pairs of vectors into vectors. Vector multiplication arises in physical situations such as fluid flow, electricity, and magnetism in physics, and leads to the study of what are called Lie algebras.)

One system of vectors over the real numbers, which is of particular interest, is the system of *complex numbers*. Motivation for the development of the complex numbers stems from the fact that there is no point on the real number line corresponding to $\sqrt{-1}$ (which we traditionally denote by i). To see this, recall that the product of a positive number with itself is always positive and that of a negative number with itself is also always positive. ($0 \cdot 0 = 0$.) These are the only possibilities, so there is no real number which, when squared, gives $^-1$.

The construction of the system of complex numbers from the real numbers is similar to our construction of the system of integers from the natural numbers but is even simpler. *Complex numbers* may be *defined* as ordered pairs (a, b), where a and b are real numbers. (The more usual notation is $a + bi$.) There are no equivalence classes involved. We *define addition* as in vector addition

$$(a, b) + (c, d) = (a + c, b + d)$$

and *multiplication* by

$$(a, b) \times (c, d) = (ac - bd, ad + bc)$$

For example, given the complex numbers $(3, 7)$ and $(2, 1)$, these definitions give $(3 + 2, 7 + 1) = (5, 8)$ as the sum and $(3 \times 2 - 7 \times 1, 3 \times 1 + 7 \times 2) = (6 - 7, 3 + 14) = (^-1, 17)$ as the product. Each of these pairs, of course, can be represented in the more usual form $a + bi$. For example, the product $(3, 7) \times (2, 1) = (^-1, 17)$ may also be represented as

$$(3 + 7i) \times (2 + i) = {}^-1 + 17i$$

In the system of complex numbers, so defined, the additive and multiplicative identities are $(0, 0)$ and $(1, 0)$, respectively. Verify this. The complex numbers have other properties in common with the reals and rationals, such as commutativity and associativity of addition and multiplication, the distributive property of multiplication over addition, and additive and multiplicative inverses. In short, the system of complex numbers is a *field*. Unlike the rationals and reals, however, the complex numbers do not admit a strict linear order. For example, there is no way to order complex numbers like $1 + i$ and $1 + {}^-i$, or $^-2 + {}^-i$ and $2 + i$ by extending the usual orderings on the number line.

Complex numbers cannot be adequately represented on a number line. For this, we need a (two-dimensional) *plane*. In plotting points it is convenient to introduce horizontal and vertical axes and to mark off a rectangular grid. Complex numbers are plotted by moving an appropriate number of units along each dimension. Thus, for example, the number $(3, 2) = 3 + 2i$ is plotted by moving 3 units to the right along the horizontal axis and 2 units up along the vertical axis. The complex numbers $(^-2, 1.4) = {}^-2 + 1.4i$; $(^-3, 0) = {}^-3 + 0i = {}^-3$; $(\sqrt{3}, 0) = \sqrt{3} + 0i = \sqrt{3}$; and $(0, {}^-2) = 0 + {}^-2i = {}^-2i$ are plotted in similar fashion (see Figure 9–2).

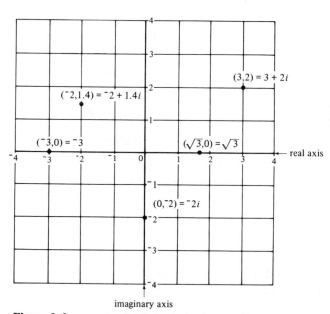

imaginary axis

Figure 9–2

To see that the reals can be embedded in the system of complex numbers, we need consider only ordered pairs of the form $(a, 0)$ where a is real. These complex numbers clearly correspond in one-to-one fashion to the reals. In other words, the real numbers correspond to those complex numbers which can be plotted along the horizontal axis. For this reason, mathematicans refer to this axis as the "real" axis. (The vertical axis is referred to as the "imaginary" axis.)

Addition and multiplication over the complex numbers correspond, respectively, to addition and multiplication over the reals. We leave the proofs as exercises and simply illustrate the correspondence with addition. For example, $(3, 0) + (2, 0) = (5, 0)$ over the complex numbers corresponds to $3 + 2 = 5$ over the reals.

Finally, we note that *all* roots (i.e., square root, cube root, etc.) are defined over the system of complex numbers and that all polynomial equations of the form

$$a_n x^n + a_{n-1} x^{n-1} + \ldots + a_1 x + a_0 = 0$$

have a solution (in fact, n solutions) over the complex numbers. (Neither statement is true over any of the other number systems we have discussed.) In particular, there is a complex number, $i (= 0 + i = (0, 1))$, which satisfies the equation $x^2 + 1 = 0$. To see this, we first compute $i^2 = (0, 1) \cdot (0, 1) = (0 \cdot 0 - 1 \cdot 1, 0 \cdot 1 + 1 \cdot 0) = (0 - 1, 0 + 0) = (^-1, 0) = ^-1$. Hence, substituting $x = i$ into $x^2 + 1 = 0$ gives $i^2 + 1 = ^-1 + 1 = 0$. Show that $x = ^-i$ also satisfies $x^2 + 1 = 0$.

EXERCISES 9-3

S–1. Perform the indicated operations with the vectors and scalers in the following:

a. $(3, ^-2, 4, 0) + (2, 1, ^-1, 8)$

d. $\left(\frac{1}{2}\right) \cdot \left[\left(1, 0, \frac{1}{2}\right) + (^-3, 2, 0)\right]$

b. $3 \cdot \left(4, \frac{1}{2}, ^-5\right)$

e. $0 \cdot (1, 3, ^-4)$

c. $2 \cdot (5, 3) + ^-3 \cdot (6, ^-1)$

f. $4 \cdot [^-2 \cdot (1, 4, 0, ^-1)]$

S—2. Compute the following sums and products of complex numbers. Represent your answers as points in the two-dimensional plane.

a. $(3, 4) + (^-5, 6)$

e. $(2 + ^-3i) \times (1 + ^-2i)$

b. $(2 + 3i) + ^-5i$

f. $(2, 5) \times (5, 2)$

c. $(1, 0) + (^-7, 0)$

g. $(a + ^-ai) + (^-a + bi)$

d. $(1, 0) \times (3, 4)$

S–3. Using the correspondence $a \rightarrow (a, 0)$ where a is real, show that the reals can be embedded in the complex numbers.

S–4. For the complex numbers show that multiplication is commutative, associative, and distributive over addition.

S–5. Find the additive and multiplicative identities of the complex-number system and show your answers are correct.

∗S–6. Since the complex-number system is a field, every non-zero complex number must have a multiplicative inverse. Find the multiplicative inverse of $(2, ^-4)$ by finding a complex number (a, b) such that $(2, ^-4) \cdot (a, b) = (1, 0)$.

E–7.   How would you define (in a natural way) equality of vectors? In particular, under what conditions would you say $(a, b, c) = (d, e, f)$?

E–8.   The *conjugate* of a complex number (a, b) is $(a, ^-b)$.

    a. What are the conjugates of $(3, 0)$, $(^-2, 5)$, and $(4, ^-7)$?

    b. Show that the sum and product of a complex number and its conjugate is "real," that is, that they are of the form $(a, 0)$ for some real number a. (*Hint:* Let (a, b) be an arbitrary complex number.)

For further thought and study: Two-dimensional vectors may be interpreted geometrically as directed line segments. The system of complex numbers under addition is isomorphic to a system of *equivalence classes of vectors* in a plane. Find a discussion of geometric numbers in an algebra II or college algebra textbook. In particular, try to determine:

    (A) how the equivalence relation is defined.

    (B) how addition of two equivalence classes of vectors is defined.

    (C) how one defines the one-to-one correspondence of complex numbers and vector classes.

    (D) if the isomorphism can be extended to include multiplication.

4. A BASIC RELATIONSHIP BETWEEN ALGEBRA AND GEOMETRY

Historically, algebra (which includes number) and geometry (the study of space) developed rather independently. The terms *algebra* and arithmetic have traditionally referred to those discrete entities and ideas (e.g., number) most naturally represented with *symbols*. *Geometry* has referred to continuously varying notions that are more easily visualized (represented) in *spatial* or iconic form.

Why these subjects went their separate ways for so long is impossible to answer definitively; it is likely, however, that any reasonable hypothesis must ultimately rest on the nature of "things," "actions," and "relations" in the real world. Observables differ in one of two fundamentally different ways. They may be *discrete* or they may vary "*continuously*."[14] Thus, each element in a set of chairs, automobiles, arrangements, or selections[15] is discrete from any other element in the respective set. The first two illustrations involve discrete things and the latter two involve discrete actions. Illustrations of entities which vary continuously, are equally profuse. Changes over time, for example, are assumed to vary continuously. The set of points on the number line or in "two space" (i.e., the plane) and even the continuous process of aging provide examples.

Although this distinction is still pedagogically convenient, it should be emphasized that the difference between algebra and geometry has become rather hazy at the frontiers of knowledge. In fact, the first major integration of algebra

---

[14] We use quotation marks to distinguish this use of the term "continuous" from its more technical meaning in mathematics.

[15] We assume that the sets are sufficiently small. If we allow enough different rearrangements or selections they may, indeed, vary "continuously."

and geometry occurred in 1637 when René Descartes published the first book on analytic geometry. Geometric motivation is still evident in many subjects, but algebraic tools (e.g., symbol manipulation) are used in almost all areas of mathematics. It is possible, for example, to use algebraic methods to solve geometric questions (and vice versa). During the twentieth century, the relationship has become so intimate at the higher reaches of mathematics that it is often difficult to distinguish algebra and geometry.

To get some idea of the basic nature of the relationship involved, we need only recall our use of the real numbers to represent the points on the number line. Because the real numbers can be defined ultimately in terms of natural numbers, we see that in this case it was possible to deal with continuously varying quantities (points) in terms of discrete entities (natural numbers). We were not able to accomplish this, however, without paying a price. Our construction of the real numbers from the rationals (and hence ultimately the naturals) involved the introduction of infinite sequences and *limiting processes*, ideas which lie at the foundations of *analysis*, more commonly known as "the calculus." It would appear, then, that the representation of continuously varying quantities, typical of geometry, requires at base reference to an infinite number of discrete (algebraic) entities.[16] Is it any wonder that algebra and geometry remained separate for so long?

---

[16] In order to more fully appreciate the importance of what is involved, it is instructive to consider a concrete illustration—for example, the problem of *measuring* the length of "continuously" varying quantities, such as sticks. In order to measure lengths, we must first choose an arbitrary *unit* length against which to compare the lengths. The problem is that a single unit will rarely suffice to determine the number of units between one end of a stick and the other (unless the stick is actually constructed to be so many units long). It would be a stroke of good fortune, indeed, if the end points were found to be *exactly n* units apart. Better approximations, of course, might be obtained by using a smaller unit or, equivalently, a number of different-sized units. For example, one unit might be meters, another, centimeters, and still another, millimeters. Unfortunately, even this will almost always still leave errors of measurement. Measuring the length of sticks with exactitude will typically require an *infinite* number of *units*.

Appendixes

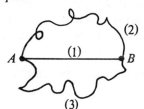

A Rule-Based Introduction to Geometry[1]

1. *Task:* Determine whether a given figure including points A and B represents a curve from A to B.

   Examples:

   a. A ⌢⌣⌢ B is not a curve from A to B.

   b. A ⌢⌣ B is a curve from A to B.

   Rule: Starting at A, trace along the figure toward B. If there are no gaps and the path ends at B, the figure is a curve from A to B. Otherwise, it is not.

2. *Task:* Given a curve from A to B, determine whether or not it represents the line segment $\overline{AB}$.

   Example:

   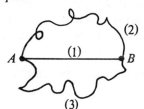 (1) is the line segment $\overline{AB}$.

   Rule: Align a standard straight edge (ruler) with both A and B. If the ruler coincides with the curve, then the curve is the line segment $\overline{AB}$; otherwise not.

[1] Many of these ideas are inherent in *S.M.S.G.*, *Studies in Mathematics: Vol. XIII: Inservice Course in Mathematics for Primary School Teachers*, Stanford, Calif., Stanford University, 1966, pp. 69–85, 183–197.

3. *Task:* Represent $\overline{AB}$ as the union of $\overline{AQ}$, $\overline{QM}$, and $\overline{MB}$.

   Example:

   a.
   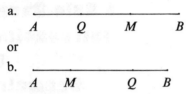

   or

   b.

   Rule: Mark points Q and M on segment $\overline{AB}$ between A and B. (Note: the union in example b involves overlapping sets (i.e., line segments).)

4. *Task:* For a given nonclosed simple curve passing through three points, determine which point is between the other two.

   Example:

   B is between points A and C.

   Rule: Start at either end point of the curve and trace a path along the curve. The second point along the path is between the other two.

5. *Task:* For a given line segment $\overline{AB}$ and points Q, M on $\overline{AB}$ between A and B (not necessarily in that order), state all possible relationships of one point being between two other points.

   Example:

   All possible relationships for line segment

   are: Q is between A and M and between A and B
   M is between A and B and between Q and B.

   Rule: Start at A, choose the first point along the segment in the direction of B. This point is between A and the second point along the segment in the direction of B; it is also between A and B.

   Next choose the second point along the segment in the direction of B. This second point is between A and B; it is also between the first point chosen along the segment in the direction of B and B itself.

6. *Task:* Given two points A and B, draw a figure representing $\overline{AB}$.

   Example:

   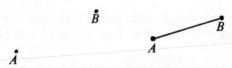

   Rule: Start at A and, using a ruler, draw the "most direct" path to B.

7. *Task:* Given two points A and B, draw a figure representing ray $\overrightarrow{AB}$.
   Example:

   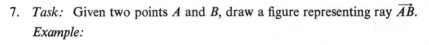

   Rule: Apply rule 6. Extend segment beyond point B. Attach arrow at end of extension pointing away from points A and B.

8. *Task:* Given two points A and B, draw a figure representing line $\overleftrightarrow{AB}$.
   Example:

   .   .

   A B A B

   Rule: Apply rule 7 and draw $\overrightarrow{AB}$ and $\overrightarrow{BA}$.

9. *Task:* For a given curve, determine whether it is simple.
   Examples:

   a.   is not a simple curve.

   b.   is a simple curve.

   Rule: Start at any point on the curve and trace along the curve in one direction. If a portion of the curve is crossed, then the curve is not simple. If an "end point" of the curve is reached, then return to the starting point and proceed in the direction opposite to the one originally chosen. If a portion of the curve is crossed, then the curve is not simple. If an "end point" of the curve or the beginning point is reached, then the curve is simple. (Do not start at an intersection.)

10. *Task:* For a given curve, determine whether it is closed.
    Examples:

    a.   is a closed curve.

    b.   is not a closed curve.

    Rule: Find a point on the curve that enables you to trace along the entire curve in one direction and return to the starting point without retracing. If you succeed, the curve is closed. If there is no such point, the curve is *not* closed.

11. *Task:* For a given curve, determine whether it is simple and closed.
    Examples:

    a. is simple and closed.

    b. is not simple.

    c. is not closed.

    Rule: Apply rules 9 and 10.

12. *Task:* For a given point in the plane and a given simple closed curve, determine whether the point is interior or exterior to the curve.
    Example:

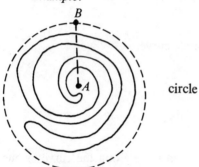

    circle

    The point *A* is exterior to the curve.

    Rule: If the given point is on the curve, then it is neither interior or exterior to the curve. If it is not on the curve, draw a circle enclosing the closed curve and select any point on the circle (e.g., *B*). Draw the line segment connecting the given point and the chosen point. Count the number of times the line segment crosses the curve. If the number is even, then the point is exterior to the curve. If the number is odd, then it is interior to the curve.

13. *Task:* Mark the interior of a given simple closed curve.
    Example:

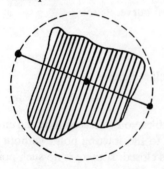

Rule: Draw a circle enclosing the curve and select any point on the circle. Draw a line segment intersecting the curve and connecting the chosen point with another point on the circle. Select any point between the first and second points of intersection of the segment and the curve. Starting at this point, shade the region containing this point, making sure not to touch the curve.

14. *Task:* Mark the exterior of a given simple closed curve.
    Example:

Rule: Apply rule 13. Mark the region not shaded and not including the curve.

15. *Task:* Given any planar figure, determine whether it is a polygon.
    Examples:

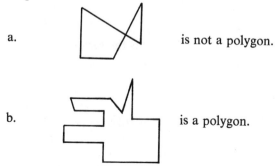

a.　is not a polygon.

b.　is a polygon.

Rule: Apply rule 11. If the curve is not "simple and closed," then it is not a polygon. If the curve is "simple and closed," apply rule 2 to each pair of consecutive vertices and determine if the curve is composed entirely of line segments. If it is, then it is a polygon; otherwise not.

16. *Task:* Given any planar figure, determine whether it is a triangle.
    Examples:

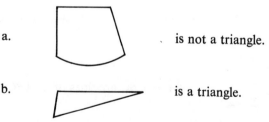

a.　is not a triangle.

b.　is a triangle.

Rule: Apply rule 15. If the figure is not a polygon, then it is not a triangle. If the figure is a polygon, then count the number of distinct segments of which it is composed. If the number is 3, then it is a triangle; otherwise not.

17. *Task:* Given any planar figure, determine whether it is a quadrilateral.

Examples:

a. is a quadrilateral.

b. 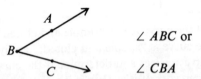 is not a quadrilateral.

Rule: Apply rule 15. If the figure is not a polygon, then it is not a quadrilateral. If the figure is a polygon, then count the number of distinct segments of which it is composed. If the number is 4, then it is a quadrilateral; otherwise not.

18. *Task:* Identify the vertex of a given representation of an angle.

Example:

Point *B* is the vertex.

Rule: Apply rule 4.

19. *Task:* Name a given representation of an angle.

Example:

∠ *ABC* or

∠ *CBA*

Rule: Starting at one of the two points other than the vertex, trace the path along the curve naming each point in the order in which it is reached.

20. *Task:* Name the sides of a given labeled representation of an angle.

Example:

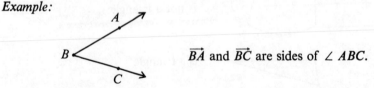

$\overrightarrow{BA}$ and $\overrightarrow{BC}$ are sides of ∠ *ABC*.

Rule: Use the vertex letter followed by one of the other letters to name one side and the vertex followed by the remaining letter to name the other side.

21. *Task:* Determine whether two given line segments are congruent.

Example:

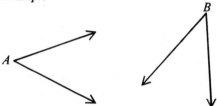

A

B   Lines *A* and *B* (*A* and *C*) are not congruent.

C   Lines *B* and *C* are congruent.

Rule: Trace one of the line segments, and place the tracing on top of the remaining line segment so that one pair of end points coincide. If the segments coincide, then they are congruent; otherwise not.

22. *Task:* Determine whether two given angles are congruent.

Example:

∠ *A* and ∠ *B* are congruent.

Rule: Trace one of the angles, and place the tracing on top of the other angle so that the vertices and one pair of sides coincide. If the other pair of sides coincide, then the angles are congruent; otherwise not.

23. *Task:* Determine whether two given triangular regions are congruent.

Example:

Triangles *ABC* and *DEF* are congruent.

Rule: Trace the triangle enclosing one of the regions. Select any angle of the remaining triangle and apply rule 22 successively using each of the three angles determined by the traced triangle. Flip the tracing over and repeat. If at any stage the triangles coincide, then they are congruent; otherwise not.

24. *Task:* Determine whether two quadragular regions are congruent.

Example:

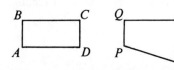

Quadrilaterals *ABCD* and *PQRS* are not congruent.

Rule: Trace the quadrilateral enclosing one of the regions. Select any angle of the remaining quadrilateral and apply rule 22 successively using each of the four angles determined by the traced quadrilateral. Flip the tracing over and repeat. If at any stage the quadrilaterals coincide, then they are congruent; otherwise not.

25. *Task:* Determine whether two given regions, determined by simple closed curves, are congruent.

Example:

a.

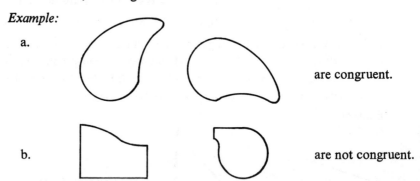

are congruent.

b.

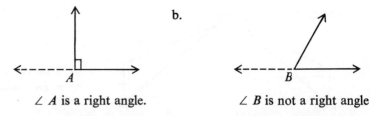

are not congruent.

Rule: Trace one of the regions, and place the tracing on top of the other region and rotate the tracing. Flip the tracing over and repeat. If the two regions can be made to coincide, then they are congruent; otherwise not.

26. *Task:* Determine whether a given angle is a right angle.

Example:

a.   b.

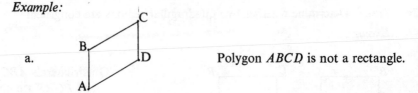

∠ *A* is a right angle.   ∠ *B* is not a right angle

Rule: Extend one of the sides through the vertex. Apply rule 22 and determine if the angle formed by the extended side, the vertex and the remaining side is congruent to the original angle. If it is, then the angle is a right angle; otherwise not.

27. *Task:* Determine whether a given polygon is a rectangle.

Example:

a.   Polygon *ABCD* is not a rectangle.

b.

Polygon *QRST* is a rectangle.

Rule: Apply rule 17 and determine if the polygon is a quadrilateral. If it is, apply rule 26 and determine if all four angles are right angles. If so, the polygon is a rectangle. If the process breaks down at any point, then the polygon is not a rectangle.

28. *Task:* Determine whether a given polygon is a square.

Example:

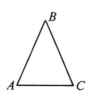

a.

Polygon *ABCD* is a square.

b.

Polygon *GEFH* is not a square.

Rule: Apply rule 27. Then apply rule 21 to each pair of sides. If the polygon is a rectangle (by rule 27) *and* the four sides are equal (by rule 21), then the polygon is a square; otherwise not.

29. *Task:* Determine whether a given triangle is isosceles.

Example:

Triangle *ABC* is an isosceles triangle.

Rule: Apply rule 21 to each pair of sides. If two (or more) line segments coincide, then the triangle is isosceles; otherwise not.

30. *Task:* Determine whether a given triangle is equilateral.

Example:

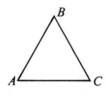

Triangle *ABC* is an equilateral triangle.

Rule: Apply rule 21 to each pair of sides. If all three lines coincide, then the triangle is equilateral; otherwise not.

31. *Task:* Determine whether a given triangle is a right triangle.

   Example:

   Triangle *ABC* is a right triangle.

   Rule: Apply rule 26 at each vertex of the triangle. If one of the angles is a right angle, then the triangle is a right triangle; otherwise not.

B

An Analysis
of 10
Elementary School
Textbook Series

INTRODUCTION

During the past 15 years, a wide variety of materials (and methods) have been developed for use in the mathematics classroom. Each set of materials has been associated with a rash of claims and other promotional devices. There has been little in the way of independent appraisal of these materials, let alone any independent attempt to identify their relative strengths and weaknesses. Without such information, and guidelines that might be drawn from them, school administrators and mathematics supervisors have had to rely entirely on their own judgments in selecting material. Although many schools have gone about this difficult task in an intelligent way, most selections have been based on inadequate information.

The general purpose of this appendix is to analyze those materials that are commercially available for use in the mathematics classroom. We have concentrated on materials prepared for use in grades K–6. The project on which this report is based was carried out in three phases: (1) identification and analysis of objectives, (2) evaluation, and (3) general appraisal.

Identification and Analysis of Objectives

The publishers (senior editors and authors) were contacted by letter and by phone to obtain a written statement of the objectives toward which the materials are directed. These statements of objectives were first analyzed to determine the degree to which they are operational (i.e., defined in behavioral terms). Second, the materials were analyzed to ascertain the extent to which one might reasonably expect them to achieve the stated objectives. Explicit attention was given to those objectives, both general and specific, that were identified by the publishers.

Evaluation

Relevant information was obtained from the publishers to determine the extent to which the materials were evaluated and the nature of these evaluations in both the pre- and post-publication periods.

General Appraisal

A subjective appraisal of the materials was also made to supplement the more objective analyses indicated above. In particular, we have commented on such things as (1) the degree to which the social applications of mathematics are emphasized and the nature and purpose of these applications; (2) the degree of emphasis placed on the study of mathematical structures, both algebraic and geometric, as opposed to a strict adherence to more traditional arithmetic; (3) the reading level and special vocabulary used in the children's texts; (4) the sort of teaching methodologies proposed by the authors in the teacher's edition; (5) the relative emphasis given to concepts and ideas as opposed to skills; (6) whether the materials make any provision for enhancing the child's ability to discover, or, to introduce him to the idea of proof, i.e., gaining knowledge by deduction; (7) evaluative devices provided for measuring changes in student achievement; and (8) the authors' style, typographical and graphic displays, and other noteworthy aspects of each series.

Scope and Content of Report

A rough estimate indicates that there are between 80 and 100 mathematics text series currently available for use in one or more of grades K–6. A complete analysis of all available materials, although an ultimate objective, is not possible at present. Instead, a review of 10 of the most widely known and used text series was undertaken.

This report consists of (1) a summary of the author-publisher's objectives and an evaluation of their mathematical objectives; (2) an attempt to determine the adequacy of the materials with respect to these objectives; (3) a summary and analysis of the extent to which the materials have been evaluated; (4) a general appraisal of the text series.

Two Views on Objectives of Mathematics Education

The past 15 years have seen much revision in the content of mathematics programs. The results have produced an impressive display of mathematics materials of better quality and more attuned to the mainstream of contemporary mathematics. Far less work has been done to clarify the changes in objectives toward which this revised content is aimed.

Particular emphasis at the elementary school level is being given not only to mathematical content but to teaching children to "think creatively," to "perceive structure," and to "make discoveries." No one questions the importance of these objectives. It is far more difficult, however, to define them satisfactorily.

This does not mean that we should abandon such general objectives; indeed, they should be at the core of curriculum planning. But the question of objectives has led to a good deal of controversy. Two fairly well-defined groups with opposing points of view have emerged.

One view supported primarily by mathematics educators can be summarized as follows: "It is useless to specify objectives since at this time only the most trivial objectives can be accurately specified." They defend this position with remarks such as, ". . . the ability to think mathematically is an important objective of mathematics education. It is also important for people to appreciate mathematics, to develop effective habits of thinking, and to appreciate the importance of deductive thought; and furthermore, we feel that we can make good judgments as to which kinds of instructional situations may be helpful in achieving these ends. Do not try to pin us down prematurely as to just what is involved."

On the other hand, many educational psychologists maintain that stating objectives is worthless unless they are stated unambiguously and in operational terms. Proponents of this view have organized curricula about such objectives as "counts orally from 1 to 10," "adds two numbers with sums up to 20 using expanded notation when required," "uses an inch ruler in measuring real objects or pictures to the nearest inch," and so on.

It is not our purpose here to resolve this problem. Our view is that complete reliance on operationally defined objectives has led some to a fragmented curriculum. On the other hand, we feel the nonobjectivists have not gone as far as possible in pinning down the vague aims of mathematics which they propose. (See Chapter 1 for further discussion.)

1. SUMMARY OF OBJECTIVES

We will first summarize the objectives as stated by the authors and/or publishers of each series. In doing this we form three classes: objectives held in common by all series, by a few series, and unique to one series. Within each of these classes we will distinguish between higher order and lower order objectives.

Higher-order objectives are those that we typically think of as being more abstract and more difficult to define. They include such things as to understand, to discover, to enjoy, to create, and are the sort of objectives cited by mathematics educators. Lower-order objectives consist of more specific, concrete kinds of tasks, which we think we understand better or at least are able to measure more easily. Examples include, to compute, to recite, to identify, to list.

These of course represent extreme points on a continuum. In a sense the fundamental difference lies in their behavioral aspects. In the lower-order objectives, a behavior is specified and thus one can give an operational definition of the objective. One can determine in an unambiguous manner what behavior is expected and whether it has been performed. There is less agreement as to the meaning of the higher order objectives, and consequently, what behavioral changes they are expected to produce.

Common Objectives—Lower-Order

It is surprising that among the common objectives in the elementary school mathematics programs under review, only one is stated in operational terms. This is the ability to compute, i.e., to perform the basic operations on the set of whole and positive fractional numbers. This objective is stated variously as "to use math skillfully," "skillful and rapid computation," "to develop and maintain arithmetic skills," and so on. Science Research Associates (Greater Cleveland Mathematics Program) is the only series that explicitly extends this objective (in grades K–6) to include the set of integers, although integers are briefly introduced in Grade 6 in several others.

Common Objectives—Higher-Order

The remaining common objectives are of the higher-order type, i.e., are more general in nature. They impress the uninitiated with their eloquence; closer analysis reveals them as being vague, intuitively defined objectives that are open to a variety of interpretations.

Most pervasive among these objectives are the ways in which students are to learn mathematics. Several terms appear with almost monotonous regularity, among which are the verbs to discover, to understand, to participate actively, to generalize, and the objects of these verbs—patterns, structure, and relationships.

Two additional objectives commonly stated are (1) to solve problems, including word and measurement problems and other applications of mathematics in the real world; and (2) to use technical terminology and symbolism. (A notable exception to this latter is Encyclopeadia Britannica Press's *Math Workshop*.)

The statements of these objectives might be summarized as follows: to develop in pupils an understanding of the structure, patterns, and relationships in mathematics; to develop this understanding by having the pupil actively participate in the learning process, discover, and make generalizations from his discoveries; to enable the pupil to express concepts using precise vocabulary, to solve problems, to relate mathematics to the real world, and to acquire arithmetic skills; and finally, to have him learn to appreciate, to enjoy, to like, and to be fascinated by mathematics.

Although such general objectives are important, for the most part they were stated in nonoperational terms. Consider as an example the objective "to understand structure." Neither the word "understand" nor "structure" was given a clear meaning.

What does "structure" mean for the authors? Are numeration systems structures? Is the commutative principle a structure? How about sets? Number sentences? Geometric figures? Equivalence classes of fractions?

What should the student learn when he studies structure? And how can we tell when he has learned it?

Mathematicians would probably agree that certain things are structures. It is far from certain, however, that the typical elementary school teacher has any idea about what a mathematical structure is. How then can we effectively teach structure?

Similar problems arise when we consider the meaning of the other general objectives cited. In all cases two crucial questions remain unanswered: What behavioral changes are expected? And how can we tell whether they have occurred?

Objectives Common to Several Series

We continue with our summary presenting next those objectives that are less frequently mentioned. In this and the next section we found that although certain objectives are not explicitly stated by a particular series, material included in their texts is directed toward them. These apparent differences reflect upon the statement of objectives—as poorly or inadequately written—or upon the lack of thought given to objectives themselves.

We list here in abbreviated form the eight kinds of objectives considered in this section (from lower-order to higher-order):

1. sets
2. measurement
3. number sentences
4. geometry
5. relationships between arithmetic and geometry
6. ways to solve problems
7. record mathematical ideas
8. nature of mathematics

Though set notions are fundamental to, and widely used by, most series in elementary school mathematics, only four series advocate their use explicitly. The clearest statements are given by Singer (*Sets and Numbers*), in which a rationale for the use of sets is given: "Sets are more concrete than numbers" and "operations on sets are more meaningful to the child. . . ." and some examples of the ways in which sets are used are discussed. Science Research Associates (*GCMP*) would have pupils learn basic set operations and ideas "in order to shed light on the basic arithmetic operations." A similar objective is voiced by Scott-Foresman (*Seeing Through Arithmetic*), "The notions of set, subset, and union and intersection of sets help unify the content and give clarity to ideas that are developed." In American Book Company (*Modern Mathematics Series*), "the language of sets is introduced when and if it is needed."

Four series state that measurement concepts are an important objective: American Book Company, Scott-Foresman, Addison-Wesley (*Elementary*

School Mathematics), and Webster (*Elementary Mathematics*). The statement from Webster is most explicit: "to solve problems involving measurement with both the metric and English systems." Addison-Wesley's is more typical: "Careful attention is given to measurement concepts;" and in American Book Company, ". . . the topic of measurement is carefully developed. Great care is taken to help the student discover and appreciate the nature and uses of measurement."

The ability to use equations or number sentences is singled out by American Book Company, Science Research Associates, Scott-Foresman, Silver Burdett (*Modern Mathematics Through Discovery*), and Singer. Singer's statement is most explicit. They point out that the use of letters as variables prepares the pupil for algebra and that "particular attention is given to number sentences, equations, and the translation of English sentences expressing quantitative relations into mathematical equations." Scott-Foresman's statement is more typical: "The use of arithmetic sentences to express mathematical ideas is a key idea throughout." A goal in Silver-Burdett is ". . . to provide a sustained use of equations and inequalities."

The inclusion of topics from geometry is cited as a major difference from traditional math programs by American Book Company, and mentioned as an objective by Addison-Wesley, Holt, Rinehart and Winston (*Elementary Mathematics—Patterns and Structures*), Scott-Foresman, Singer, and Webster. In Holt, Rinehart and Winston, "A consistent, intuitive development of basic geometric topics is provided;" in Singer, "The inclusion of a substantial body of content from geometry is introduced by work with simple geometric constructions in the first, second, and third grades." In Addison-Wesley a key feature is "a unique geometry of the real world. . . ." Similar statements are given by the remaining series.

A related objective in Silver-Burdett, American Book Company, and Holt, Rinehart and Winston is to show pupils the interrelationships between geometry, arithmetic, and algebra. By this they mean an introduction to coordinate geometry.

The ability to record mathematical ideas is cited by Harcourt Brace Jovanovich (*Elementary Mathematics*), Scott-Foresman, and is a major objective in Encyclopedia Britannica Press. This, along with the ability to see that there are many ways to solve a problem (Webster, Science Research Associates, Encyclopedia Britannica Press) are objectives which we feel reflect an essentially different emphasis from other series.

Finally, three series suggest that students should have a conception of the nature of arithmetic. Students should see mathematics as "a body of related concepts" (Addison-Wesley); or "as a study rather than a collection of separate topics" (Silver-Burdett); Encyclopedia Britannica Press gives the clearest answer to the question, "What is arithmetic?" They hope that children learn ". . . that everything that is said in arithmetic is a report of an experiment with things that either has been or could be carried out." They consider arithmetic to be essentially . . . "(1) one-by-one counting, (2) recording results, and (3) developing strategies and shortcuts to avoid one-by-one counting."

Objectives Unique to a Given Series

We conclude this summary by noting those general objectives stated by a single series only. We will include here only those unique objectives which are of a general nature, i.e., not highly specific like solve three-step word problems, although we again find that material directed toward these objectives may be included in other series. We will present these objectives by series in alphabetical order.

Only five of the 10 series being considered state unique general objectives. Addison-Wesley (*Elementary School Mathematics*) is concerned with an approach to problem solving. They hope that children "develop an attitude that each time they face a new problem, they do not feel the need to be taught a method of solving this problem, but that they can attack the problem bringing together a number of things they already know."

A related objective in Encyclopedia Britannica Press (EBP—*Math Workshop*) is to develop problem-solving ability but with a different interpretation of the word "problem." We cite at relative length these objectives, because EBP's approach is substantially different from all other series. They use the word "problem" in its "mathematical sense." "The question $8 + 5 = \underline{?}$ is an 'example' to anyone who knows that $8 + 5 = 13$; it is a 'problem' to a child who can count and understands the question as calling for an experiment he can carry out. It is nonsense if he does not understand what addition is about." They then hope to "replace the unmotivated drill of classical arithmetic by problems which illustrate new mathematical concepts" and to produce arithmetic skill as a "side effect" of the study of such mathematical concepts.

The child will study such concepts and problems through a discovery approach, which in EBP means with a minimum of verbalization. Pages in the text include enough (visual) clues so that "many children can work with full independence—once they realize that working from clues is the basic game we play in arithmetic . . . At the lower levels, in particular, it is best to rely almost not at all on verbal directions, oral or written . . . Math Workshop tries to present mathematical ideas with the fewest words possible. . . ."

Holt, Rinehart and Winston (*Elementary Mathematics—Patterns and Structures*) would have the pupil learn to be "self-reliant" and to be able to "uncover his own errors."

"Studying math materials" is an objective cited by Scott-Foresman (*Seeing Through Arithmetic*). We note parenthetically that they distinguish between two types of objectives, mathematical content and mathematical abilities. This is the only series to make this distinction.

A unique objective in Singer (*Sets and Numbers*) is for children to be able to give a precise answer to the question, "What is a number?" They say that "students can learn a clear, simple, and meaningful characterization of number as a property of sets."

Among the more specific objectives of Webster (*Elementary Mathematics*) we find the following: Pupils should have the ability "to explain the different situations which a fraction can suggest." They also encourage the ability "to think independently, not merely as the author thinks," "to check and verify answers," and "to select and evaluate good procedures."

This concludes our summary of what we feel to be the objectives stated by the authors and publishers. We wish to point out that each series states in its teacher's editions specific objectives for each lesson. Also available are scope and sequence charts which enumerate the topics covered. What we have been concerned with here are the over-all objectives which a series hopes to achieve in elementary school mathematics.

Following a brief evaluation of the statements themselves we will attempt to assess the degree to which the objectives are in fact realized by the materials.

Evaluation of Statements of Objectives

We find that most objectives are stated in vague, general terms, which require the reader's intuition to interpret them. Such statements reflect what we believe to be a general lack of understanding as to exactly what the objectives are.

Ambiguously stated objectives may tend to confuse the typical teacher about her role in mathematics education. Instead of clarifying and helping to focus attention on the desired goals, they obscure and often mislead. One consequence is that teachers faced with the task of teaching the "new math" have little idea of what it consists of and what its aims are.

We find ourselves in agreement with this description offered by Francis J. Mueller (*Mathematics Teacher*, November, 1967, p. 705):

> We, the missionaries and distributors of a product that knows no peer in precision and succinctness do indeed state our objectives—our goals—in the vaguest sort of rhetoric, couched in soaring platitudes, rich in fervor and zeal, *but utterly devoid of any measurable criteria* [emphasis ours].

He continued,

> Inevitably, we are going to be forced to decide just what it is the student is expected to learn . . . and how we can tell whether or not he has learned it.

This task has not as yet been performed.

2. DEGREE OF CORRESPONDENCE BETWEEN OBJECTIVES AND MATERIALS

To what extent might one reasonably expect the materials to achieve the stated objectives? The answer to this question depends very much on what the objectives mean and who is using the material.

Unless otherwise specified, we will assume that an "average" teacher is using the materials with an "average" class. We will further assume that this teacher is following the suggestions offered in the teacher's guide—though we feel that the ultimate results depend very much upon the teacher. A "good" teacher can get results even with "poor" materials.

Determining what the objectives really are is a difficult problem. Most objectives are not operationally defined, and we are in turn forced to interpret them according to our *own* intuitive conceptions. In effect, then, our conclusions consist of intuitive interpretations of intuitively stated objectives.

Correspondence Between Materials and Common Objectives

Because most series differ little in (essential) content, their primary differences lie in mode of presentation and degree of emphasis.

Under the conditions stated, computational skills with whole and fractional numbers should be achieved in all series. The treatment and attention given to computation in Science Research Associates (SRA) might be slightly more effective than in other series—especially with respect to rapidity and accuracy in computation, and Encyclopedia Britannica Press's (EBP) emphasis upon discovery and problem solving may be bought at the expense of developing less efficient computational skills.

On the other hand, one of EBP's strengths is its effectiveness in having pupils participate actively and discover patterns and relationships. This is not as successfully achieved in other series where there tend to be fewer real opportunities to discover and to generalize from discoveries.

Understanding structure is a particularly difficult ability to assess. A pupil who is able to state the commutative and associative laws but who cannot solve the problem $8\frac{2}{3} + 5\frac{4}{5} + 6\frac{1}{3} = $ _____ mentally is certainly not using these principles to shorten computation—and we seriously question whether he "understands" these principles. Insofar as structural properties are emphasized, however, we feel that Singer's *Sets and Numbers* presentation is most likely to encourage such understanding.

No series is particularly effective in *teaching* problem solving, although EBP definitely provides more opportunities to solve problems than most. At best a pupil will learn a technique for solving a certain class of problems—those which can be translated into or described by an equation with one unknown. Addison-Wesley (*Elementary School Mathematics*) best does the job of integrating mathematics with other subject matters and in this way, of relating mathematics to the real world.

Singer (*Sets and Numbers*) is most conscious of, and effective in, developing vocabulary, i.e., having pupils understand and use technical terminology to express mathematical ideas. On the other hand, we feel that excessive use of words (verbiage) in some series may tend to get in the way of ideas. This is especially true for those series in which much reading ability is called for. We note SRA and EBP for the lack of extensive verbal instructions in the early grades and Silver-Burdett (*Modern Mathematics Through Discovery*) for too much required reading.

We feel that objectives such as to appreciate or to enjoy are so vague as to make it impossible to judge the extent to which they are achieved.

We have singled out those series which we feel are more successful in achieving some of the general objectives stated. The remaining series realize these objectives to a lesser and largely indeterminate degree. Pupils using their material may learn "fancy" words and may be able to state mathematical properties, but they have relatively few actual opportunities to discover relationships, solve problems, and generalize patterns.

Objectives Common to Several Series

We turn now to those objectives stated by a few series. Here, too, objectives are vaguely stated and we cannot be certain what abilities students are expected to acquire.

Singer introduces the idea of a set and set operations earlier than most series and emphasizes their use at all grade levels. The tendency in the other series citing this objective is to introduce set ideas in the primary grades but to ignore them in the intermediate grades. Only in Singer do we feel that sets have become, to a limited extent, an object of study in and of themselves although, to be sure, their primary function is to lead to number concepts.

The objective concerning measurement is attained about equally by those series listing it. Addison-Wesley and Webster pay more attention to measurement—and to applications in general—so we feel they may have a slight advantage in this respect. American Book Company (*Modern Mathematics Series*) discusses with students the nature of measurement, but no more so than other series. These discussions center about the approximate nature of measurement, which we think students will grasp.

The use of number sentences receives much attention in solving word problems and in studying structural properties and relations between operations. Particular attention is given in Singer, Science Research Associates (SRA), and Scott-Foresman (*Seeing Through Arithmetic*) to translating English sentences into arithmetic.

Beyond the ability to identify figures and recognize some of their properties, it is difficult to determine what is expected of students with respect to geometry. Students will learn these things and in the case of Addison-Wesley and Singer, students will also learn how to perform some Euclidean constructions—but not why they work. We do not feel that pupils will appreciate the difference between metric and nonmetric geometry nor will they see geometry as an abstract model of physical space.

Several series, including Silver-Burdett, American Book Company, and Holt, Rinehart and Winston, introduce coordinate systems and the graphing of some linear equations by Grade 6. However, the relationship between geometry and arithmetic is not emphasized in these activities, and we conclude that this objective may not be realized.

What is a "mathematical idea" and how do we "record" it? This ability is emphasized in Harcourt Brace Jovanovich and Scott-Foresman to the extent that pupils are required to read graphs and use number sentences in solving word problems. Students using Encyclopedia Britannica Press (EBP) materials record data from numerous mathematical "experiments"—and we think that EBP is the only series that really develops this ability in students.

The ability to see that there are many ways to solve problems is achieved by those series stating it as an objective. In SRA, students are encouraged to do problems in any way they can. A similar emphasis is inherent in the discovery approach used by EBP. Cartoons are often used in Webster to show different ways of solving problems.

Children using EBP materials cannot avoid being impressed with arithmetic as a real "science" concerned with experimenting, recording results, and devising ways of avoiding one-by-one counting. This conception of arithmetic is quite different from those implied by other series. EBP is the only series which we feel will convey to pupils an over-all impression of the nature of arithmetic.

Correspondence with Respect to Unique Objectives

As we turn to the unique objectives we still find ourselves in the position of having to interpret intuitively stated objectives.

We do not see that Addison-Wesley develops in children an attitude toward problem solving that is essentially different from most other series. It is not clear what they mean by "a new problem" and we feel that "bringing together things that they already know" is necessary for all problem solving.

Perhaps Encyclopedia Britannica Press (EBP) is most successful in achieving its objectives. The format of their text pages forces one to take the attitude they suggest—namely, here is a problem, think about it, what can you say? They emphasize mathematical concepts in the context of problem solving far more than other series; they discourage verbalizing discoveries, because they feel this may hinder generalizing; and they provide little in the way of explicit drill but present problems that may require a great deal of computation to find the solution.

There are no special provisions in Holt, Rinehart and Winston (HRW) to teach students to "be self-reliant" nor to "uncover their own errors." As in other series, checking work is taught but we do not feel that it receives particular emphasis.

It is interesting to note that Scott-Foresman (STA) singles out the ability to study mathematical materials as an objective, which we interpret to mean the ability to read mathematics. It is less fortunate that this objective is not promoted in the text materials.

We think that children will be able to answer the question, "What is a number?" in the way suggested by Singer. This is accomplished by the development of number concepts, working from concrete sets to the N-notation to symbolic representation. This development is not atypical of other series but is emphasized more in Singer.

Webster's objectives are stated in relatively operational terms, i.e., (occasionally) they specify the behavior expected. One example of this is their objective "to explain different situations which a fraction can represent," which we feel students will be able to do.

We also feel that they accomplish the other goals specified. They do this largely through illustrations on many student pages in which different approaches to solving problems and ways of verifying answers are presented. They do not insist that any one way be followed but encourage discussion of the relative merits of each possible method.

We conclude this section with some general remarks. First, our estimation of whether particular objectives are achieved is based mainly on our interpretation of the statements of objectives. Second, we have not discussed every objective but only those which we feel reflect the over-all objectives of elementary school mathematics as cited by the authors. Finally, we wish to note that probably all of the objectives cited for one series or another would be acceptable to most authors. And, in fact, they tend to be achieved about equally in each series, with the relatively few exceptions noted above.

3. EVALUATION BY PUBLISHERS AND AUTHORS

We now examine the methods employed to evaluate the elementary school mathematics materials. As already indicated, most objectives are stated vaguely and nonoperationally. They refer to higher-order learning and do not specify the changes in behavior that are to be expected, i.e., what a pupil should be able to do as a result of "understanding mathematics."

Under these conditions no program of evaluation can hope to determine successfully the extent to which such objectives are being achieved.

Summary of Evaluations

Information about evaluation is available for six series. They show that two basic methods of evaluation were employed. Four of the six have used informal techniques that rely largely on subjective teacher comments. The other two have conducted formal testing programs with achievement tests.

Informal Evaluation

Addison-Wesley—*Elementary School Mathematics*

The Addison-Wesley series has its roots in experimental work performed by the authors at the Burris Laboratory School of Ball State University. Some of the material incorporated into the series was tested in this school. Other material used was tested under the Greater Cleveland Mathematics Program with which three of its authors were associated. No other prepublication evaluation was undertaken.

The chief source of evaluation at present is the subjective judgments of teachers, whom the authors met at workshops and conferences. Based largely on these judgments, a revised series was published in 1968.

Encyclopedia Britannica Press—*Math Workshop*

The Encyclopedia Britannica Press is primarily the outgrowth of an in-service program taught by the authors. Classes in this program were built around worksheets which each participating teacher tried out with his class. Teachers reported their results, modifications were made, and new activities were developed. A complete six-grade program, based largely on these materials, was prepared. The authors visited classes where the program was being tried; a revised program was published in 1962.

On the basis of direct classroom experience and teacher feedback, a further revision was begun, resulting in the publication of a revised program in 1967. Teachers who had used the program materials for at least two years were asked to answer the following questions: (1) What pages are excellent and why? (2) Which pages should be discarded; (3) What learnings need further reinforcement; (4) How could pages better provide for individual differences? The answers to these questions and additional suggestions from salesmen were taken into account in making this revision.

Silver-Burdett—*Modern Mathematics Through Discovery*

Plans for the present Silver-Burdett edition were conceived in 1958. At that time a research project was undertaken to evaluate the strengths and weaknesses of the older S-B series, *Making Sure of Arithmetic*. This project involved eight schools in New York, New Jersey, and Indiana and took place over the academic year 1958–1959.

Work was then begun on texts for grades 1 and 2 with experimental teaching of selected lessons and key ideas in schools in New Jersey and Illinois. These informal tryouts were conducted by the regular classroom teachers who had traditional training. Ideas found effective in the traditional series were retained in the new one. Based largely on the teachers' reports, revisions were made and the texts published.

Texts for Grades 3–6 also incorporated successful ideas from the older series and contained new ideas that had been evaluated in experimental programs such as the Greater Cleveland Mathematics Program.

No standardized tests have been used to evaluate the series since publication. Reports requesting information as to the performance of pupils on standardized tests have been solicited. Teacher comments and an analysis of student workbooks collected from selected schools formed the basis upon which a revision of grades 1 and 2 was made in 1964.

Consideration is being given to the preparation of standardized tests with norms and to a more formal approach to evaluation. At the time this report was prepared, the informal procedures described above were being continued.

Webster—*Elementary Mathematics—Concepts, Properties, and Operations*

Webster's present series was conceived in 1962 during a revision of *Exploring Arithmetic*, the forerunner of the series. Most of the manuscript was completed by early 1966. The present series incorporates the pedagogy used in the earlier series.

No formal controlled experimentation was conducted. Class tryouts, however, were performed by the authors. Some testing of ideas was also conducted at the University Elementary School, University of Iowa.

No objective evaluation has been conducted since publication. Subjective evaluation has been obtained from the authors and from some elementary school teachers. The teachers involved were selected because they had some knowledge of mathematics. The question of primary interest to the publishers was, "Could the pupil page stand on its own as far as clarity was concerned?"

Analysis of Informal Evaluations

The informal evaluations rely on (subjective) teacher comments as their major source of information. Although these comments are often obtained from large numbers of teachers who are distributed over large geographical regions, we feel that this method is open to serious criticism.

The fundamental flaw is the uncertainty concerning what is being measured. Student achievement, both in skills and concepts, is measured indirectly at best.

Such methods measure the "teachability" of a series or what might be called teacher likes and dislikes. These are important factors but we have no indication that teachers are competent judges of mathematical content, of teaching methodology, or of the extent and nature of the mathematics learned even in their own classes. There is, in fact, incomplete agreement on these points among professional mathematics educators.

Our purpose is not to be critical of teachers but to point out that they are being called upon to perform tasks for which they are inadequately prepared and not trained.

To be sure, evaluations by teachers provide useful information. It may be, as some contend, that they are the best means available today to measure such things as understanding. But we feel they often do not succeed even here. Teachers are obliged to make subjective judgments that are necessarily based on varying concepts of what constitutes understanding. The evaluations are further confounded by the possibility that the teachers themselves may not fully understand the material they are teaching.

We note too, that these methods have essentially ignored recent developments in the field of mathematics education which are aimed at clarifying such issues.

Summary of Formal Evaluations

The basic formula employed in the two formal programs of evaluation was to administer a specially prepared achievement test and a standardized achievement test to a sample of pupils who have used the materials being evaluated. The achievement test measures student mastery of all the material and the standardized test measures achievement on traditional content. The standardized test also enables comparisons to be made with traditional or other programs.

Singer—*Sets and Numbers*

During 1962–1963 the Singer evaluation program (for Grades 1–3) involved approximately 300 classes at various grade levels in the San Francisco Bay area.

Each of these experimental classes was given an achievement test at the completion of each book in the series.

In addition, classes were matched by administrators within each district on the basis of known variables such as student ability, socioeconomic level of neighborhood, and others. Children in control classes had been in traditional programs. The Metropolitan Achievement Test, a standardized achievement test, was given with the result that children in the experimental classes did as well or better on traditional content as those in the traditional programs.

Science Research Associates—*Greater Cleveland Mathematics Program*

The extensive evaluation of the materials published by Science Research Associates was all undertaken by the Educational Research Council of Greater Cleveland as part of its role in developing these materials. The tests used were the Greater Cleveland Mathematics tests, designed to measure achievement on the SRA materials; The Stanford Arithmetic tests (standardized tests which focus on traditional material), and the Lorge-Thorndike Intelligence test which is used to measure IQ.

Approximately 3000 pupils in over 22 school districts in the Cleveland area participated in the testing program in 1962–1963 and again in 1963–1964.

Among the major questions to be answered based on the results of these tests were:

1. How well do pupils perform on various tests after they have gone through the GCMP materials?
2. In what way is IQ associated with pupils' performance?
3. How well do the pupils perform in areas emphasized in the conventional standardized tests?

The general conclusions reached were (1) most of the materials covered by the tests were well mastered by the majority of the pupils; (2) the GCMP materials can be taught reasonably well to pupils of various intelligence levels; (3) the grade equivalents of participating pupils were all above the grade placement at the time of testing.

Analysis of Formal Evaluations

Formal evaluation programs have two distinct advantages. First, they employ objective tests which, provided they are adequate and properly administered, guarantee that the specified objectives will be measured. Second, both the achievement and standardized tests focus attention on the pupil and what he has learned, as opposed to teachers' opinions of the same.

Conclusions with respect to specific skills can be obtained with a high degree of confidence. The standardized tests, which have been in use over a long period of time, provide a reliable basis from which comparisons with traditional programs can be made—at least on a normative basis.

The same cannot as yet be said for these tests when used to measure such intangibles as pupil understanding. This is due not only to inadequacies in the tests themselves but also because such concepts are not yet well defined. In this area, we must still rely, for the most part, on subjective judgments rather than empirical facts.

4. GENERAL APPRAISAL OF TEXT SERIES

With the realization that pupils are capable of learning more mathematics and more sophisticated mathematics earlier than was formerly believed possible has come a host of revisions in the elementary school mathematics curriculum. What criteria should be used to determine whether a topic should be included or omitted? What choices (decisions) have, in fact, been made? The purpose of this analysis is to answer the latter question.

The particular aims of this section are to indicate significant character-istics of each series; to note those areas in which major or no differences occur; and to provide you with guidelines and information from which you can proceed to examine in greater detail those materials which seem likely to fulfill your requirements.

To accomplish these aims each series has been analyzed with respect to eight issues which reflect the recent curriculum revisions and which are recog-nized as being crucial in mathematics learning today.

The Issues

Social Applications: What is the role of social applications in elementary school mathematics? How much emphasis is placed on applications and what is the nature and purpose of these applications? Positions on this issue may vary from, "Social applications are the only reason for learning mathematics (they should accordingly occupy a major role)," to "Applications tend to obscure the structure and logical nature of the subject and should be considered as of secondary importance."

Structure: What is the role of mathematical structures in elementary school mathematics? To what extent does the study of mathematical structures increase understanding and appreciation of the subject? To what degree is the study of mathematical structures emphasized as opposed to the more traditional arith-metic?

Geometry: What kind of and how much geometry is presented? What is the purpose of geometry in the curriculum? Topics in geometry have gained recent admission in elementary school programs but there does not as yet appear to be a clear consensus as to what its nature or role should be. Positions vary from, "It is a 'fun' topic and consequently serves primarily as a diversion from arithmetic," to "It is an integral part of the program, though not on an equal footing with arithmetic." This is an area which clearly requires further investigation and clarification on the part of mathematics educators.

Vocabulary: To what extent is technical vocabulary and notation used in elementary school texts? Are pupils required to use technical vocabulary as well as to recognize it? To what extent does successful use of the text depend upon reading ability? The fundamental question here is, "Do words get in the way of ideas? Do words hinder students or enhance their ability to learn? And closely related to these questions is another: At what level should terminology be introduced and to what extent?

Methods and Teacher's Edition: What sorts of teaching methodologies are proposed? Are particular topics or techniques emphasized? Is there a recommended sequence of topics or time schedule? Is special teacher training necessary? It is not clear if one learns best from lectures and demonstrations or from exploration and experimentation. The appropriateness of each technique probably depends on the particular context. The organization of text materials, however, may be oriented to make one method more efficient than others.

Although the basic question is concerned with methodology, it was felt that it would be helpful to also provide information as to the nature of the teacher's edition and the extent to which the teacher is dependent on it. What information is provided and in what format? Is it useful? To what extent is the pupil's text independent of the teacher's edition from the point of view of the teacher?

Concepts versus Skills: To what extent is arithmetic viewed as a tool subject? To what extent are concepts emphasized as opposed to skills? Some feel that the ability to learn concepts and ideas depends on the prior mastery of skills. Others suggest that understanding must precede use. Related to this issue is the question: What are the minimal acceptable levels of competency in arithmetic in today's society?

Proof: Are students encouraged to generalize? Is there an attempt to teach logic formally? Is it made clear that induction is not proof, i.e., that demonstration of many instances of a mathematical generalization does not prove it? Are counterexamples used? Formal proof plays a small role in the elementary school curriculum. To what extent are pupils prepared for this notion which they will face in their later studies?

Evaluation: Are materials provided to measure changes in pupil behavior, i.e., the extent to which skills and concepts have been acquired? What is the nature of such materials? Are there provisions for students to acquire the habit of self-evaluation? Equally important but beyond the range of this analysis is the problem of evaluating long-term changes, i.e., the cumulative effects that take place over periods of years rather than those that occur within a year or semester.

General Findings

Social applications appear in two basic forms in elementary school mathematics: word or story problems and measurement. The tendency is for the word problems to be brief and to consist mainly of practice of particular arithmetic skills. A common technique for solving word problems is to identify the action taking place (as "putting together" or "separating") and to form a number sentence that describes this action. Many series pay particular attention to translating from English to arithmetic. This approach to problem solving is limited in scope; such techniques are applicable to a restricted class of problems. They are not, for example, applicable to geometry or to more complicated situations. In addition, most of the problems tend to be artificial and do not really relate arithmetic skills to ordinary daily situations. They are also unrealistic in that they are not open-ended, as are most important problems in real life.

All series cover essentially the same content with respect to structure—including such things as the commutative, associative, distributive laws and properties of zero and one. These topics are typically introduced and formally defined by grade 3. In the primary grades the emphasis is on the set of whole numbers; in the intermediate grades it shifts to the set of fractional numbers. Integers (negative numbers) are typically briefly introduced in grade 6. Additional structural properties include numeration systems, functions, relations, and number theory.

Structural properties are often presented in a somewhat fragmented setting. For example, operations are related to each other (inverses) but are not compared with respect to a given structural property, e.g., addition is commutative but subtraction is not. There is a tendency to "unify" these properties in grade 6.

Geometry usually begins with the study of points and lines in grade K and works up to plane and three-dimensional figures. Extensive use of concrete objects is recommended in the early grades. Constructions appear at various grade levels. Both nonmetric properties (e.g., interior-exterior, open-closed) and metric properties (length, area) are included. A major emphasis is placed on recognition and classification of polygons and many diversionary topics are included, e.g., symmetry, paper cutting and paper folding. There does not seem to be any unified approach or emphasis. This subject tends to be presented in a somewhat fragmented way.

Technical vocabulary appears in all series but there is considerable variability in emphasis. At one extreme all concepts are defined formally from the start and children are expected to use these terms; at the other end, all verbalization is minimized. Reading ability, especially in the intermediate grades, is usually a necessary skill. Most series encourage students to express ideas, although not always in technical language, and this ability is often used as a measure of understanding. There is great variability in this area, and the results probably depend more upon the teacher than on any other factor.

Most series claim to teach by discovery, which usually means by presenting a carefully composed sequence of questions which lead the student to the main idea. Teachers are advised to lead, to guide, to discuss—but never to lecture. Four standard techniques are used to learn computational skill and concepts: sets, number sentences, the number line, and arrays of various sorts. Sets tend to be discarded in the intermediate grades. The trend is to use number sentences including inequalities at all levels and in many different situations.

Provisions for slow learners consist mainly of graded problem sets, additional drill problems, and the omission of topics. Supplementary activities are all enrichment for faster learners. With a few notable exceptions there are few observable differences in recommended methodology.

Teacher's editions usually contain, in addition to some philosophical remarks and an overview of the program, lesson-by-lesson notes that indicate the purpose of the lesson, necessary mathematical background, and suggestions (often quite general) for teaching the lesson. Supplementary problems may be included, as well as prebook activities for each lesson, a glossary of terms, tables of measures, and other miscellaneous information. Most are easy to read and provide useful mathematical information, but often the methodology proposed is not incorporated into the pupil's text.

Most series emphasize concepts (i.e., understanding) as prerequisite to the acquisition of skills. They are all oriented toward concepts in this sense, although a few come close to achieving something of a balance between skills and concepts. A common practice is to include supplementary practice sheets (of drill problems) at the end of the text (the teacher refers students who need more practice to them) and fact tests interspersed throughout the text to keep children in "shape."

Few situations are presented in which students are called upon to make independent generalizations or to draw conclusions. These occur mainly in often optional enrichment activities. On the other hand, many concepts are taught by the presentation of examples of a concept prior to naming or explaining the concept. In this way pupils may generalize an idea before they are told it, but there is no way of telling whether or not this occurs. It is often forestalled by having the teacher state the concept immediately after working the examples.

In a few series elementary formal logic is taught and most series include "reasoning exercises" of various sorts that call for making simple deductions. These, however, appear infrequently and there is no systematic approach to the presentation.

Chapter or unit tests are provided by most series, usually containing problems similar to those in the text. These tests measure mainly computational ability. Diagnostic aids to evaluate them usually consist of page references indicating where the topic was originally taught. The ability to discover, to understand, to reason, to solve problems, so highly touted as objectives, are hardly ever measured—and their achievement is open to question.

Analysis of Individual Series

The preceding is a brief summary of the general findings over all series. In the discussion that follows, an entry of "standard" indicates that the series in question does not differ in any significant way from the description presented above. We shall concentrate in this discussion on features which tend to "stand out." The results of this analysis are presented in abbreviated form in the table on pages 384–391.

Addison-Wesley—*Elementary School Mathematics*

Addison-Wesley contains more than the average amount of social applications, with special emphasis on integrating mathematics with other areas in science. One chapter in each text is devoted to measurement, and the metric system is introduced in grade 1.

Structural principles are emphasized by comparing different number systems (e.g., positive integers and rationals), and by summarizing these principles in tabular form. Operations are not, however, compared with respect to structural principles. An interesting feature is the introduction of the function concept in grade 3 (which is much earlier than usual).

The aim with regard to geometry is to develop a good general feeling for the subject, which is thought of as a model based on the real world. The approach is largely intuitive and informal, with one geometry activity at the conclusion of each chapter. Many paper-folding activities are included. Mastery of facts is not demanded.

The use of technical terminology is not strongly emphasized, although students are encouraged to express their ideas when possible. Required reading is somewhat less than average and should not handicap pupils. Short story problems are very short and the longer ones contain many interesting (and usually scientific) facts.

The teacher's edition is standard in format. In addition to the mathematics background provided in the lesson notes, an in-service manual in elementary school mathematics is appended to each teacher's edition. No particular teaching methodology is strongly advocated. The notes offering teaching suggestions are brief general remarks and are devoted to giving the teacher the necessary mathematical background. They serve only as rough guides. A typical suggestion is to have the students do some problems and discuss what they did. In general, the teacher leads the pupils to the ideas, except in exercises designated for discovery enrichment.

One chapter in each of the intermediate grades is devoted exclusively to computational work. This series is concept-oriented, however, in the sense that concepts are considered prerequisites to the learning of skills.

Discussion exercises in the texts ask students to make simple deductions. No explicit attempt is made to study formal logic.

No pages are designated as test pages. The authors feel that daily subjective evaluation on the part of the teacher is the best available means of evaluating student performance. Review pages are provided, however, which may be used as tests. Exercises on these pages are similar in format to problems on the regular pupil pages.

American Book Company—*Modern Mathematics Series*

As in several other series the basic tool used in problem solving is the equation. Problems aimed at increasing skill in translation are presented. Most applications consist of routine word problems in which an arithmetic skill is practiced. The metric system is introduced briefly in grade 6 but an entire chapter in that grade is devoted to ways of representing data, e.g., graphs, tables, and so on.

Structural properties are introduced and used in the primary grades but not named until grades 3 and 4. Extensive use is made of the number line and equations. The emphasis on unifying knowledge in grade 6 includes a good review of the structural properties of the sets of whole numbers and positive rational numbers, and a brief introduction to the set of integers. Structural properties are used mainly to increase computational ability.

Geometry is approached intuitively with little emphasis on technical terminology. A spiral approach is used so that topics are reviewed and restudied at each grade level. This results in covering somewhat less material than usual.

Technical vocabulary is not stressed but is used. Few situations are presented which require students to use technical terms, although they are encouraged to express ideas. The amount of required reading increases with grade level but should pose no obstacle to the average student.

The teacher's edition contains overprinted notes on each student's page and more detailed remarks in a 120-page insert at the end. In addition to an overview of the program, lesson notes contain the major emphasis for each lesson, a vocabulary list, and suggested teaching procedures. In striking contrast to other series, comparatively little material designed to give the teacher additional mathematical background is provided. Remarks in the teaching notes tend to be somewhat general, e.g., "build understanding of. . . ." In the primary grades the teacher leads the children to the main ideas; in the intermediate grades text questions perform this job. Enrichment activities are included for faster learners and exercises are graded in difficulty for slower learners.

"Do you remember?" pages interspersed throughout the text and "Do you need more practice?" pages at the end of the text provide opportunities for extra drill in computational skills. Concepts tend to be emphasized more than skills.

Enrichment problems in the text provide opportunities for children to discover patterns and some problem sets in the intermediate grades are explicitly designed to encourage making generalizations. Formal logic is not taught.

Standard tests are provided. They are similar in format to exercises in the pupils text. No diagnostic aids are provided. The authors recommend that observations made by the teacher be used to supplement these tests in evaluating student performance.

Encyclopedia Britannica Press—*Math Workshop*

Few social applications appear in standard format in this series. Concepts of measurement are taught but short story problems rarely occur. Applications are used in several nonroutine ways. "Headline for the day" activities present the student with a number sentence and ask *him* to give a situation described by the sentence, a "reversal" of the usual procedure. This is the closest the series comes to the usual pattern of problem solving. The stories used tend to run for several pages, may be open-ended, and usually require a variety of problem solving abilities. This sort of problem solving is often used to introduce new concepts.

Structural properties are also introduced through such stories and other gamelike activities (e.g., magic squares). Much of the work is presented in the form of problems to solve, and structural properties (and computational skills) are learned as a by-product of these activities. For example, a base 7 system is studied in the setting of a candy factory that packages its candies singly, in boxes of 7, and in cartons of 7 boxes. Or the commutative principle of addition is studied by adding magic squares. Through these and similar activities the authors manage to include and emphasize all the usual structural properties with little attention paid to vocabulary or symbolism. In the last part of grade 6 arithmetic is described from an abstract, axiomatic approach, where properties are formally defined and sets of numbers contrasted.

Geometry forms an integral part of the program with major emphasis on classification of figures, area and perimeter, and a study of reflections (symmetry). All activities are conducted with an emphasis on exploration and experimentation, with much drawing on graph paper recommended.

Few concepts are formally defined until the end of grade 6 and in general there is great emphasis on minimizing verbalization—both in the text materials and on the part of the student. Expressing ideas out loud is not encouraged and in fact is frowned upon. Visual clues and patterns are used extensively to replace written or verbal directions. The format of the texts changes in grades 5 and 6 where the student is called upon to do more reading, although technical vocabulary is still not used.

Each teacher's edition begins with 40–60 pages of introductory material that indicate the authors' emphasis on discovery as their main pedagogical objective. Lesson notes in levels A through D run 2–3 pages per pupils page but are briefer in levels E and F. They indicate the purpose of the page (stated behaviorally), mathematical background, and have many explicit and detailed suggestions and supplementary activities for teaching the page. For most teachers this is *required* reading.

The major pedagogical emphasis is on teaching for discovery, which means several things: Foremost is the minimum of verbalization; second is the widespread use of open-ended mathematical problems as the vehicle for learning; visual clues are given to replace written instructions where possible; and a spiral approach is used in which mastery upon the initial presentation of a topic is not demanded. Induction is widely encouraged, and pupils are encouraged to do problems as they can and not according to some prescribed routine. The attitude on many pages is "Here is a problem—think about it—what can you say (or do)?" This approach clearly places a greater burden on the teacher, which the notes attempt to ease. It calls for great flexibility on the part of the teacher and may place excessive demands on older teachers used to a particular way of doing things.

The gamelike activities and problems used require the students to perform many computations and are used to replace routine drill. Supplementary pages containing only drill problems are available. The main emphasis is clearly on understanding concepts.

Many of the problems presented offer opportunities to generalize and most call for some reasoning (making deductions). The authors are careful to point out, however, that generalizing from a few examples does not prove a result. A game called "fewest questions" is specifically designed to help develop reasoning ability and this ability is essential to many of the discovery situations presented.

Test pages are scattered throughout each text. Questions are often multiple-choice and are designed mainly to measure computational ability. Measurement concepts are frequently used in these tests so that knowledge of addition facts is tested by having the pupils add 5 pints and 3 pints rather than 5 plus 3. Self-evaluation is encouraged indirectly when a student makes a mistake discovering. He is required to use his discovery and when he finds it does not work, he must rethink the problem and find out where he went wrong.

Harcourt Brace Jovanovich—*Elementary Mathematics*

In the primary grades, students are taught to solve problems by finding an appropriate "mathematical model." This involves recognizing the action taking place and forming a number sentence that describes it. In the intermediate grades the focus shifts to solving more complicated problems. The amount and types of problems are standard.

Sets are used informally in the primary grades, but number sentences are the major tool used to study structural properties. In the first five grades their main purpose is to extend computational ability. Operations are related to each other but are not compared with respect to structural properties. In grade 6 there is an attempt to unify these properties by studying relationships between different sets of numbers (e.g., the positive integers and rationals). Other numeration systems are studied and the Cartesian coordinate system is introduced.

The distinction between metric and nonmetric geometry is made, with attention divided about equally between them. Included are classifications of figures and computation of perimeters, areas, and volumes. Two interesting features are the "inquiry into geometry" pages where nonroutine geometrical activities are found and a technique of relating geometry to real situations by superimposing polygons on photographs of real objects.

In the primary grades, vocabulary is not strongly emphasized, although most technical terminology is introduced by grade 4. In grades 5 and 6 symbolic statements of properties appear. Precision of expression is expected. The amount of reading material is about average and increases with grade level. In grades 5 and 6 reading ability is crucial.

In addition to the standard material, prebook activities and a discussion of some psychological and pedagogical principles are included in the teacher's edition. The teaching suggestions are brief and easy to follow.

Many lessons begin by having the pupils do or discuss some examples. Oral work is important and verbalization of ideas is encouraged. Independent work is encouraged on "be your own teacher" pages and in enrichment workbooks for grades 3–6. When the enrichment materials are used the teacher is freed to work with slower learners.

The authors aim for and achieve a rough balance between skills and concepts. Making generalizations is explicitly encouraged on enrichment pages scattered throughout each text. Logic is not formally taught and counterexamples are not used.

Separate test booklets, each containing seven tests, supplement the checkup and review pages included in the text. The authors suggest a simple sort of statistical analysis as a diagnostic aid. Self-evaluation is encouraged on "self-help" tests. Following each test item is a page number indicating where the topic was taught in the text. When a pupil makes an error, he refers to these pages for the help he needs.

Holt, Rinehart and Winston—*Elementary Mathematics*

Standard topics of measurement and work problems are covered with particular emphasis on the use of number sentences in problem solving.

A particularly thorough coverage of structural properties is made including such things as closure and binary operations. Little is done to unify these topics. More attention than average is devoted to number patterns and non-decimal numeration systems.

Geometric topics are approached intuitively with an attempt to show how geometry is a "model" of physical space. The major goals appear to be recognizing and classifying various figures, with emphasis on some geometric constructions in the intermediate grades.

The use of vocabulary by pupils is not stressed though all technical terminology is introduced somewhat earlier than usual. Vocabulary tests are provided at the end of each chapter. The amount of required reading increases with grade level and is generally somewhat above average.

The teacher's edition is standard in format. The notes become briefer in the intermediate grades and stress mathematical content rather than a particular method for teaching each lesson. This is possible, in part, because the pupil's pages are designed to lead the pupil directly to the main idea with little assistance from the teacher. This works fine provided pupils can read. The teacher's main purpose is to explain unclear points. This method places a greater burden of reading on the pupil.

One chapter in each text entitled "Practice Exercises," and "Keep in Practice" pages dispersed throughout each text are aimed at maintaining computational ability. In general, however, concepts are emphasized more than skills.

Some informal practice in proofs is given by having students supply reasons to justify algebraic manipulations. Formal logic is not taught nor are there many opportunities to make deductions or generalizations.

The testing program is standard.

Science Research Associates—*Greater Cleveland Mathematics Program*

Story problems are used in the primary grades as applications of arithmetic skills. Understanding the role of number sentences in problem solving is emphasized. In some word-problem sections numerical answers are not demanded but only finding the appropriate number sentence. The usual systems of measurement are covered including the metric system. The emphasis on applications is average.

Structural properties tend to be formally introduced earlier than usual. Number sentences and the number line are the major tools used, with sets being used informally. Structural properties as an aid in computation are emphasized in the primary grades. In grades 5 and 6 a review, including symbolic statements of principles and comparisons between different number systems (e.g., the integers and rationals), emphasizes the structural nature of arithmetic.

Geometry is treated as a diversionary part of the program. It is studied briefly in grade K and picked up again in grade 4. Approximately two units in each of the intermediate grades are devoted to geometry. This study is characterized by more-than-average emphasis on three-dimensional figures and planes. The aim is to expose the student to geometrical concepts, and mastery of particular facts is not demanded.

Formal terms are present in grade 1 but are not stressed until grade 3. Understanding is measured in part by the ability of pupils to explain concepts but it is not required that they do so using technical terminology. The text is written in simple language with less-than-average required reading. One is impressed with the "uncluttered" appearance of many pages. Reading ability should not handicap students.

An outstanding feature of this series is the teacher's edition. The section on prerequisite skills is particularly helpful as a diagnostic aid in detecting student difficulties. Materials that provide broad mathematical background for the teacher form a supplement to the text. These notes are clearly written, easy to understand, and provide many specific teaching suggestions and examples. Teaching this series should pose few problems for even the inexperienced teacher.

As a general method, the teacher does problems with the students and guides them to the main concepts. Extensive use of physical objects is recommended in the primary grades.

In sections where exploration of a concept is the objective, mastery is not demanded. Problems in such sections usually call for verbal explanations and few numerical answers.

Mastery of computational skills is demanded in sections that consist entirely of drill problems. In these sections understanding is assumed, and speed and accuracy are emphasized. Computational skills are explicitly emphasized more in this series than in the others but not at the expense of concepts.

Informal attempts at formal proof are introduced by having students justify multistep algebraic manipulations by using structural properties. Although no specific portions of the text are devoted to teaching how to make generalizations, this ability is required on some test items in which the pupil is asked to extend already learned concepts.

Unit tests are provided for all grade levels. They are multiple-choice tests, accompanied by scoring keys. Diagnostic aids accompanying them indicate the specific objective which each test item is designed to measure.

Scott-Foresman—*Seeing Through Arithmetic*

A detailed analysis of problem situations is given in which problems are classified as additive or subtractive, rate situations, or comparison situations. These problem types, solvable by number sentences, are systematically presented to the students. Attention is paid to interpreting answers as well as getting them. In the intermediate grades extensive use is made of "rate pairs" (ratios) in solving word problems.

The treatment of structure and geometry is standard. Few comparisons between different number systems are made (e.g., the positive integers and rationals), and operations are not compared with respect to structural properties. Geometry plays a relatively small role in the series; emphasis is on recognizing figures and their properties.

Although technical terminology is used in the text, there is little emphasis on its use by pupils. This is an unusual situation in which students see and read technical terms but are not required to use them.

The teacher's edition is standard in format, containing general remarks for each lesson and "keyed notes" that refer to specific exercises on the pupil page. The notes are quite detailed and many suggestions are presented.

A four-step teaching sequence is proposed: see, think, try, do. See and think are performed under the guidance of the teacher; try and do are performed independently by the pupil. This teaching sequence consists basically of an expository technique.

One chapter in each text is devoted exclusively to developing computational skills. Concepts are emphasized but mainly as explanations of skills that can already be performed.

Some concepts of logic are taught—mainly the logical connectives "and" and "or"—but there are few opportunities for generalizing or making deductions.

The beginning chapters in several grades consist of review materials, accompanied by an achievement test. This test is designed to evaluate the student's beginning level of achievement. Review materials are to be used to bring the slower pupils up to class level, and so the first five chapters of a text might be located at the end of the previous book.

In addition to chapter tests and cumulative tests, the authors recommend that teachers keep an evaluation record, based on daily observations, for each pupil.

Silver-Burdett—*Modern Mathematics*

A special attempt is made to include many applications with particular attention to measurement. An above-average number of word problems is included and the use of number sentences in problem solving is emphasized.

Structual properties are studied primarily through the use of number sentences. Operations are compared with respect to these properties and different number systems are compared with each other. Functional relationships are studied as number sentences in two variables and graphs of linear functions are plotted.

The major emphasis in geometry is on recognition and classification of figures. Some applications are made—mainly using properties of the sphere (i.e., globe).

Although the use of technical vocabulary is not strongly stressed, much more required reading is called for than average. This large amount of reading places an unusual burden upon the student.

The teacher's edition format is standard. The main efforts of the authors are concentrated in remarks on mathematical content and teaching suggestions tend to be somewhat vague and general. This is possible, because many leading questions and explanations are included on the pupil's page. No distinct methodology was observed.

No attempt is made to teach formal logic and the only opportunities for discovery or making generalizations are in supplementary enrichment problems. Concepts tend to be emphasized but computational ability is viewed as basic to the program.

The tests provided are standard. Self-evaluation is encouraged by having the students grade their own tests occasionally.

Singer/Random House—*Sets and Numbers*

The student is kept constantly aware of applications as the author has included an average of two to three word problems on most pupil pages. These are short problems that usually involve a direct application of the skill just learned. Particular attention is paid to solving mathematical sentences and translating from English into arithmetic. In grade 3, students are taught to solve problems involving two equations with two unknowns—this is much earlier than usual.

Structural properties are strongly emphasized. Sets and set properties are used to introduce the concept of number and the arithmetic operations. This is the only series in which the use of sets is carried through all grade levels and in a variety of ways: solution sets of problems, descriptions of geometrical figures, and so on.

Structural properties are studied in separate chapters. All (binary) operations are compared with respect to a given principle. Relationships between operations are indicated; different number systems are compared (e.g., the positive integers and rationals). Functions are defined as certain sets of ordered pairs and graphs of lines and some curves (parabolas) are plotted.

Geometry is an integral part of the program. About one chapter in 10 is devoted to it. More emphasis than average is placed on nonmetric concepts such as concave and convex figures, open and closed figures, interior and exterior.

This series emphasizes vocabulary more strongly than any other. Concepts are defined formally, accompanied by symbolic statements upon their first presentation. Students are expected to use these words as well as recognize them. Set terminology forms a basic part of the child's vocabulary and is used repeatedly throughout the series. The N-notation for cardinality of a set is used.

The teacher's edition is standard in format and content. Prelesson activities are included and consist mainly of practicing prerequisite skills. Teaching notes tend to highlight the main points of the lesson and indicate errors that should be avoided, as well as provide specific teaching suggestions.

Ideas in the text are presented in precise but simple language, so that many pages can be worked independently by average students. Concepts are developed by working from the concrete to the semiconcrete to the abstract (symbolic) level. Relatively little emphasis is given to discovery. Concepts are emphasized as prerequisite to computational skill. Many drill problems are included in the text.

This is the only series that includes chapters devoted to formal logic. Pupils are taught the meaning of the logical connectives and given the opportunity to make deductions. They also do exercises in which they are required to supply the reasons which justify algebraic manipulations.

The testing program is standard, i.e., chapter and cumulative tests. However, the chapter tests are also cumulative and serve for review as well as testing.

Webster/McGraw-Hill—*Elementary Mathematics*

In addition to standard topics of measurement and word problems, a serious effort is made to relate mathematics to the real world. Use of applied situations motivate concepts, and inclusion of "reasoning problems" require the pupil to select appropriate procedures for finding the solution.

Structural properties are approached primarily through the use of sets and the number line. These properties are used mainly to extend computational ability. Structural properties are emphasized by comparing operations with respect to them and by comparing different number systems with each other.

More space than average is devoted to geometry. Some extra topics included are the Cartesian coordinate system, symmetry, and finding the volume of many three-dimensional figures. No unified approach was observed.

There seems to be no consistent pattern in the use of vocabulary. In general, the use of technical terminology is not emphasized. Required reading is about average.

The teacher's edition is standard in format. Teaching suggestions are presented by describing an "illustrative classroom experience." The remarks tend to be brief and easily understood.

A favorite teaching (and motivational) device is the use of colorful illustrations in which students discuss the central idea of each lesson. Through this technique different methods of doing a problem are suggested, and the student is encouraged to do problems in whatever way he can. It is suggested that some procedures are more efficient than others and, hence, should be preferred. The general methodology employed is slightly closer to guided discovery than straight exposition, although there are few places where the student really makes independent discoveries.

Few opportunities to make generalizations are present. However, counterexamples are occasionally used to disprove (false) generalizations made in the text. Making simple deductions is encouraged in "reasoning problems," but no formal study of logic is attempted.

Evaluation materials are standard. However, there is more emphasis than average on students verifying answers by doing the problem in different ways.

SUMMARY, CONCLUSIONS, RECOMMENDATIONS

Ten widely used and commercially available textbook series in elementary school mathematics were analyzed. The objectives of these series were identified, the extent of evaluation of these materials by the publishers and authors was reported, and an estimate was made of the degree to which the materials might achieve the stated objectives. In addition, the materials were carefully analyzed with particular attention given to eight crucial dimensions.

Our findings can be summarized as follows: Most objectives are stated in nonoperational terms and consequently convey little usable information to the teacher or educator. Statements of this form, we believe, suggest either that the objectives are not clearly understood or that insufficient thought has been given to them. Evaluations have for the most part been informal and their results are open to question. Even the formal evaluation programs do not come close to measuring such higher-order objectives as understanding or discovering.

It is almost impossible, except in very general terms, to tell whether the materials can achieve the stated objectives, except where computational ability is concerned. (They *do* do this.)

Our own subjective analysis of the materials suggests that most series emphasize structure and concepts more than computational skills; that technical vocabulary is used with varying degrees of emphasis; that geometry and social applications constitute comparatively small portions of each series (though many "word problems" may be present); that little attention has been given to evaluating student performance except insofar as computational ability is concerned; and that teaching methodologies vary considerably, with a tendency to use guided discovery most often.

We feel that the question of objectives is a particularly crucial one. With poorly defined goals, it is difficult to design effective materials and impossible to determine whether the goals are being achieved—or for that matter, exactly what *is* being achieved.

Little can be said about producing more effective or efficient materials or techniques of instruction until this problem has been resolved. We strongly urge that steps be taken in this direction.

Addison-Wesley
Elementary School Mathematics

Social Applications

A serious effort is made at relating math to everyday life. Measurement is emphasized in realistic word problems. Short story problems are really short and translating into arithmetic is emphasized.

above-average emphasis

Structure

Grades 1–3: counting numbers. Grades 4–6: positive rational numbers. Function concept introduced in grade 3. Structural properties are formally named and summarized in tables in grade 6.

Geometry

Two kinds of geometry are distinguished: physical and mathematical. The emphasis is on physical geometry. The major objectives are to expose the pupils to some of the notation and notions of geometry and to give them a good feeling for the study of geometry.

Vocabulary

Students are encouraged to express ideas but precision of expression is not demanded. The text does not rely heavily on the pupil's reading ability.

American Book Company (ABC)
Meeting Mathematics / Exploring Mathematics / Developing Mathematics / Understanding Mathematics / Learning Mathematics / Unifying Mathematics

---

Social Applications

Number sentences are a basic tool in problem solving. Practice in translating from English to arithmetic is provided. One or two chapters in each book are devoted to measurement.

average amount of material

Structure

Structural properties are used before they are named. Arithmetic laws are studied by "reverse facts." In grade 6 emphasis is on unifying knowledge. Equations and the number line are used extensively. Brief treatment of other numeration systems.

Geometry

The main ideas are recognition and classification of figures, and measurement of area and volume. Constructions are included as enrichment activities. Geometry plays a small role in the program.

less material than average

Vocabulary

Children learn new words mainly by imitating teachers. Technical terminology is not stressed. Reading level increases with grade level but is kept simple.

Encyclopedia Britannica Press (EBP)
Math Workshop

Social Applications

Applications are used as tests of computational ability and skill in measurement. Many problems are open-ended, with several answers or incomplete information. There are few "conventional" problems.

less material than average

Structure

Structural properties are learned through game-like activities. Most properties are learned by the end of level D. Operations are related to each other and compared with respect to properties.

above-average emphasis

Geometry

The major emphasis is on classification of figures, comparison of areas and perimeters, partitioning of regions, and similarity transformations (i.e., symmetry). Geometry activities are generally of an exploratory nature.

Vocabulary

Few concepts are formally defined. Directions are minimized with the use of visual patterns and clues. Some lessons are nonverbal. In levels E and F reading is not difficult but is crucial. Problem situations are used rather than directions.

Teacher's Edition	Methods	Concepts vs. Skills; Proof	Student Evaluation
A five-page introduction includes the philosophy of the authors. Notes face each pupil's page and state purpose of the lesson, preparation activities, math content, and directions for use of the page. Coded exercise answers enable pupils to check their own work.	Directions brief and general. Teachers advised to treat notes as suggestions. Emphasis is on content rather than technique. Format suggests a guided discovery approach, but many decisions are left to the teacher.	Concept-oriented. Proof: Few examples are used for generalization; no counterexamples were observed. Some discussion exercises called for simple deductions.	No pages are designated as test pages. There are review pages that may be used as tests. The authors feel that teacher observations give the best evaluation.

Teacher's Edition	Methods	Concepts vs. Skills; Proof	Student Evaluation
Bound at back of teacher's edition is 120-page guide. Included are an overview of the program and notes that indicate major emphasis, new vocabulary, and teaching procedures for each lesson. More comments are overprinted on pupil's page. Provides little extra math background for teacher.	Teaching comments are vague and general. No distinct method was observed. In early grades teacher does most of the talking. In upper grades questions in the text guide pupils to the main ideas.	Concepts are emphasized more than skills, though skills are maintained. No portion of the text explicitly devoted to proof or logic. Some opportunities to make generalizations are provided in enrichment activities.	Chapter tests are provided. Cumulative midyear and year-end tests are also provided. Teachers are also encouraged to keep records based on daily observations.

Teacher's Edition	Methods	Concepts vs. Skills; Proof	Student Evaluation
A 60-page introductory section includes authors' philosophy and description of materials. Lesson notes concentrate on teaching techniques, including detailed examples and suggestions. Math objectives for each page are explained.	Emphasis is on teaching for discovery; on *doing* more than on explaining what is being done. Pupils often work problems independently at their own rates and using their own methods. A spiral approach technique places a heavy responsibility on the teacher. The notes are essential.	Emphasis is on concepts. Skills are acquired as a by-product of problem solving. No formal efforts to teach proof. Counterexamples are used to warn pupils about making generalizations too quickly.	Teacher judgment is considered the best measurement of pupil achievement. Pages suggested as tests are used more for investigation than grading. Student self-evaluation is built into the discovery process.

Harcourt Brace Jovanovich
Elementary Mathematics

Social Applications

The crucial step is finding a "mathematical model" that fits a physical situation, and then to form a number sentence. Problems with nonnumerical answers and problems in which pupils make up the question are included. Interpreting answers is emphasized and estimating answers is encouraged.

Structure

Sets and set operations are used in the early grades, number sentences in the later grades. Structure is used mainly to explain computational algorithms. Relationships between number systems are emphasized.

Geometry

The text distinguishes between metric and nonmetric geometry. The main emphasis is on nonmetric, i.e., recognition and classification of figures. Three-dimensional figures are introduced in later grades and nonroutine enrichment problems with some applications are presented.

Vocabulary

Concepts are used first and then defined informally. Most technical terminology is introduced by grade 4. Pupils are expected to verbalize understandings. Reading ability is crucial in intermediate grades.

Holt, Rinehart and Winston (HRW)
Elementary Mathematics, Patterns and Structures

Social Applications

Word problems at the end of each chapter. One chapter in each grade devoted to measurement. Basic tool is number sentences. Attention paid to translating from English to arithmetic. Some applications to science.

Structure

Grades 1–4: whole numbers. Grades 5–6: positive rationals. Extensive use of number sentences. Operations related to each other but not compared with respect to structural properties. Much attention is paid to different numeration systems.

Geometry

Concepts approached intuitively. Emphasis on plane figures in primary grades; on constructions in later grades. Concept of space introduced in grade 3—earlier than usual. Applications are not emphasized.

Vocabulary

Students not pressured to use technical terms. Definitions are informal. Symbolic statements used in grade 6. Usage is stressed in grades 5 and 6. Reading ability is crucial in these grades. Vocabulary lists are provided.

more-than-average

Science Research Associates (SRA)
Greater Cleveland Mathematics Program (GCMP)

Social Applications

Applications are used as additional drill; not used to motivate topics. Number sentences are emphasized but pupils may do problems any way they can. Problems with nonnumerical answers are used. Much attention to translating from English to arithmetic.

Structure

Grades K–4: emphasis on whole and fractional numbers. Negative numbers in grade 5. The main tool is number sentences. Structural properties often introduced one to two grades earlier than average. Structure emphasized in grades 5 and 6.

above-average emphasis

Geometry

Grade K: brief taste of geometry, points and lines. Geometry reintroduced in grades 4–6. Above-average emphasis on three-dimensional figures and planes. Aim is for exposure and mastery is not demanded. Geometry viewed as "fun" and as a break from regular subject matter.

Vocabulary

Formal terms are introduced in grade 3, at which time pupils are expected to use them. The emphasis is mainly on recognition of terms. Vocabulary is kept to a minimum and reading ability is not crucial.

Teacher's Edition
A 30-page introduction gives an overview of the program and a discussion of teaching principles. Each lesson is faced with notes giving the purpose, math content, prebook teaching, and procedures for teaching the page.

Methods
Typical lesson begins with pupil doing something. Oral work is very important. A spiral approach is used. Enrichment books are available for grades 3–6. Some independent work is encouraged.

Concepts vs. Skills; Proof
A rough balance between concepts and skills. Skills maintained in practice pages at end of text. Pupils encouraged to make generalizations in enrichment problems. No attempt to teach proof or formal logic.

Student Evaluation
Separate test booklets contain seven tests per year: inventory test, four achievement tests, midterm and end-of-year exam. Diagnostic aids are provided. Review pages and checkup pages are also provided in the pupil's text.

Teacher's Edition
Brief introduction followed by scope and sequence charts and suggested time schedules. Notes for each lesson include Main ideas, math background, prebook teaching, and procedures for using the page. Notes are readable and helpful.

Methods
The bulk of the comments are devoted to math content. Questions in the text lead pupils to the main ideas. An accelerated sequence is available for grades 4–6.

Concepts vs. Skills; Proof
Concepts more than skills. Drill pages at the back of the text are used to maintain skills. No special provisions for teaching proof or logic. Few places for generalizing. Some "give reasons" problems.

Student Evaluation
Chapter tests at the end of each chapter. Four cumulative tests included in the teacher's edition. No diagnostic aids are provided.

Teacher's Edition
Specific objectives cited for each lesson. Discussion of math concepts and readiness for the lesson. Material for increasing the teacher's math background presented. Particularly well written with detailed comments—useful and easy to read.

above average

Methods
Topics presented in a straightforward manner. Students are led by the teacher to a statement of the main ideas. Except in chapters exploring concepts, mastery is required on each chapter. Students must be able to explain material covered. The texts for the primary grades are very close to being "programmed" material.

Concepts vs. Skills; Proof
Concepts are emphasized more than skills. Computation is emphasized more than average. No portion of the text is devoted to teaching logic or proofs. Pupils are asked to "justify steps" by stating the appropriate properties.

Student Evaluation
A packet consisting of four tests is administered upon completion of a specified portion of the text. A multiple choice test for each unit is provided in the intermediate grades. Diagnostic aids and additional quizzes are also provided.

Scott-Foresman
Seeing Through Arithmetic (STA)

Social Application

Three-step problem-solving technique: (1) recognize action; (2) describe action in symbols; (3) find solution; interpret it. Emphasis on verbal interpretation of solutions. Extensive use of "rate pairs."

Structure

Grades 1–4: whole numbers. Grades 4–6: positive rationals. Grade 6: negative numbers. Few comparisons of operations with respect to structural properties. Structure studied in separate chapters. Different numeration systems in grade 5.

Geometry

Emphasis on recognition of plane figures in primary grades; on three-dimensional figures in grades 5–6. Constructions on enrichment pages. Some concepts of symmetry and attempts to relate geometry to daily situations.

less-than-average emphasis

Vocabulary

Students are not required to use much technical terminology, although page headings are often technically worded. Concepts are defined informally. Little emphasis on expressing ideas. Small number of words used; reading level is average.

Silver-Burdett (S-B)
Modern Mathematics Through Discovery

Social Application

Particular attention to measurement. Number sentences, a basic tool. Estimating answers encouraged.

above-average amount of material

Structure

Concepts formally defined in grade 4, with symbolic statements in grade 6. Operations related to each other and compared with respect to structural properties. Functional relations, negative numbers, different numeration systems studied in grade 6.

above-average emphasis

Geometry

Topics approached intuitively. Emphasis is mainly on plane figures. Few constructions. Three-dimensional figures in upper grades. Some applications. Geometry plays a small role in the program.

Vocabulary

Little technical terminology until grade 4. Concepts are defined informally. Use of technical terms not stressed. Heavy demands at all levels upon reading ability.

more-than-average reading

Singer/Random House
Sets and Numbers

Social Application

Number sentences are widely used with special attention on translating from English to arithmetic. Word problems are present on many pages, including multistep problems in the later grades.

above-average amount of material

Structure

Structure studied in chapters called "Laws of Arithmetic." Formal verbal and symbolic statements of properties given in grade 3. Heavy use of sets and set notions. Negative numbers in grade 5.

above-average emphasis

Geometry

Approximately one chapter in 10 is devoted to geometry, with a rough balance between constructions and descriptions. Above-average emphasis on topological concepts. Language of sets used to describe geometrical figures. An emphasis on three-dimensional figures in intermediate grades.

Vocabulary

Use of precise language heavily emphasized. Concepts are defined formally, with symbolic statements. Pupils are expected to incorporate technical terminology into their vocabularies (beginning in grade 1).

above-average emphasis

Teacher's Edition	*Methods*	*Concepts vs. Skills; Proof*	*Student Evaluation*
Thirty pages of introductory material includes overview of program. Lesson notes give: math background, lesson objectives, teaching procedures, vocabulary list, and keyed notes. Time schedule and scope and sequence chart included. Texts are long—grade 4 is nearly 400 pages.	A four-step teaching sequence is proposed. "See and think" are done with the teacher; "Try and do" are done independently by the pupils. These comments are quite detailed and provide specific suggestions. The methodology is basically one of exposition.	Concepts are emphasized but skills are thought of as forming the foundation for them. One chapter in each text devoted to drill; also, "Keeping Skillful" pages. Grades 5–6: some formal logic. Few opportunities for making generalizations.	Chapter tests, mid-year and year-end exams are provided in pupil's text. Inventory tests in grades 3 and 4. The authors recommend that an Evaluation Record for each student be kept by the teacher.

Teacher's Edition	*Methods*	*Concepts vs. Skills; Proof*	*Student Evaluation*
Fifty to 60 pages of guide materials are devoted largely to math content. Lesson notes include the purpose of the lesson, list of terms, comments on teaching the page, discussion of math concepts. Comments are brief and indicate main ideas.	Remarks concerning methods are somewhat vague and general. Main effort is devoted to math content. Questions that guide pupils are often located in pupil's text.	Computational ability is viewed as basic to the program. Many drill problems are given. Concepts are emphasized more than skills. Few opportunities to generalize; no specific attention to proof or logic.	Learning stage tests, mid-year and end-of-year exams provided in pupil's text. Diagnostic aids are provided. Self-evaluation encouraged by having students occasionally check their own work.

Teacher's Edition	*Methods*	*Concepts vs. Skills; Proof*	*Student Evaluation*
Five-page introduction to each text is supplemented by comments for each chapter. The lesson notes indicate objectives, suggestions for introducing and teaching the page, and supplementary activities. Notes are helpful and easy to read.	Methodology is close to exposition. Mastery of each lesson is desired before proceeding. Advanced students are encouraged to work independently. Most formal presentation of all series but not austere.	Main emphasis is on concepts but many drill type exercises are provided. Beginning in grade 3, one chapter in each text is devoted to formal logic—included are many opportunities to make deductions.	Cumulative chapter tests and an end-of-the-year exam are included in the pupil texts. Test items are grouped under different headings which enables the teacher to pinpoint areas of difficulty.

above average

Social Applications

Applications are used to introduce and motivate new topics. Cartoons depict "real" situations. "Reasoning" problems include problems with irrelevant information and problems where pupils make up the question.

above-average
emphasis

Structure

Widespread use of sets, set notation, and the number line. Concepts used before they are defined. Negative numbers and other numeration systems in grade 5.

above-average
emphasis

Geometry

Main emphasis is on recognition and classification of figures. Some constructions and concepts of symmetry are included in grade 3. Geometry forms an integral part of the program.

above-average
amount of material

Vocabulary

Some technical terms used in grade 1, but no consistent pattern of usage was detected. Use of technical terms not strongly emphasized. Students are encouraged to express ideas. Reading level is about average.

Teacher's Edition	Methods	Concepts vs. Skills; Proof	Student Evaluation
One-page introduction mentions the objectives of the program. Comments for each lesson include math concepts, purposes, orientation and background, an illustrative classroom experience, and supplementary teaching suggestions.	A guided discovery approach in which pupils are led by questions posed by teacher and text. Colorful cartoons serve as an interesting motivational device. The format encourages active participation. Pupils are encouraged to see that there is more than one way to do a problem.	Concepts are emphasized in texts but correlated workbooks (grades 3–6) consist mainly of drill problems. Few places for making generalizations but pupils do use counterexamples and are asked to justify their work. "If . . . then" statements in grade 5 but no formal logic.	Chapter tests are provided in the pupil texts. Some diagnostic aids accompany them.

Answers to
Odd-Numbered
Exercises

EXERCISES 1–1.2

S–1. a. 36, 49, 64

Square the successive natural numbers 1, 2, 3, ...

    b. ↓, ←, ↑

Rotate the arrow clockwise 90° to get the next element in the pattern

    c. 64, 128, 256

Multiply the number by 2 to get the next number

    d. ⬡ , ⬡ , ⬡

Each figure has one more side than the preceding figure

  *e. 34, 55, 89

Each number is the sum of the two previous numbers.

    f. 1000001, 10000001, 100000001

Each numeral is the same as the previous one with an extra 0 added in the middle

S–3. a. meaning of digits 0, 1, 2, ..., 9, divisibility, understanding of place-value system

    b. addition of fractions

    c. ability to multiply

    d. ability to add and multiply

    e. ability to add and multiply

393

S–5. Elements in n row increase by 1

Elements in e row increase by 2

Elements in o row increase by 2

In each column, $e = 2n$

$$o = 2n - 1$$

$$o = e - 1$$

Each upper left to lower right diagonal sums to a multiple of 5, e.g., $1 + 4 + 5 = 10$, $3 + 8 + 9 = 20$, $7 + 16 + 17 = 40$.

S–7. a. i. Any element in column 1 added to any element in column 2 gives an answer in column 3.
   ii. Any element in column 2 added to any element in column 4 gives an answer in column 1.
   iii. Any element in column 3 added to any element in column 4 gives an answer in column 2.
  b. If any number in column 5 is added to any number in column n ($n = 1, 2, 3, 4, 5$), the answer will be in column n.
  c. Yes. If any number in column m ($m = 1, 2, 3, 4, 5$) is multiplied by any number in column n ($n = 1, 2, 3, 4, 5$), the answer is always in the same column as $m \times n$.
  d. The regularities will be similar, except that there will be six columns.

M–9. Addition: Each upper right to lower left diagonal has all the same numbers; the numbers in each upper left to lower right diagonal jump by twos; the part of the table above the main diagonal is the mirror image of the part below, and so on. Multiplication: The elements in each row and column jump by a common multiple; the elements down the main diagonal are all squares; in any row or column headed by an odd number, the entries in the row or column alternate odd and even; in any row or column headed by an even number, the entries in the row or column are all even.

EXERCISES 1–1.3

S–1. a. a traffic light
  b. the verses of a song such as "Old MacDonald's Farm"

EXERCISES 1–1.4

S–1. a. $2 + 4 + 6 + \ldots + 2n = n \times (n + 1)$
  $2 + 4 + 6 + \ldots + 20 = 10 \times 11 = 110$
  $2 + 4 + 6 + \ldots + 100 = 50 \times 51 = 2550$

b. If $n \rightarrow p$, then $n + 1 \rightarrow n + p$
    $8 \rightarrow 7 + 24 = 31$
    $9 \rightarrow 8 + 31 = 39$

c. $Y = 72 \div X$
    If $X = 1$, then $Y = 72$; if $X = 12$, then $Y = 6$; if $X = 18$, then $Y = 4$.

d. Each group of three numbers is sent into the largest one in the group.
    $13 \rightarrow 15$; $21 \rightarrow 21$; $35 \rightarrow 36$.

E–3. Rotate around the indicated diagonal.

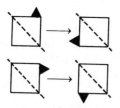

EXERCISES 1–2

S–1. a. a waltz or an oom-pah-pah on a tuba

b. a game of jacks: bounce the ball; pick up one jack; bounce the ball; pick up two jacks; ...

c. a sequence of lights on a computer where a light on stands for 1 and a light off stands for 0

S–3. a. i. $3^2 + 3 = 12$; $5^2 + 5 = 30$. The assertion must be false.

ii. 2, 3, 4 (3 is divisible by 3); 18, 19, 20 (18 is divisible by 3). The assertion seems reasonable.

iii. $1^2 - 1 + 41 = 41$; $4^2 - 4 + 41 = 53$. The assertion seems reasonable. However; $41^2 - 41 + 41 = 41 \times 41$. The assertion is *not* true.

iv.

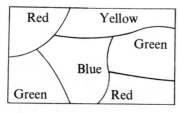 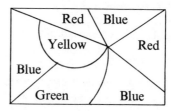

The assertion seems reasonable. In fact, it has never been proved to be true in all cases.

v. For 39747, $(3 + 7 + 7) - (9 + 4) = 4$.
    When 39747 is divided by 11, there is a remainder of 4.

For 48906, $(4 + 9 + 6) - (8 + 10) = 11$. 48906 is divisible by 11. The assertion seems reasonable.

b.   *ii, iii, iv, v* seem reasonable; *i* must be false. In fact, *ii* and *v* are true, *iii* is false, and it is not known whether or not *iv* is true.

M–5.   One possible story is as follows: Johnny has four bags with five marbles in each one. On the way to school he finds two more marbles. How many marbles does Johnny have? (Be sure to make up your own stories.)

EXERCISES 1–3.2

S–1.   a.   stop
b.   go
c.   treble clef
d.   yield right of way
e.   half-note
f.   sixteenth-note

g.   peace symbol
(or "ban-the-bomb" symbol)
h.   SOS (send help)
i.   Olympic Games symbol
j.   and
k.   null (empty) set

S–3.   The larger number goes at the larger end of the symbol and the smaller number goes at the smaller end of the symbol, e.g., $3 < 7$, $12 < 29$.

E–5.   a.   15
b.   39
c. $\dfrac{25}{12}$

d.   55
e.   84
f.   36

EXERCISES 1–3.3

S–1.   a.   symbol
b.   icon
c.   symbol
d.   icon

e.   icon
f.   symbol
g.   icon

EXERCISES 1–3.4

S–1.   a.   crossroad ahead
b.   side-road on the right ahead
c.   road curves to the right with a side road on the left
d.   grade crossing ahead
e.   pavement narrows from 3 lanes to 2

M–3.   Four possible symbols may be: □ for home, ⌒ for dog, Λ for boy and ⊓ for school. One story using these symbols could be about the ⌒ following the Λ from □ to ⊓ .

EXERCISES 1–4.2

S–1. a. 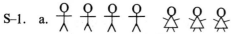

     b. 

     c. 

     d. 

S–3. a. 

     b. 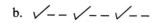   or

     c. SOS or ••• – – –•••

     d.

     e. H_2O or 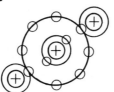

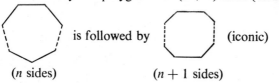

S–5. a. The nth term is the square of n (verbal)
       $a_n = n^2$ (symbolic)

     b. ↑ is followed by →
       → is followed by ↓
       ↓ is followed by ←
       ← is followed by ↑ (iconic)
       rotate the arrow clockwise 90° to get to the next entry in the list
       (verbal)

     c. The nth term is 2 raised to the nth power (verbal)
       $a_n = 2^n$ (symbolic)

     d. The nth entry is a polygon with $(n + 2)$ sides (verbal)

       (n sides)   is followed by   ($n + 1$ sides)   (iconic)

     e. Each entry is the sum of the previous two entries (verbal)

       $a_{n+2} = a_n + a_{n+1}$ (symbolic)

    f.   Each entry is the same as the previous one with an extra 0 in the center (verbal)

       1 0 0 ... 0 1 is followed by 1 0 0 ... 0 0 1 (symbolic)

$$\underbrace{1\ 0\ 0\ ...\ 0\ 1}_{n} \qquad \underbrace{1\ 0\ 0\ ...\ 0\ 0\ 1}_{n+1}$$

S–7.   The display is symmetric about the center vertical line. For any two adjacent entries on one line, the sum is written between them on the next line, e.g.,

$$1\quad 4\quad 6\quad 4\quad 1$$
$$1\quad 5\quad 10\quad 10\quad 5\quad 1$$

E–9. a. $\displaystyle\sum_{i=1}^{7} i$ d. $\displaystyle\sum_{i=1}^{5} (i-1)^2$

       b. $\displaystyle\sum_{i=1}^{8} \frac{1}{i}$ e. $\displaystyle\sum_{i=1}^{11} \frac{1}{i^2+1}$

       c. $\displaystyle\sum_{i=1}^{9} 2^i$ f. $\displaystyle\sum_{i=1}^{27} X_i$

EXERCISES 1–4.4

S–1. a. $x = y + 3$; or $x - 3 = y$

      b. $1 + 3 + ... + (2n - 1) = n^2$; or $\displaystyle\sum_{i=1}^{n} (2i - 1) = n^2$

      c. $c^2 = a^2 + b^2$
      d. $\square = \triangle$ or $\triangle = \square$
      e. $A = b \times h$
      f. $4 \times 6 = 24$
      g. $10 - 4 = 6$; or $10 - 6 = 4$; or $4 + 6 = 10$.
      h. $\square = \dfrac{\triangle + \lozenge}{2}$

S–3. a. Defense Department

other departments

      b. tuition

other income

c.

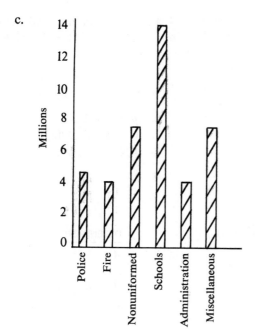

EXERCISES 1–5.1

S–1. Discovery and exposition

EXERCISES 1–5.2

S–1. a. Alice is a sophomore.
  b. If the teacher is sick, the class is noisy.
  c. No queeks have red hair.
  d. Santa Claus did not come down the chimney.
  e. a is larger than b.

S–3. a. $7 + 8 = 7 + (3 + 5)$
    $= (7 + 3) + 5$
    $= 10 + 5$
    $= 15$
  b. $6 + 7 = 6 + (4 + 3)$
    $= (6 + 4) + 3$
    $= 10 + 3$
    $= 13$

S–5. a. Let $2m$ and $2n$ be any even numbers. Then $(2m) \times (2n) = 2(2mn)$, which is an even number because it is a multiple of 2.
  b. Even $\times$ Odd = Even. Let $2m$ be any even number and $2n + 1$ be any odd number. Then $(2m) \times (2n + 1) = 2[m(2n + 1)]$, which is an even number because it is a multiple of 2.

c. Odd $\times$ Odd = Odd. Let $2m + 1$ and $2n + 1$ be any odd numbers. Then $(2m + 1) \times (2n + 1) = 4mn + 2m + 2n + 1 = 2(2mn + m + n) + 1$, which is an odd number.

EXERCISES 1–5.3

E–1. a. $(p \vee q) \supset r$ e. $p \supset (s \vee t)$
 b. $\backsim p \vee \backsim q$ f. $(\forall x)(r(x) \supset t(x))$
 c. $q \supset \backsim r$ g. $t(y)$
 d. $\backsim p \supset s$

EXERCISES 1–6

S–1.

	INDEPENDENT	DEPENDENT
a.	B	A
b.	A, B	——
c.	A or B	A or B
d.	A	B
e.	A, B	——
f.	A, B	——
g.	A	B
h.	B	A

E–3. a. For example

$$(a * b) * c = e = a * (b * e)$$

$$(c * d) * f = a = c * (d * f)$$

b. There are several examples. For instance

$$b * d = f \quad \text{but} \quad d * b = e$$

$$c * f = b \quad \text{but} \quad f * c = d$$

EXERCISES 1–6.1

S–1. (1) and (4)

EXERCISES 1–7.1

S–1. a. $x =$ amount Tommy started with

$$x - \$1.37 = \$2.14$$

b. $\square =$ age of furnace

$$2 \times \square = \text{age of house}$$

$$\square + 2 \times \square = 48$$

c. $\triangle$ = number of marbles Tommy started with

$$(\triangle - 11 - 13) = \frac{\triangle}{2}$$

d. y = number of dolls Susie has

$2y$ = number of dolls Patty has

$2y + y = 4 \times 3$

S–3.

	2	4	6	8	10	12
2	4	6	8	10	12	2
4	6	8	10	12	2	4
6	8	10	12	2	4	6
8	10	12	2	4	6	8
10	12	2	4	6	8	10
12	2	4	6	8	10	12

S–5.

	O	A	R	B	L	T
O	O	A	R	B	L	T
A	A	R	T	O	B	L
R	R	T	L	A	O	B
B	B	O	A	L	T	R
L	L	B	O	T	R	A
T	T	L	B	R	A	O

EXERCISES 1–8

M–1. Detect regularities; particularize; interpret descriptions; describe; make logical inferences; axiomatize. The illustrations chosen will depend on the text-book used.

S–3. a. Detecting a regularity

b. i. Detecting a regularity

ii. Interpreting a description

S–5. Write the percent as the numerator and 100 as the denominator (verbal); $\square\% = \square/100$ (symbolic).

EXERCISES 2–1

S–1. a. operation

b. relation

c. thing

d. relation

e. operation

[M–3.] The numerals for 11 and 17 have the same number of digits, 11 is less than 17, 17 is greater than 11, both numbers are between 10 and 20, the numerals for 11 and 17 have the same "tens" digit. Ask children to compare the numbers and the digits in the numerals.

EXERCISES 2–2

S–1. a. {January, February, March}
      b. {words, phrase, which, letters}
      c. {6, 8, 10, 12}

S–3. A collection of objects from which elements can be drawn.

      a. months of the year
      b. all the words in the phrase
      c. all the numbers between 5 and 13 inclusive

S–5. a. is a day of the week which begins with the letter T
      b. is one of the first 3 letters of the alphabet
      c. is one of the 3 colors on a traffic light
      d. is an even number between 1 and 11
      e. is one of the primary colors

S–7. a. $\{x \mid x$ is one of the last four letters in the alphabet$\}$
      b. $\{x \mid x$ is a polygon of three, four, or five sides$\}$
      c. $\{x \mid x \in N$ where N = natural numbers and $x \leq 6\}$
      d. $\{x \mid x$ is divisible by 2 and $2 \leq x \leq 10\}$
      e. $\{x \mid x$ is a primary color$\}$

E–9. a, b, e, g

E–11. a. {Wednesday, Thursday, Saturday, Sunday}
       b. {odd natural numbers}
       c. {all natural numbers not divisible by 5}
       d. {3, 6, 7, 8}

S–13. a. {4, 6, 2, 8}
       b. {5, 3, 7}
       c. e.g., {3}, {5, 8}, {4, 5, 6} and so on.

EXERCISES 2–3

S–1. a.   b. c.

d.

S–3. a.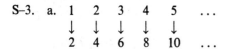

b. 10
c. 2
d. 480
e. 120
f. one-to-one correspondence

S–5. There are many correct answers for each part. Sample answers are:

a. {right, left} and {on, off}
b. {button, zipper, snap} and {?, !}

S–7. *b, c, d, e*

S–9. One possible set is $S =$ {is an uncle of, lives next door to, is equal to}. Another is $T =$ {is older than, is the sister of, has the same color eyes as}.

∗S–11. a. express each number in the first set as a product of factors of 2 and pair it with the polygon whose number of sides is 2 more than the number of factors of 2.
b. verbal
c. $2^n \to n + 2$, where $n + 2$ is the number of sides of the polygon.

EXERCISES 2–4

S–1. Some possible answers are:

a. weighs the same as, is older than, is as tall as, is the cousin of.
b. is a teacher of, is not a member of, is president of.

S–3. Some possible answers are: flip the switch, turn on the faucet, tie your shoe, button your coat, wash the dishes, feed the dog, open the drawer.

EXERCISES 2–5

S–1. Some possible answers are: "hear" and "here," which are equivalent in sound but not in meaning; two similar triangles that are equivalent in shape but not in size (area); "couldn't" and "could not," which are equivalent in meaning, but not in structure; 22 and 43, which are equivalent in the number of digits but not in value; cat and boy, which are equivalent in the sense that they are living beings but are not equivalent in the sense of same number of legs; $\frac{1}{2}$ and $\frac{2}{4}$, which are equivalent in the sense that they represent the same rational number but not in the sense of having the same numerator.

EXERCISES 2–5.1

S–1.   a.   Some possible properties are: the property of "writing with your left hand," the property of "being a girl," "having blue eyes," "being able to play the piano."

      b.   One example is: $S =$ the set of checkers on a checker board. Equivalence classes can be formed according to color (black or red) or according to kings or singles. Another example is $T = \{4, 5, 11, 12, 13\}$. The relation "has the same value" and the relation "has the same number of digits" each partitions T into different equivalence classes.

S–3.   Each of the following is only one of many possible ways:

      a. $E_1 = \{1, 3, 5, 7, 9\}$ and $E_2 = \{2, 4, 6, 8, 10\}$ according to odd or even.

      b. $E_1 = \{$red, white, blue$\}$ and $E_2 = \{$large, small$\}$ according to color or size.

      c. $E_1 = \{\ \triangle\ ,\ \square\ ,\ \lozenge\ \}$ and $E_2 = \{2, 10, 3, 9\}$ according to polygon or number.

      d. $E_1 = \{1, 3, 5, \ldots\}$ and $E_2 = \{2, 4, 6, \ldots\}$ according to odd or even.

      e. $E_1 = \{$January, June, July$\}$ and $E_2 = \{$March, May$\}$ according to an initial letter of J or M.

S–5.   Reflexive: For any a, a is related to a.

Symmetric: For any a and b, if a is related to b, then b is related to a.

Transitive: For any a, b, and c, if a is related to b, and b is related to c, then a is related to c.

S–7. $\{0, 5, 10, \ldots\}$, $\{2, 7, 12, \ldots\}$, $\{3, 8, 13, \ldots\}$, $\{4, 9, 14, \ldots\}$

S–9.

	R	S	T
a.	yes	yes	yes
b.	no	no	no
c.	no	yes	no
d.	no	no	yes
e.	yes	no	yes
f.	no	no	yes
g.	yes	no	yes
h.	yes	no	yes
i.	yes	yes	yes

[M–11.]   Have them define and give examples of an equivalence relation. Another method: Determine if a given relation is an equivalence relation. Another method: Select the equivalence relation out of a set of relations.

*E–13.   a.   reflexive and transitive only
         b.   symmetric and transitive only

EXERCISES 2–6

S–1.   a.   for $a \in S$, $a\,R\,a$.
       b.   for $a, b \in S$, if $a\,R\,b$, then $b\,R\,a$.
       c.   for $a, b \in S$, if $a\,R\,b$, then $b\,\not R\,a$.
       d.   for $a, b, c \in S$, if $a\,R\,b$ and $b\,R\,c$, then $a\,R\,c$.
       e.   for $a \neq b$, $a, b \in S$, either $a\,R\,b$ or $b\,R\,a$, but not both.

S–3.   The strict linear order relation (e.g., "comes after" in time) possesses the property of linearity while the strict partial order relation (e.g., "is a descendant of") does not. Both are antisymmetric, transitive, and irreflexive.

S–5.   Antisymmetric and transitive

S–7.

	R	I	S	A	T	L
a.	no	yes	no	yes	yes	no
b.	yes	no	no	yes	yes	no
c.	no	yes	no	yes	yes	yes
d.	no	yes	no	yes	no	no
e.	no	yes	no	yes	yes	no
f.	yes	no	no	yes	yes	yes
g.	yes	no	yes	no	yes	no
h.	yes	no	yes	no	yes	no
i.	yes	no	yes	no	no	no

S–9. $a - 4, b - 2, c - 3, d - 5, e - 1$

S–11.   a.   If you multiply both sides of an inequality by a positive number, the inequality sign is preserved.
        b.   If you multiply both sides of an inequality by a negative number, the inequality sign is reversed.

EXERCISES 2–7

S–1.   a.   binary
       b. $5 \circ 7 = 6, 8 \circ 10 = 9, 12 \circ 14 = 13$
       c. $7 \circ 5 = 6, 10 \circ 8 = 9, 14 \circ 12 = 13$
       d.   equal, equal
       e.   7
       f.   7
       g.   Nothing definite. Examples do not *prove* a general statement. But one would guess that they are equal.
       h. $(5 \circ 7) \circ 12 = 9, 5 \circ (7 \circ 12) = 7\frac{1}{4}$. They are not equal.

S–3.   a.   ternary operation
       b.   10
       c. $15\frac{1}{3}$

S–5. a. ternary
　　 b. unary
　　 c. 4-ary

S–7. $T = \$1.50m + \$.75n$; binary operation

E–9. a. binary operation
　　 b. 19, 17
　　 c. no
　　 d. If $A = 7$, $B = 5$ and $C = 2$, then $(A * B) * C = 40$ but $A * (B * C) = 26$.

EXERCISES 3–1

S–1. a. true or false
　　 b. always true
　　 c. true or false
　　 d. always false
　　 e. always false
　　 f. always false
　　 g. always true
　　 h. true or false

E–3. a. true with usual interpretation
　　 b. logically true
　　 c. logically true
　　 d. true with usual interpretation
　　 e. logically true

EXERCISES 3–2

S–1. a. observation　　　　　　 e. deduced logically
　　 b. deduced logically　　　 f. deduced logically
　　 c. observation　　　　　　 g. deduced logically
　　 d. observation

M–3. All boys in school are on the football team.

　　　Tom Smith is a boy.

　　　Therefore Tom Smith is on the football team.

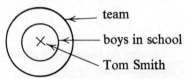

team
boys in school
Tom Smith

M–5. Observation: Give the rabbit a carrot and see if it eats it.

　　　Deduction: All rabbits like carrots.

　　　　　　　　 This pet is a rabbit.

　　　　　　　　 Therefore, this pet likes carrots.

EXERCISES 3–3

S–1. a. You should buy the automobile that is the best seller.
 b. A longer, wider, and heavier car is a better car.
 c. The better a boy is, the more Christmas presents he will receive.
 d. The suburban schools are better because they are decentralized.
 e. The more a model appliance costs, the better it must be.

EXERCISES 3–4

S–1. a. All pro football players are good athletes.
   My cousin is a pro football player.
 b. Any two of the three statements will serve as axioms.
 c. One third of the people in my class are boys.
 d. All good students get 3.5 averages or better.
   I am a good student.
 e. There are two possible sets of axioms.
   1. Sam is older than Tom.
     Tom and Elizabeth are twins.
     Sam is younger than Susan.
   2. Tom and Elizabeth are twins.
     Sam is older than Elizabeth.
     Sam is younger than Susan.
 f. $A + B = B + A$
   $A \times B = B \times A$
   $A \times (B + C) = (A \times B) + (A \times C)$
 g. In the National League, only the Phillies, Dodgers, Giants, Pirates, Cardinals, and Mets played yesterday.
   One sportswriter picked the Dodgers, Pirates, and Cardinals to win their games.
   One sportswriter picked the Dodgers, Mets, and Giants to win their games.
 h. $A + B = B + A$
   $A \times (B + C) = (A \times B) + (A \times C)$
   If $A + B = A + C$, then $B = C$
   $A + 0 = A$
   $A \times 0 = 0$ (This statement is not really basic; with a little effort it can be deduced logically from the other four basic statements. *Hint:* Let $B = C = 0$ in the second statement.)

M–3. Any time students are given a general rule together with a list of exemplars, they can reduce the list to the general rule. For example, if boys have assembly on Monday and girls have assembly on Tuesday, followed by a list of boys and girls with the days they go to assembly, then the list of names can be deleted and only the general rule kept.

EXERCISES 3–5

S–1. a. Mr. Smith is a Rotarian.
   b. Susie has blonde hair.
   c. Dave is missing one mitten.
   d. Trudy is very placid.
   e. My pet is not a golguk.

S–3. a. Tommy is a boy or a girl.
   b. John is either 6 or 7 years old.
   c. My automobile is not red.

E–5. a. The hypothesis "All multiples of 3 are odd" is false. 6, 12, 18, 24, 30, etc., are all multiples of 3 which are even.
   b. This is reasoning from the converse. No assumption is made about all even numbers; even numbers may or may not be multiples of 6.
   c. The hypothesis "Senator Mansfield is not a Democrat" is false.

EXERCISES 3–6.1

S–1. a. {5, 10, 15, 20, 25}
   b. {2}
   c. {4, 5, 6, 7, 8, 9, 10}
   d. {1000, 2000, . . ., 9000}
   [e.] {5, 10, 15, . . .}. There are an infinite number of elements.
       $\{x \mid x = 5n \text{ for } n = 1, 2, 3, \ldots\}$

S–3. a. {1, 2}, {1}, {2}, ∅
   b. Some possible answers are: the set of all polygons
                    the set of all triangles
                    the set of all right triangles
   c. Some possible answers are: the set of multiples of 4
                    the set of even numbers between 5 and 25
                    {2, 4, 6, 10, 14}
                    {2, 102, 202, 302, 402}
                    the set of all even numbers

EXERCISES 3–6.2

S–1. a. {2}
   b. {1, 2, 3, 5, 7, 9}
   c. { }
   d. {1, 2, 3, 4, 5, 6, 7, 8, 9}
   e. {4, 6, 8, 9, 10}
   f. {1, 3, 5, 7, 9, 10}
   g. {1, 3, 5, 7}
   h. {4, 6, 8}
   i. {2, 4, 6, 8}
   j. {1, 3, 4, 5, 6, 7, 8, 9, 10}

k. {1, 3, 4, 5, 6, 7, 8, 9, 10}
l. {4, 6, 8, 10}
m. {4, 6, 8, 10}

S–3. a.

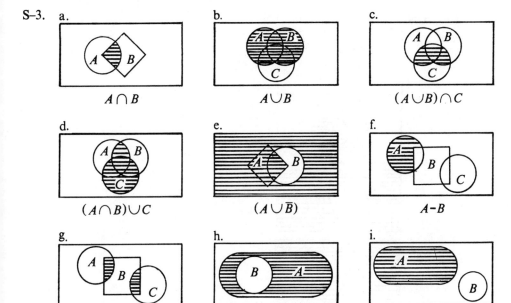

$A \cap B$

b.

$A \cup B$

c.

$(A \cup B) \cap C$

d.

$(A \cap B) \cup C$

e.

$\overline{(A \cup \overline{B})}$

f.

$A - B$

g.

$B \cap (A \cup C)$

h.

$A - B$

i.

$A - B$

j.

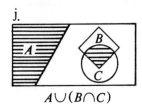

$A \cup (B \cap C)$

k.

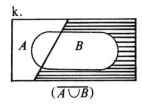

$(\overline{A \cup B})$

l.

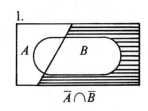

$\overline{A} \cap \overline{B}$

EXERCISES 3–7

S–1. a. true   g. true
      b. false   h. false
      c. false   i. true
      d. false   j. false
      e. true   k. true
      f. true   l. true

S–3. a. Tomorrow will be clear.
      b. One coin is a nickel.
      c. No conclusion is possible.
      d. No conclusion is possible.
      e. Today is not December 31.
      f. Mr. Jones owns a sports car.
      g. No conclusion is possible.

[M–5.]   Present the class with numerous examples of the type given on page 123 and Problem S–4. Show them that $p \rightarrow q$ can be considered false only if p is true but q is false.

EXERCISES 3–8

S–1.   a.

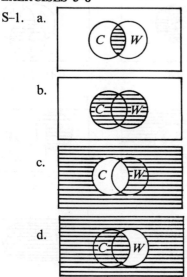

b.

c.

d.

[S–3.]   In each of the following, the premise set intersection is shaded.

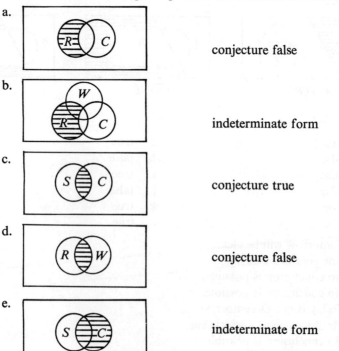

a.   conjecture false

b.   indeterminate form

c.   conjecture true

d.   conjecture false

e.   indeterminate form f.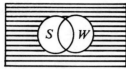

conjecture true

g.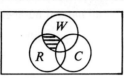

conjecture true

EXERCISES 3–9

S–1. a. Some babies are cute.
     All babies are cute.

   b. Some multiples of 4 are even.
     All multiples of 4 are even.

   c. All of my debts are large.
     Some of my debts are large.

E–3. a.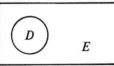

D = set of distinguished members
E = set of even numbers

   b.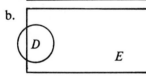

   c.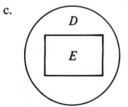

   d.

EXERCISES 3–10

S–1. a. $\forall\, x \in N \,|\, x + x \neq x$

   b. $\sim \exists\, x \in N \,|\, x + x = 0$

   c. $\exists\, x \in N \,|\, x + 5 = 9$

   d. $\sim \forall\, x \in N \,|\, 3x = 12$

S-3. a.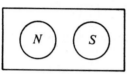

N = all natural numbers

$S = \{x \mid x + x = x\}$

    b.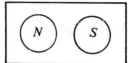

N = all natural numbers

$S = \{x \mid x + x = 0\}$

    c.

N = all natural numbers

$S = \{x \mid x + 5 = 9\}$
$N \cap S = \{4\}$

The shaded intersection includes the single point 4

    d.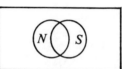

N = all natural numbers

$S = \{x \mid 3x \neq 12\}$

where $N - S$ is the single point 4

EXERCISES 4–1

S-1.

$\oplus$	0	1	2	3
0	0	1	2	3
1	1	2	3	0
2	2	3	0	1
3	3	0	1	2

S-3. a. rotations of 0°, 72°, 144°, 216°, 288°

    b. followed by

    c.

	0°	72°	144°	216°	288°
0°	0°	72°	144°	216°	288°
72°	72°	144°	216°	288°	0°
144°	144°	216°	288°	0°	72°
216°	216°	288°	0°	72°	144°
288°	288°	0°	72°	144°	216°

[S-5.] a. The letter in the first position is moved to the second position; the letter in the second position to the third; the letter in the third to the first. $\varnothing$ takes $\langle b, c, a \rangle$ into $\langle a, b, c \rangle$ and $\langle c, a, b \rangle$ into $\langle b, c, a \rangle$.

    b. $\varnothing * T = N; T * T = \varnothing$

    c.

*	N	$\varnothing$	T
N	N	$\varnothing$	T
$\varnothing$	$\varnothing$	T	N
T	T	N	$\varnothing$

[M–7.] One possible concrete embodiment consists of three figures (dolls, different colored objects, etc.) in a row. N leaves the figures alone; $\varnothing$ moves the figure at the right end of the line to the left end; T moves the figure at the right end of the line to the left end of the line, and then repeats the process a second time. Any three different figures may be used.

EXERCISES 4–1.1

[S–1.]

×	−1	1
−1	1	−1
1	−1	1

E–3.

⊕	0	1	2	3	4	5
0	0	1	2	3	4	5
1	1	2	3	4	5	0
2	2	3	4	5	0	1
3	3	4	5	0	1	2
4	4	5	0	1	2	3
5	5	0	1	2	3	4

EXERCISES 4–1.2

E–1.

×	1	2	3	4
1	1	2	3	4
2	2	4	1	3
3	3	1	4	2
4	4	3	2	1

E–3. a. infinitely many
 b. yes
 c. no. $abbab + bab = abbabbbab$
  $bab + abbab = bababbab$

EXERCISES 4–2

S–1. a. $(1 \oplus 3) \oplus 4 = 4 \oplus 4 = 3$
  $1 \oplus (3 \oplus 4) = 1 \oplus 2 = 3$
 b. $(180° \circ 90°) \circ 270° = 270° \circ 270° = 180°$
  $180° \circ (90° \circ 270°) = 180° \circ 0° = 180°$
 c. $(P_1 * P_4) * P_2 = P_3 * P_2 = P_5$
  $P_1 * (P_4 * P_2) = P_1 * P_3 = P_5$

S–3. a. 240° b. 2 c. (0, 1)

S–5. There is no single element that will generate all the others.

S–7. a. has closure property
 b. not associative because $(e * c) * d \neq e * (c * d)$
 c. b is an identity
 d. has the commutative property

S-9.

∪	{ }	{a}	{b}	{a, b}
{ }	{ }	{a}	{b}	{a, b}
{a}	{a}	{a}	{a, b}	{a, b}
{b}	{b}	{a, b}	{b}	{a, b}
{a, b}	{a, b}	{a, b}	{a, b}	{a, b}

There is no generator.

[E–11.] The table is symmetric about the diagonal given in the hint. That is, the part below the diagonal is the mirror image of the part above the diagonal.

E–13. Because only *one counterexample* is necessary to show that a general property does not hold, but a *proof* is necessary to show that the property holds for *all* of the elements in the embodiment.

M–15. 12

EXERCISES 4–3.1

S–1. Two examples are: Example (7) is a generalization of Example (4); and Example (2) is a restriction of Example (3).

S–3.

*	(0,0,0)	(0,0,1)	(0,1,0)	(1,0,0)	(0,1,1)	(1,1,0)	(1,0,1)	(1,1,1)
(0,0,0)	(0,0,0)	(0,0,1)	(0,1,0)	(1,0,0)	(0,1,1)	(1,1,0)	(1,0,1)	(1,1,1)
(0,0,1)	(0,0,1)	(0,0,0)	(0,1,1)	(1,0,1)	(0,1,0)	(1,1,1)	(1,0,0)	(1,1,0)
(0,1,0)	(0,1,0)	(0,1,1)	(0,0,0)	(1,1,0)	(0,0,1)	(1,0,0)	(1,1,1)	(1,0,1)
(1,0,0)	(1,0,0)	(1,0,1)	(1,1,0)	(0,0,0)	(1,1,1)	(0,1,0)	(0,0,1)	(0,1,1)
(0,1,1)	(0,1,1)	(0,1,0)	(0,0,1)	(1,1,1)	(0,0,0)	(1,0,1)	(1,1,0)	(1,0,0)
(1,1,0)	(1,1,0)	(1,1,1)	(1,0,0)	(0,1,0)	(1,0,1)	(0,0,0)	(0,1,1)	(0,0,1)
(1,0,1)	(1,0,1)	(1,0,0)	(1,1,1)	(0,0,1)	(1,1,0)	(0,1,1)	(0,0,0)	(0,1,0)
(1,1,1)	(1,1,1)	(1,1,0)	(1,0,1)	(0,1,1)	(1,0,0)	(0,0,1)	(0,1,0)	(0,0,0)

{(1, 0, 0), (0, 1, 0), (0, 0, 1)}

serves as a set of generators. There are other possible sets of generators.

*S–5. a. 0 b. 1 c. $n - 2$ d. $4, 5, \ldots, n - 1, 0, 1, 2, 3$

EXERCISES 4–3.2

S–1. a. $240° \circ 240° = 120°$

$$\updownarrow \qquad \updownarrow \qquad \updownarrow$$

$$\underline{2} + \underline{2} = \underline{1}$$

  b. $120° \circ 0° = 120°$

$$\updownarrow \qquad \updownarrow \qquad \updownarrow$$

$$\underline{1} \oplus \underline{0} = \underline{1}$$

  c. $240° \circ 120° = 0°$

$$\updownarrow \qquad \updownarrow \qquad \updownarrow$$

$$\underline{2} \oplus \underline{1} = \underline{0}$$

S–3. There is no one-to-one correspondence between the basic sets.

S–5. $0 \leftrightarrow a$ or $0 \leftrightarrow a$

   $1 \leftrightarrow c$ $1 \leftrightarrow d$

   $2 \leftrightarrow b$ $2 \leftrightarrow b$

   $3 \leftrightarrow d$ $3 \leftrightarrow c$

*E–7. $m + n \leftrightarrow \dfrac{m + n}{1} = \dfrac{m}{1} + \dfrac{n}{1}$

[M–9.] Have the child describe what an isomorphism is, or better, have him determine whether given pairs of simple systems are isomorphic or not. Also have him find the isomorphisms (correspondences).

EXERCISES 4–3.3

S–1. a. no
   b. yes
   c. no
   d. yes
   e. no

S–3. $N \leftrightarrow P_0$

   $\emptyset \leftrightarrow P_1$

   $T \leftrightarrow P_2$

EXERCISES 4–3.4

S–1. $1 + 5 = 6$ or $3 + 7 = 10$
    $\updownarrow$ $\updownarrow$ $\not\updownarrow$ $\updownarrow$ $\updownarrow$ $\not\updownarrow$
    $2 + 6 = 8$ $4 + 8 = 12$

[S–3.] An isomorphism must be a one-to-one correspondence, but a homomorphism need not be; no; yes.

*E–5. Let $a = 3 \cdot m \overset{h}{\leftrightarrow} m$

     $a' = 3 \cdot n \overset{h}{\leftrightarrow} n$

     Then $a + a' = 3m + 3n = 3(m + n) \overset{h}{\leftrightarrow} m + n.$

EXERCISES 5–1

S–1. a. 3   f. 9
   b. 17 (1900 was *not* a leap year)   g. 4
   c. 50   h. 8
   d. 2   i. answer varies
   e. 7

S–3. Some possible answers are:

$S = \{a, b, c\}$ and $T = \{0, 1\}$

$a \leftrightarrow 0$
$b \leftrightarrow 1$
c

$S = \{!, *, ?, \#, \$\}$ and $T = \{w, x, y\}$

$! \searrow w$
$*$
$? \leftrightarrow x$
$\#$
$\$ \nearrow y$

S–5. Draw a piece of chalk from box No. 1 and then a piece of chalk from box No. 2. Lay the pieces aside. Repeat the above process until you cannot draw any more pieces of chalk from one of the boxes. If the other box likewise contains no chalk, then the sets have the same cardinal number. If the other box still contains some chalk, then the sets do not have the same cardinal number.

[S–7.] a. 4   d. 0
     b. 5   e. yes
     c. 9

[S–9.] a. 5   d. 0
     b. 2   e. yes
     c. 7

[S–11.] S T a. 5
     b. 3
     c. 6
     d. 2
     e. no

(Venn diagram: S and T overlapping circles containing a, b, d in S; c, e in the intersection; f in T)

S–13. No; the important thing is the existence of the matching, not the particular pairing.

EXERCISES 5–2

S–1. a. is below _____ in the rainbow; 3
     b. comes before (in time); 17 (*Note:* 1900 was *not* a leap year.)
     c. comes after (alphabetically); 50
     d. comes after (alphabetically); 7
     e. sounds higher than; 4
     f. is less than; 8
     g. is older than; answer varies

S–3. Some possible answers are:

     a. the set of members in Jeffrey's immediate family; is older than the set of children in Jeffrey's class; comes after (alphabetically)
     b. the set of cities in the United States; has a larger population than
     c. the set of toothpastes on the market; sells more than

S–5. No; every set has a cardinal number but only ordered sets have ordinal numbers.

S–7. a. geography   history

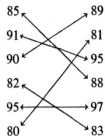

b. 6, 6, 6
c. geography 3, history 2, alphabetical order 3

EXERCISES 5–3

S–1. a. ordinal   d. cardinal
     b. cardinal   e. ordinal
     c. ordinal

S–3. Some possible answers are $S = \{1, 2, 3, 4, 5, 6\}$; 6; the set of days in the week; 7.

S–5. a. 5   b. 5   c. 5

S–7. a. 8   b. 8   c. 3

EXERCISES 5–4

S–1. a. 1
     b. Some possible answers are:
         S = set of even natural numbers; 2
         $T = \{9, 13, 21, 64\}$; 9
         W = set of non-negative (whole) numbers; 0
     c. Some possible answers are:
         S = set of fractions larger than 0 and less than 1
         T = set of negative integers; $\{\ldots, ^-3, ^-2, ^-1\}$

S–3. 211

∗S–5. a. (1), (3)
      b. (2) will represent a natural number if a is greater than b
          (4) will represent a natural number if $a = b \times c$ for some natural number c.

EXERCISES 5–5

S–1. Some possible answers are:
  a. Ruth gives her pebbles to David.
  b. David gives his pebbles to Ruth.
  c. The person with fewer pebbles gives them to the other person.
  d. Ruth and David each give their pebbles to a third person. Find how many pebbles this third person has.

E–3. Yes, because if $a \neq b$, then $a + c \neq b + c$.

S–5. Commute: to give in exchange for another, exchange, change, alter

Associate: to join or connect together

Yes; The commutative property was so named because the order of performing the operation on two elements, a and b, does not change the result, i.e., we can exchange a for b and b for a and get the same result.

The associative property was so named because the manner by which we join or group three elements does not change the result.

S–7. Some possible examples are: opening the window and putting your head out the window, boiling water to get steam and condensing steam to get water, opening a book and reading it.

*E–9. a. $10 > 7$ because $10 = 7 + 3$ and 3 is a natural number
  b. $10 + 3 > 7 + 3$
  c. $c > d$ implies $c = d + n$ where $n \in N$

Therefore $c + e = (d + n) + e$

$c + e = d + (n + e)$ (associative property)

$c + e = d + (e + n)$ (commutative property)

$c + e = (d + e) + n$ (associative property)

Therefore $c + e > d + e$ because $n \in N$

We know there is such a number n because we were told $c > d$.
  d. For any natural numbers, if the first is greater than the second, then the first plus a third one is greater than the second plus the third one.

EXERCISES 5–6

S–1. $13 - 7 = 6$ because $13 = 6 + 7$.

S–3. a. no   b. no   c. no   d. no

[S–5.] Some possible examples are:

$7 - (3 - 2) \neq (7 - 3) - 2$ because $6 \neq 2$,

$9 - (2 - 1) \neq (9 - 2) - 1$ because $8 \neq 6$,

$12 - (8 - 3) \neq (12 - 8) - 3$ because $7 \neq 1$.

[S–7.] a. (1) $\{?, *\}$ (2) $\{10, 7, 6, 9, 14\}$
          (3) $\{m, n, o, p, q\}$ (4) $\{\ \ \}$

b. (1) *A* / *B* / ? / ! / # / * ← *A–B*   (2) A / 10 / B / 7 / 14 / 4 / 6 / 9 ← *A–B*

(3) A / *m* / *n* / *o* / *q* / *p* $=A-B$ (4) A $=B;\ A-B=\{\ \}$

c. $(A - B) \cap B = \{\ \ \}$
d. $(A - B) \cup B = A \cup B = A$ when $B \subset A$
e. (1) 4, 2, 2;   (2) 6, 1, 5;   (3) 5, 0, 5;   (4) 4, 4, 0
f. cardinal number of A − cardinal number of B = cardinal number of $(A - B)$

EXERCISES 5–7

S–1. a. $\{(\square, a), (\square, b), (\square, c), (\square, d), (\triangle, a), (\triangle, b), (\triangle, c), (\triangle, d),$
$(\bigcirc, a), (\bigcirc, b), (\bigcirc, c), (\bigcirc, d)\}$
b. $\{(1, r), (1, s), (2, r), (2, s), (3, r), (3, s), (4, r), (4, s), (5, r), (5, s)\}$
c. $\{(\cap, \times), (\cap, +), (\cap, \div), (\cap, -), (\cup, \times), (\cup, +), (\cup, \div), (\cup, -)\}$
d. $\{(!, w), (!, x), (!, y)\}$
e. $\{(w, 10), (w, 9), (m, 10), (m, 9), (t, 10), (t, 9)\}$

S–3. a. $A \times B = \{(a, 1), (a, 2), (b, 1), (b, 2), (c, 1), (c, 2)\}$
$B \times A = \{(1, a), (2, a), (1, b), (2, b), (1, c), (2, c)\}$
b. $A \times B = \{(1, 1), (1, 2), (2, 1), (2, 2)\}$
$B \times A = \{(1, 1), (1, 2), (2, 1), (2, 2)\}$
c. $A \times B = \{(w, y), (w, w), (w, x), (x, y), (x, w), (x, x), (y, y), (y, w), (y, x)\}$
$B \times A = \{(w, y), (w, w), (w, x), (x, y), (x, w), (x, x), (y, y), (y, w), (y, x)\}$

*[S–5.] B contains only one element.

S–7. a. commutative property of multiplication
b. closure of the natural numbers under multiplication
c. associative property of multiplication
d. associative property of multiplication
e. commutative property of multiplication

E–9. Either $a = 0$ or $b = 0$ or both a and $b = 0$.

E–11. a. yes
b. No; because if $c = 0$ and a and b are unequal, then $a \times c = b \times c$ is true since $0 = 0$ but $a = b$ is false. For example, $3 \times 0 = 7 \times 0$ but $3 \neq 7$.

EXERCISES 5–8

S–1. c, f, h, k

S–3. $s = 1, t = 11; s = 2, t = 8; s = 3, t = 5; s = 4, t = 2.$

[S–5.] Some possible examples are: $6 \div 2 \neq 2 \div 6$ because $6 \div 2 = 3$ and $2 \div 6$ is undefined, $10 \div 2 \neq 2 \div 10$ because $10 \div 2 = 5$ and $2 \div 10$ is undefined.

S–7.   a.   Turn off the television.
     b.   Untie your shoe.
     c.   Condense the steam to change it to water.
     d.   does not have an inverse operation

EXERCISES 5–9

S–1. $79 \cdot 37 + 79 \cdot 63 = 79 \cdot (37 + 63) = 79 \cdot 100 = 7900$

S–3. $7 \times 13 = 7 \times (10 + 3) = (7 \times 10) + (7 \times 3) = 70 + 21 = 91$

S–5.   a. $7 \times 13 = 7 \times (10 + 3) = (7 \times 10) + (7 \times 3) = 70 + 21 = 91$
     b. $8 \times 15 = 8 \times (10 + 5) = 80 + 40 = 120$
     c. $6 \times 18 = 6 \times (10 + 8) = 60 + 48 = 108$
     d. $9 \times 14 = 9 \times (10 + 4) = 90 + 36 = 126$
     e. $8 \times 19 = 8 \times (10 + 9) = 80 + 72 = 152$

EXERCISES 5–10

S–1. a, c, d, e, f, h, j

S–3.

a.	$98 = 2 \cdot 7^2$		f.	173 is prime
b.	$40 = 2^3 \cdot 5$		g.	$726 = 2 \cdot 3 \cdot 11^2$
c.	$72 = 2^3 \cdot 3^2$		h.	$371 = 7 \cdot 53$
d.	$360 = 2^3 \cdot 3^2 \cdot 5$		i.	$713 = 23 \cdot 31$
e.	$150 = 2 \cdot 3 \cdot 5^2$		j.	$2592 = 2^5 \cdot 3^4$

S–5.

a.	6		f.	5
b.	6		g.	36
c.	12		h.	10
d.	5		i.	12
e.	8		j.	4

S–7.

		GCF	LCM
a.	7, 9	1	63
b.	6, 8	2	24
c.	5, 25	5	25
d.	12, 15	3	60
e.	18, 22	2	198
f.	126, 140	14	1260
g.	220, 600	20	6600
h.	9, 12, 15	3	180

EXERCISES 5–11

S–1. a. 0, 1, 2, 3
       b. yes
       c. yes
       d. yes
       e. yes, 0 is the identity.
       f. yes, 2 is the inverse of 3.
       g. yes, 0 is the inverse of 0.
                 4 is the inverse of 1.
                 3 is the inverse of 2.
                 1 is the inverse of 4.

[S–3.] a. yes. Any element in S operated on by any other element in S will result in an element of S.
        b. yes
        c. yes, # is the identity.
        d. yes, #
        e. yes, $
        f. yes, !

*E–5. a. yes
       b.

$\circ$	E	R	R^2
E	E	R	R^2
R	R	R^2	E
R^2	R^2	E	R

       c. yes
       d. yes
       e. yes because $(R) \circ R^2 = E$ and $E \circ (E) = E$
       f. yes, E
       g. yes, E
       h. yes, R^2
       i. yes, R

EXERCISES 6–1

S–1. a. iconic   d. iconic
       b. symbolic   e. verbal
       c. symbolic   f. iconic

E–3. a. number   e. number
       b. numeral   f. number
       c. numeral   g. numeral
       d. number

EXERCISES 6–2.1

E–1. a. 8   b. 24   c. 59
       d. 700   e. 900   f. 42

E–3. a. DCCXXVIII
       b. XXV
       c. LXV
       d. LXXXVI

EXERCISES 6–2.2

[E–1.] We need zero as a "place holder." For example, the numeral "15" means "five ones and one ten," while "105" means "five ones, no tens, and one hundred." Without the symbol "0" we would not be able to distinguish the two.

[M–3.] 23 = 2 girls and 3 boys

         24 = 2 girls and 4 boys

         .   .

         .   .

         .   .

         30 = 3 girls and no boys

         31 = 3 girls and 1 boy

EXERCISES 6–2.3

S–1. 331_5

S–3. a. 1100100_2
       b. 10201_3
       c. 144_8
       d. 84_{twelve}

S–5. a.

	dollars	halves	quarters	nickels	pennies
57¢ =	0	1	0	1	2
185¢ =	1	1	1	2	0
378¢ =	3	1	1	0	3

      b.

	dollars	halves	dimes	nickels	pennies
57¢ =	0	1	0	1	2
185¢ =	1	1	3	1	0
378¢ =	3	1	2	1	3

E–7. base 2

EXERCISES 6–3

S–1. a.

12
+23
35

b. 38
  +27
   65

c.| 17
   29
  +35
   81

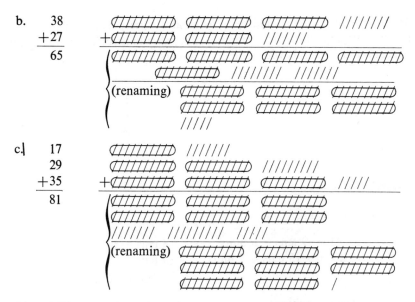

E–3.   The addition algorithm for three-digit numbers may be stated as follows: Add the units column; write the units digit of the sum and "carry" the tens digit to the tens column. Then add the tens column; write the units digit of the sum in the tens place in the answer and "carry" the tens digit to the hundreds column. Then add the hundreds column and write the sum in the hundreds place in the answer.

The addition algorithm for four- and five-digit numbers can be stated similarly.

An example using the algorithm for five-digit numbers is

```
  7 3 2 1 5        1←       1←       1←       1←
 +8 7 9 6          7        3        2        1        5
 ─────────                  8        7        9        6
  8 2 0 1 1        ──────────────────────────────────────
                   8       2        0        1        1
```

Iconically we could group into bundles of 10s, 100s (bundles of ten 10s), 1000s (bundles of ten 100s), etc., but this would be tedious. The implicit regularity can be obtained by understanding the addition algorithm for two-digit numbers, as exemplified in the text.

M–5.   a.   Represent 36 by 3 bundles of 10 pencils each, held by rubber bands, together with six loose pencils. Represent 17 by 1 bundle of 10 pencils held by a rubber band, together with seven loose pencils. When all of the loose pencils are combined, they form one bundle with 3 left over. When the bundles are combined there are 5 bundles. The sum then is 5 bundles of ten pencils each and 3 loose pencils, or 53.

     b.   paintbrushes, beads strung 10 per string, any kind of object which can be grouped and held together

M–7. The examples should be of as great a variety as possible. Some should involve carrying and some should not. They should involve different numbers of addends and numerals with different numbers of digits. Four possible examples are:

$$
\begin{array}{r} 235 \\ +364 \\ \hline \end{array}
\qquad
\begin{array}{r} 390 \\ +872 \\ \hline \end{array}
\qquad
\begin{array}{r} 27 \\ 43 \\ 79 \\ +84 \\ \hline \end{array}
\qquad
\begin{array}{r} 47 \\ 3 \\ +14 \\ \hline \end{array}
$$

EXERCISES 6–4

S–1.   a.   Borrowing

$$
\begin{array}{rll}
93 = & 9 \text{ tens} + 3 \text{ units} & = \quad 8 \text{ tens} + 13 \text{ units} \\
-57 = & -(5 \text{ tens} + 7 \text{ units}) & = -(5 \text{ tens} + \;\, 7 \text{ units}) \\
\hline
36 & & \quad 3 \text{ tens} + \;\; 6 \text{ units}
\end{array}
$$

Equal additions

$$
\begin{array}{rll}
93 = & 9 \text{ tens} + 3 \text{ units} & \xrightarrow{+10} \quad 9 \text{ tens} + 13 \text{ units} \\
-57 = & -(5 \text{ tens} + 7 \text{ units}) & \xrightarrow{+10} -(6 \text{ tens} + \;\, 7 \text{ units}) \\
\hline
36 & & \quad 3 \text{ tens} + \;\; 6 \text{ units}
\end{array}
$$

Complement

$$
\begin{array}{rll}
93 \longrightarrow & 93 + 43 & = \quad 136 \\
-57 \longrightarrow & -(57 + 43) & = \quad -100 \\
\hline
36 & & \quad 36
\end{array}
$$

b.   Borrowing

$$
\begin{array}{rl}
476 = & 4 \text{ hundreds} + 7 \text{ tens} + 6 \text{ units} \\
-239 = & -(2 \text{ hundreds} + 3 \text{ tens} + 9 \text{ units}) \\
\hline
237 &
\end{array}
$$

$$
\begin{array}{rl}
= & 4 \text{ hundreds} + 6 \text{ tens} + 16 \text{ units} \\
= & -(2 \text{ hundreds} + 3 \text{ tens} + \;\, 9 \text{ units}) \\
\hline
& 2 \text{ hundreds} + 3 \text{ tens} + \;\; 7 \text{ units}
\end{array}
$$

Equal additions

$$
\begin{array}{rl}
476 = & 4 \text{ hundreds} + 7 \text{ tens} + 6 \text{ units} \\
-239 = & -(2 \text{ hundreds} + 3 \text{ tens} + 9 \text{ units}) \\
\hline
237 &
\end{array}
$$

$$
\begin{array}{rl}
\xrightarrow{+10} & 4 \text{ hundreds} + 7 \text{ tens} + 16 \text{ units} \\
\xrightarrow{+10} & -(2 \text{ hundreds} + 4 \text{ tens} + \;\, 9 \text{ units}) \\
\hline
& 2 \text{ hundreds} + 3 \text{ tens} + \;\; 7 \text{ units}
\end{array}
$$

Complement

$$476 \longrightarrow \quad 476 + 761 = \quad 1237$$
$$-239 \longrightarrow -(239 + 761) = -1000$$
$$\overline{237} \qquad \qquad \qquad \qquad \overline{237}$$

c. **Borrowing**

$$921 \quad = \qquad 9 \text{ hundreds} + 2 \text{ tens} + 1 \text{ unit}$$
$$-737 \quad = \quad -(7 \text{ hundreds} + 3 \text{ tens} + 7 \text{ units})$$
$$\overline{184}$$

$$= \qquad 9 \text{ hundreds} + 1 \text{ ten} \ + 11 \text{ units}$$
$$\longrightarrow = \quad -(7 \text{ hundreds} + 3 \text{ tens} + \ 7 \text{ units})$$

$$\longrightarrow = \qquad 8 \text{ hundreds} + 11 \text{ tens} + 11 \text{ units}$$
$$= \quad -(7 \text{ hundreds} + \ 3 \text{ tens} + \ 7 \text{ units})$$
$$\overline{1 \text{ hundred} + \ 8 \text{ tens} + \ 4 \text{ units}}$$

Equal additions

$$921 \quad = \qquad 9 \text{ hundreds} + 2 \text{ tens} + 1 \text{ unit}$$
$$-737 \quad = \quad -(7 \text{ hundreds} + 3 \text{ tens} + 7 \text{ units})$$
$$\overline{184}$$

$$\xrightarrow{+10} \ 9 \text{ hundreds} + 2 \text{ tens} + 11 \text{ units}$$
$$\xrightarrow{+10} -(7 \text{ hundreds} + 4 \text{ tens} + \ 7 \text{ units})$$

$$\xrightarrow{+100} \ 9 \text{ hundreds} + 12 \text{ tens} + 11 \text{ units}$$
$$\xrightarrow{+100} -(8 \text{ hundreds} + \ 4 \text{ tens} + \ 7 \text{ units})$$
$$\overline{1 \text{ hundred} + \ 8 \text{ tens} + \ 4 \text{ units}}$$

Complement

$$921 \longrightarrow \quad 921 + 263 = \quad 1184$$
$$-737 \longrightarrow -(737 + 263) = -1000$$
$$\overline{184} \qquad \qquad \qquad \qquad \overline{184}$$

[S–3.]   a. $53 - 38 = (50 + 3) - (30 + 8)$ renaming

$$= (40 + 13) - (30 + 8) \qquad \text{renaming}$$

$$= (40 - 30) + (13 - 8) \qquad \text{by the property on page 199 of Chapter 5}$$

$$= \quad 10 \ + \ 5 \qquad \text{renaming}$$

$$= \quad 15 \qquad \text{renaming}$$

b. $81 - 33 = (80 + 1) - (30 + 3)$ renaming

$$= (70 + 11) - (30 + 3) \qquad \text{renaming}$$

$$= (70 - 30) + (11 - 3) \qquad \text{by the property on page 199 of Chapter 5}$$

$$= 40 + 8 \qquad \text{renaming}$$

$$= 48 \qquad \text{renaming}$$

[S–5.] a. $53 - 38 = (53 - 38) + 0$ adding 0 to any number does not change the sum

$\qquad\qquad\quad = (53 - 38) + (62 - 62)$ renaming 0 as $62 - 62$

$\qquad\qquad\quad = (53 + 62) - (38 + 62)$ by the property on page 199 of Chapter 5

$\qquad\qquad\quad = 115 - 100$ renaming

$\qquad\qquad\quad = 15$ renaming

b. $81 - 33 = (81 - 33) + 0$ adding 0 to any number does not change the sum.

$\qquad\qquad\quad = (81 - 33) + (67 - 67)$ renaming 0 as $67 - 67$

$\qquad\qquad\quad = (81 + 67) - (33 + 67)$ by the property on page 199 of Chapter 5

$\qquad\qquad\quad = 148 - 100$ renaming

$\qquad\qquad\quad = 48$ renaming

M–7. Represent 35 by 3 bundles of 10 pencils each, held by rubber bands, together with 5 loose pencils. Represent 18 by 1 bundle of 10 pencils together with 8 loose pencils.

     a. For the borrowing algorithm, take the rubber band off 1 bundle, so that 35 is represented by 2 bundles and 15 loose pencils. Then take away 1 bundle and 8 loose pencils.

     b. For the "equal additions" algorithm, add 10 loose pencils to 35 and 1 bundle of 10 to 18. Then 35 is changed to 3 bundles and 15 loose pencils, and 18 is changed to 2 bundles and 8 loose pencils. Then take 2 bundles and 8 pencils away from 3 bundles and 15 pencils.

     c. For the complement algorithm, add 8 bundles and 2 loose pencils to both collections. Then 35 is changed to 11 bundles and 7 pencils and 18 is changed to 10 bundles. Then take 10 bundles away from 11 bundles and 7 pencils.

M–9. No. Children need to be able to *do* subtraction problems; but which algorithm a student uses makes little difference. It is important, however, that they have the proper foundations for understanding them. Teachers should know all three algorithms, however, in order to deal with any textbook and to be able to present alternative algorithms to children who are having trouble with any particular algorithm.

EXERCISES 6–5

S–1. a.   3   10

8	24	80

$24 + 80 = 102$

b.

	3	10
2	6	20
10	30	100

$$6 + 20 + 30 + 100 = 156$$

c.

	3	20
1	3	20
20	60	400

$$3 + 20 + 60 + 400 = 483$$

S–3. a. $(39 \times 71) + (39 \times 29) = 39(71 + 29) = 3{,}900$
  b. $(41 \times 76) + (59 \times 76) = (41 + 59) \times 76 = 7{,}600$
  c. $(146 \times 468) + (146 \times 532) = 146 \times (468 + 532) = 146{,}000$

E–5. To multiply by 25, multiply by 100 and then divide by 4. To multiply by 50, multiply by 100 and then divide by 2.

  a. 400
  b. 6825

  c. 950
  d. 36800

M–7. a. 37 is written under the last 2 digits of 296 instead of under the first 2 digits.
  b. When multiplying 18×37, the student wrote down the product of 7×8, instead of writing down 6 and carrying 5.
  c. The product of 8 and 37 is 296.

M–9. Students should demonstrate their knowledge of the multiplication algorithm by using it to solve a wide variety of multiplication problems. It is not enough to be able to state the algorithm verbally; children must be able to use it.

EXERCISES 6–6

S–1. a.

```
      15
   5)76
      5
     ___
     26
     25
     ___
      1 R
```

There are (at least) 10 5's in 76.
This accounts for 50 out of 76.
There are 26 left over.
Five 5's account for 25 more.
There is a remainder of 1.

$$\begin{array}{r} 12 \\ 24\overline{)293} \\ 24 \\ \hline 53 \\ 48 \\ \hline 5\,R \end{array}$$

b.   There are (at least) 10 24's in 293.
This accounts for 240.
There are 53 left over.
Two 24's account for 48 more.
There is a remainder of 5.

$$\begin{array}{r} 96 \\ 34\overline{)3296} \\ 306 \\ \hline 236 \\ 204 \\ \hline 32\,R \end{array}$$

c.   There are (at least) 90 34's in 3296.
This accounts for 3060.
There are 236 left over.
Six 34's account for 204 more.
There is a remainder of 32.

E–3. a. divisible by 2, 3, 6, 9
      b. divisible by 2, 3, 4, 5, 6, 8, 10
      c. divisible by 2, 3, 4, 6, 8, 9
      d. divisible by 2, 3, 4, 6, 8, 9
      e. divisible only by 1 and 977 (a prime)

M–5. Children might be expected to detect the rules for divisibility by 2, 5, and 10 without help. To detect the others the teacher should present a number of exemplars and nonexemplars of each and have the children consider the digits, sums of digits, last few digits, and so on.

EXERCISES 7–1

S–1. There are two elements in the first set and four in the second; the property can be represented by the fraction $\frac{2}{4}$.

S–3. Three; a and c represented by $\frac{1}{3}$, b represented by $\frac{3}{1}$, d represented by $\frac{2}{2}$

S–5. $\frac{2}{3}, \frac{2}{5}, \frac{3}{2}, \frac{3}{5}$ are all possible answers.

E–7. Some possible answers are

      a. $(\{\ \square\ ,\ \triangle\ ,\ \diagup\ ,\ \bigcirc\ ,\ \square\ \}, \{x, y, z\})$
      b. dividing a 5-foot piece of string into three equivalent parts
      c. taking five pieces of pie from (at least) two pies which have been divided into thirds.
      d. cutting each of several (at least two) pies into thirds and serving five of the resulting pieces.

M–9. Some possible questions are:

What part of a yard is a foot? What part of a foot is an inch?
Can you draw a picture of the fraction $\frac{3}{4}$?
What fraction is represented by *aabbb*?
How would you compare the size of the set of vowels with the set of consonants?

EXERCISES 7–2

S–1. $\dfrac{212}{318} \cong \dfrac{2}{3}$ because $(212) \times (3) = (318) \times (2)$

$\dfrac{4}{7} \not\cong \dfrac{2}{3}$ because $(4) \times (3) \neq (7) \times (2)$

S–3. $\left\{ \dfrac{7}{3}, \dfrac{14}{6}, \cdots, \dfrac{7n}{3n}, \cdots \right\}$

[E–5.] One possible answer is

M–7. Some things you might expect your students to do are: (1) accept equivalent parts in trade (e.g., if he gives you $\frac{3}{4}$ of a pie, he will accept $\frac{6}{8}$ in return but he will not accept $\frac{4}{8}$); and (2) be able to recognize and tell you when parts of a measure are equivalent such as $\frac{1}{2}$ a foot and $\frac{3}{6}$ of a foot.

EXERCISES 7–3.1

S–1. a.. Some possible names are $\frac{5}{12}$, $\frac{15}{36}$, $\frac{20}{48}$, $\frac{25}{60}$, and $\frac{30}{72}$

b. $\left\{ \dfrac{5}{12}, \dfrac{10}{24}, \cdots, \dfrac{5n}{12n}, \cdots \right\}$

S–3. a. $\dfrac{2}{11}, \dfrac{7}{11}, \dfrac{12}{11}, \dfrac{19}{11}$

b. $\dfrac{7}{12}, \dfrac{2}{3}, \dfrac{3}{4}$

c. $\dfrac{1}{5}, \dfrac{2}{9}, \dfrac{4}{15}$

M–5. a. The cancellation is incorrect; to cancel the 5 in the denominator, the numerator must be written as the product of 5 and some other natural number(s).

b. The cancellation would be allowed if there were no 4 present because of the rule $a/b \cong na/nb$

E–7. Yes. Because $\dfrac{a}{b} \cong \dfrac{n \cdot a}{n \cdot b}$ for any natural number n. This is still true if n is the greatest common divisor of the numerator and denominator of the fraction $\dfrac{n \cdot a}{n \cdot b}$.

EXERCISES 7–3.2

S–1. a. .75   c. 3.2

b. .625   d. 2.0

S–3. a, d, e, f, h

S–5.   a.   .1666 ...
       b.   .888 ...
       c.   .307692$\overline{307692}$

M–7.   One possible example is

.10110111011110111110 ... where an additional "1" follows each "0."

Another possible example is

.01001000100001000001....

M–9.   Several questions might be:

1. Find decimal representations of the fractions $\frac{7}{9}$, $\frac{2}{3}$, $\frac{3}{5}$ ....
2. Find fractional representations of decimals
   .25, .37$\overline{37}$, 2.3925$\overline{739257}$ ....
3. Are decimal representations of a particular rational unique?
4. Are fraction representations of a particular rational unique?
5. Write a decimal that does not represent a rational number.

EXERCISES 7–4

S–1.   a. $\dfrac{1}{6} + \dfrac{3}{10} = \dfrac{1 \cdot 10 + 6 \cdot 3}{6 \cdot 10} = \dfrac{28}{60}$

       b. $\dfrac{1}{6} + \dfrac{3}{10} = \dfrac{5}{30} + \dfrac{9}{30} = \dfrac{14}{30}$

       $\dfrac{28}{60} \cong \dfrac{14}{30}$ since $28 \cdot 30 = 840 = 60 \cdot 14$.

M–3. $\frac{1}{4}$ of all 20 paper clips is 5 clips.

$\frac{2}{5}$ of all 20 paper clips is 8 clips.

$\frac{1}{4} + \frac{2}{5}$ of all 20 paper clips is 13 clips, but 13 clips is $\frac{13}{20}$ of all clips.

Hence, $\frac{1}{4} + \frac{2}{5} = \frac{13}{20}$.

M–5.   a. $\dfrac{a}{b} + \dfrac{c}{d} = \dfrac{a+c}{b+d}$

       b.   By using "pieces of a pie." Two-thirds of a pie plus four-fifths of a pie must be more than a whole pie and not six-eights of a pie. This "algorithm" also leads to such nonsensical results as $\frac{1}{2} + \frac{1}{2} = \frac{2}{4}$, $\frac{1}{3} + \frac{0}{2} = \frac{1}{5}$, $\frac{1}{4} + \frac{3}{4} = \frac{4}{8}$, and so on.

E–7. $\dfrac{a}{b} + \dfrac{c}{b} = \dfrac{a \cdot b + b \cdot c}{b \cdot b}$ by definition of addition

       $= \dfrac{b \cdot (a + c)}{b \cdot b}$ by distributive and commutative laws for natural numbers

       $\cong \dfrac{a + c}{b}$ by definition of equivalence of fractions

EXERCISES 7–5

S–1. a. $\dfrac{7}{8} - \dfrac{7}{10} = \dfrac{7 \cdot 10 - 8 \cdot 7}{8 \cdot 10} = \dfrac{70 - 56}{80} = \dfrac{14}{80}$

b. $\dfrac{7}{8} - \dfrac{7}{10} = \dfrac{7 \cdot 5}{8 \cdot 5} - \dfrac{7 \cdot 4}{10 \cdot 4} = \dfrac{35}{40} - \dfrac{28}{40} = \dfrac{35 - 28}{40} = \dfrac{7}{40}$

Note that $\dfrac{14}{80} \cong \dfrac{7}{40}$ because $14 \cdot 40 = 7 \cdot 80$.

S–3. To show that subtraction is well defined, we must show that no matter what pairs of representative fractions are chosen from the equivalence classes of two rational numbers, the difference always results in equivalent fractions. Let the equivalence classes of the two rational numbers be:

$$\left\{ \frac{a}{b}, \frac{2a}{2b}, \frac{3a}{3b}, \cdots, \frac{ma}{mb}, \cdots \right\} \text{ where } m \text{ is a natural number}$$

$$\left\{ \frac{c}{d}, \frac{2c}{2d}, \frac{3c}{3d}, \cdots, \frac{nc}{nd}, \cdots \right\} \text{ where } n \text{ is a natural number}$$

We must show that $\dfrac{a}{b} - \dfrac{c}{d}$ is equivalent to $\dfrac{ma}{mb} - \dfrac{nc}{nd}$.

Now $\dfrac{a}{b} - \dfrac{c}{d} = \dfrac{ad - bc}{bd}$ by definition

And $\dfrac{ma}{mb} - \dfrac{nc}{nd} = \dfrac{ma \cdot nd - mb \cdot nc}{mb \cdot nd}$ by definition

$\qquad\qquad = \dfrac{mn \cdot ad - mn \cdot bc}{mn \cdot bd}$ by commutative and associative properties of multiplication

$\qquad\qquad = \dfrac{mn \cdot [ad - bc]}{mn \cdot bd}$ by distributive property

But $\dfrac{mn \cdot [ad - bc]}{mn \cdot bd} \cong \dfrac{ad - bc}{bd}$ because

$mn \cdot [ad - bc] \cdot bd = mn \cdot bd \cdot [ad - bc]$ by the commutative property of multiplication

E–5. Yes. Some particular cases are:

$$\frac{1}{4} = \frac{3}{4} - \frac{1}{2} \text{ if and only if } \frac{3}{4} = \frac{1}{2} + \frac{1}{4}$$

$$\frac{1}{4} = \frac{7}{8} - \frac{5}{8} \text{ if and only if } \frac{7}{8} = \frac{5}{8} + \frac{1}{4}$$

$$\frac{7}{13} = \frac{11}{13} - \frac{4}{13} \text{ if and only if } \frac{11}{13} = \frac{4}{13} + \frac{7}{13}$$

M–7. Divide a circle into 18 equal slices. $\frac{8}{9}$ of these sections is 16 sections (shaded horizontally); $\frac{5}{6}$ of these is 15 sections (shaded vertically). The number of sections shaded horizontally but *not* vertically is 1. Hence, $\frac{8}{9} - \frac{5}{6} = \frac{1}{18}$.

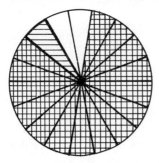

EXERCISES 7–6

[S–1.] $\frac{a}{b} \times \frac{c}{d} = \frac{ac}{bd}$ by definition.

Now ac is a natural number because the natural numbers are closed under multiplication.

And bd is a natural number for the same reason.

Therefore, $\frac{ac}{bd}$ is a fraction representative of a rational number.

S–3. a. state-by-state interpretation

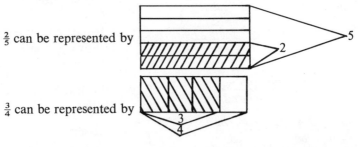

$\frac{2}{5}$ can be represented by

$\frac{3}{4}$ can be represented by

$\frac{2}{5} \times \frac{3}{4}$ can be represented by the cross-hatched area $= \frac{6}{20}$

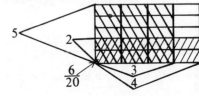

state-by-operator interpretation

$\frac{2}{5}$ can be represented by

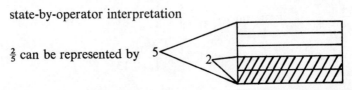

Partition the state represented by $\frac{2}{5}$ into fourths.

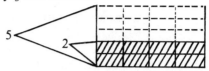

The answer is represented by that part of the original shading included in three of the fourths.

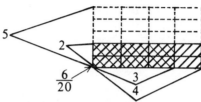

b.   state-by-state interpretation

$\frac{5}{3}$ can be represented by

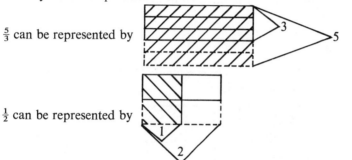

$\frac{1}{2}$ can be represented by

$\frac{5}{3} \times \frac{1}{2}$ is represented by the ratio of the cross-hatched area (five sections) to the area contained entirely in the *unit* rectangle (six sections).

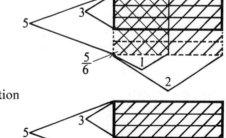

state-by-operator interpretation

$\frac{5}{3}$ can be represented by

Partition the state represented by $\frac{5}{3}$ into halves. Then $\frac{5}{3} \times \frac{1}{2}$ is represented by that part of the unit rectangle (6 sections) that is included in one of the halves.

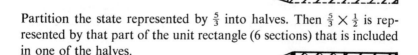

M–5. Four possible test questions are:

    1. Construct a state-by-state interpretation of $\frac{2}{3} \times \frac{1}{4}$ using the unit rectangle and show that your answer gives the same result as the definition of multipication.

    2. Construct a state-by-operator interpretation of $\frac{3}{4} \times \frac{1}{2}$, using the unit rectangle.

    3. Find the products $\frac{1}{4} \times \frac{8}{9}$, $\frac{7}{8} \times \frac{1}{3}$, and $\frac{3}{4} \times \frac{6}{11}$.

    4. Find the products $.76 \times 1.2$, $10.6 \times .0024$ and $\frac{1}{4} \times .60$.

E–7. a. $\dfrac{x}{y} = \dfrac{1}{1}$ or any fraction equivalent to $\dfrac{1}{1}$

      b. $\dfrac{x}{y} = \dfrac{9}{7}$ or any fraction equivalent to $\dfrac{9}{7}$

      c. $\dfrac{x}{y} = \dfrac{21}{8}$ or any fraction equivalent to $\dfrac{21}{8}$

EXERCISES 7–7

S–1. $\frac{3}{4} \div \frac{1}{2}$. The least common multiple of the denominator is 4. Thus $\frac{3}{4}$ is represented by $\frac{3}{4}$ and $\frac{1}{2}$ by $\frac{2}{4}$.

$\frac{3}{4}\longleftrightarrow$ 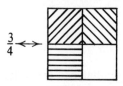 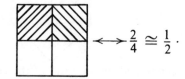 $\longleftrightarrow \frac{2}{4} \cong \frac{1}{2}$.

Hence $\frac{3}{4} \div \frac{1}{2}$ equals $\frac{3}{2}$.

[S–3.] $r_1 = \dfrac{x}{y}$ and $r_2 = \dfrac{z}{w}$. Then $r_1 \div r_2$ is represented by $\dfrac{x}{y} \div \dfrac{z}{w}$, which by the definition of division, is equal to $\dfrac{xw}{yz}$. Because x, w, y, and z are natural numbers, xw and yz are also natural numbers, because the naturals are closed under multiplication. Thus, $\dfrac{xw}{yz}$ represents a positive rational number. This property is the closure property for division of positive rational numbers.

S–5. 2.325

[M–7.] $1.23\frac{4}{7}$. This is not good notation because the meaning of the $\frac{4}{7}$ is ambiguous. Does it represent $\frac{4}{7}$ of a whole or $\frac{4}{7}$ of .01? In this case it represents the latter. It is better to "round off" such answers to, say, 1.24.

EXERCISES 7–8.1

S–1. a. yes   d. yes

       b. no   e. yes

       c. yes   f. yes

E–3. a. closed
      b. not closed
      c. closed
      d. not closed—consider $\frac{3}{4} \div \frac{0}{3}$

EXERCISES 7–8.2

S–1. $\frac{3}{4} + \frac{5}{2} = \frac{3}{4} + \frac{10}{4} = \frac{3 + 10}{4} = \frac{10 + 3}{4} = \frac{10}{4} + \frac{3}{4} = \frac{5}{2} + \frac{3}{4}.$

This illustrates the commutative property of addition of positive rational numbers.

S–3. 1. $\dfrac{a}{b} \cdot \dfrac{c}{d} = \dfrac{a \cdot c}{b \cdot d}$ by definition

      2. $= \dfrac{c \cdot a}{d \cdot b}$ by commutative property of multiplication of natural numbers

      3. $\dfrac{c}{d} \cdot \dfrac{a}{b}$ by definition of multiplication of positive rationals

EXERCISES 7–8.3

S–1. $r_I \div \dfrac{3}{4} = r_I \times \dfrac{4}{3} = \dfrac{4}{3} \neq \dfrac{3}{4}. \left(\text{Any rational not equivalent to } r_I \text{ will work in place of } \dfrac{3}{4}.\right)$

E–3. $r_o \times r = r_o$ because $\dfrac{0}{n} \times \dfrac{ma}{mb} \cong \dfrac{0}{n}.$

EXERCISES 7–8.4

S–1. $\dfrac{a}{b} \times \dfrac{b}{a} = \dfrac{a \times b}{b \times a} = \dfrac{ab}{ba} = \dfrac{ab}{ab} \cong \dfrac{1}{1}$ (the multiplicative identity)

EXERCISES 7–8.5

S–1. Both are equal to $\frac{19}{60}$

S–3. a. $\dfrac{4}{3}$ b. $\dfrac{11}{17}$ c. $\dfrac{14}{15}$

EXERCISES 7–8.6

S–1. The rational number m/n must appear in the mth row (down from the top) and the nth column (from the left) for any and all natural numbers m and n.

S–3. a. There is none.
      b. $\frac{0}{1}$, that is, r_o.

EXERCISES 8–1

S–1. See the first paragraph of this section for three possible reasons.

S–3. One example of each is:

a. ($\{a, b, c\}, \{d, e\}$)   e. ({Chris, Wendy, Sandi}, {Lou})

b. ($\{1\}, \{2, 3, 4, 5\}$)   f. ($\{\ \ \}, \{\triangle, \bigcirc\}$)

c. ($\{\square, \square\}, \{,\square\ \square\}$)   g. ($\{\square\ , \square\ , \square\ \}, \{\ \ \}$)

d. ($\{\ \text{♀}\ \}, \{\ \text{♀}\ \}$)

[E–5.] $^{-}2$ (see Section 8.3). Can you think of any other good notations for the property "2-less"?

M–7. For example, some questions might be:

(1) Which tower contains more blocks?

(2) How many blocks would have to be removed from the larger tower to make it the same height as the smaller tower?

(3) The tower of three blocks is how many less than the tower of five blocks?

M–9. Some methods for doing this might be:

(1) Consider the game of "giant steps" (see Section 2), where a leader gives the players a pair of numbers, say (2, 4), meaning "take two steps forward and four steps backward." Thus, (2, 4) and (4, 2) result in different final positions.

(2) Let the left-hand position be the number of toys Janie gets and the right-hand position the number of toys Joey gets. Thus, (2, 4) and (4, 2) represent quite different situations for Janie and Joey.

(3) Can you find other methods?

EXERCISES 8–2

S–1. Some examples are:

a. (2, 0), (5, 3), (17, 15), (27, 25)

b. (1, 4), (8, 11), (0, 3), (31, 34)

c. (5, 5), (7, 7), (0, 0), (6, 6)

d. (6, 5), (1, 0), (10, 9), (8, 7)

e. same as "no-moreness." (answer c above)

S–3. a. 2-lessness

b. 0-moreness = 0-lessness = no-moreness

c. 3-lessness

d. 7-moreness

S–5.   Some examples are:

    a.   (6, 0), (12, 6), (74, 68), (50, 44)
    b.   (4, 0), (31, 27), (12, 8), (23, 19)
    c.   (0, 3), (12, 15), (6, 9), (72, 75)
    d.   (0, 0), (15, 15), (23, 23), (61, 61)
    e.   (4, 7), (3, 6), (7, 10), (12, 15)
    f.   (5, 5), (51, 51), (8, 8), (64, 64)
    g.   (1, 0), (6, 5), (4, 3), (32, 31)

M–7.   Suppose the child is given the command (5, 2). Then you might say to him, "Try to get where you are going in the shortest way possible." Repeat this with several other ordered-pair commands.

EXERCISES 8–3

[S–1.]   a.

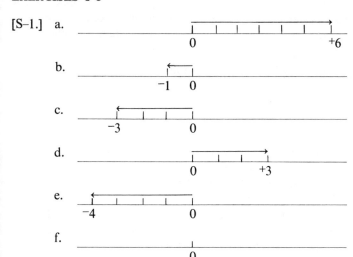

[S–3.]   a.   (0, 2)   e.   (5, 0)
       b.   (12, 0)   f.   (0, 7)
       c.   (0, 0)   g.   (0, 2)
       d.   (10, 0)   h.   (0, 0)

E–5.   It is not possible to find an integer between *any* two different integers, for example, between $^{+}4$ and $^{+}5$. Hence, the integers are not dense.

EXERCISES 8–4

S–1.   a.   (7, 2) + (3, 6) = (10, 8) which represents $^{+}2$.
      b.   (3, 0) + (16, 10) = (19, 10) which represents $^{+}9$.
      c.   (9, 12) + (4, 6) = (13, 18) which represents $^{-}5$.
      d.   (7, 1) + (3, 3) = (10, 4) which represents $^{+}6$.

S–3. $(3, 8) + (10, 4) = (13, 12)$

$(1, 6) + (8, 2) = (9, 8)$

$(13, 12) \doteq (9, 8)$, because $13 + 8 = 12 + 9$

S–5. $(m + 5, m) + (n, n + 2) = (m + 5 + n, m + n + 2)$.

$(m + 5 + n, m + n + 2) \doteq (5, 2)$ because

$m + 5 + n + 2 = m + n + 2 + 5$.

(We could prove that these are equal by repeatedly applying the associative and commutative properties of addition of whole numbers.)

[S–7.] Let ^-m and ^+n be represented by $(0, m)$ and $(n, 0)$, respectively, where $m \geq n$. Then $^-m + {}^+n$ is represented by $(0, m) + (n, 0) = (n, m) = (0, m - n)$, which represents $^-(m - n)$. Algorithm 4 gives the same result.

E–9. Simply map $^+0 \leftrightarrow 0$, $^+1 \leftrightarrow 1$, $^+2 \leftrightarrow 2$, $^+3 \leftrightarrow 3$, etc. This mapping is one to one and onto. Also, addition is preserved by the mapping, because if m and n are any natural numbers, then $^+m + {}^+n = {}^+(m + n)$ corresponds to the whole number $m + n$.

M–11. See Problem S–8. Can you construct similar situations?

EXERCISES 8–5

S–1. a. $(7, 6) - (1, 4) = (6, 2)$, which represents $^+4$.

b. $(7, 3) - (3, 2) = (4, 1)$, which represents $^+3$.

c. $(3, 5) - (0, 1) = (3, 4)$, which represents $^-1$.

d. $(5, 7) - (1, 3) = (4, 4)$, which represents 0.

e. $(7, 3) - (6, 5)$ cannot be done, since $3-5$ is undefined for whole numbers. We need to choose a different representative of $^+4$ than $(7, 3)$, say $(10, 6)$. Then $(10, 6) - (6, 5) = (4, 1)$, which represents $^+3$.

S–3. By the definition of subtraction, $(3, 0) - (0, 2) = (3 - 0, 0 - 2)$, but $0 - 2$ is undefined in the whole numbers.

E–5. a. $(5, 2) - (0, 2) = (5, 0)$

b. $(10, 7) - (4, 6) = (6, 1)$

b. $(5, 0)$ is more-less equivalent to $(6, 1)$, because $5 + 1 = 0 + 6$.

S–7. a.

b.

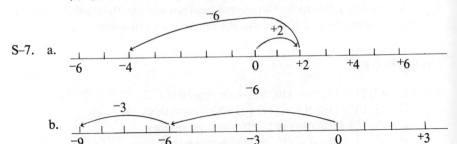

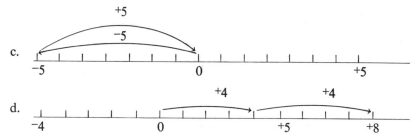

S–9. a. $^+212 - {}^+32$ can be represented by $(212, 0) - (32, 0) = (180, 0)$, which represents $^+180$. So the difference between the boiling and freezing points is $180°$.

b. $^+313 - {}^-13$ can be represented by $(326, 13) - (0, 13) = (326, 0)$, which represents $^+326$. Thus, the difference between the highest and lowest points is 326 feet.

c. $^+21 - {}^-9$ can be represented by $(30, 9) - (0, 9) = (30, 0)$ which represents $^+30$. Thus, the temperature rose $30°$.

[M–11.] Pair $^+3 + {}^+2 = {}^+5$ with $^+5 - {}^+2 = {}^+3$

$^+8 + {}^-6 = {}^+2$ with $^+2 - {}^-6 = {}^+8$

$^-5 + {}^+3 = {}^-2$ with $^-2 - {}^+3 = {}^-5$

EXERCISES 8–6

S–1. a. $(5, 3) \times (3, 2) = (21, 19)$, which represents $^+2$.

b. $(6, 8) \times (4, 1) = (32, 38)$, which represents $^-6$.

c. $(3, 6) \times (1, 5) = (33, 21)$, which represents $^+12$.

d. $(7, 7) \times (6, 2) = (56, 56)$, which represents $^+0$.

S–3. a. $^+3 \times {}^+2 = {}^+6$

b. $^-4 \times {}^-1 = {}^+4$

c. $^+5 \times {}^-2 = {}^-10$

d. $^-3 \times {}^+3 = {}^-9$

e. $^-16 \times {}^+13 = {}^-208$

f. $^-18 \times {}^+0 = {}^+0$

g. $^+63 \times {}^-11 = {}^-693$

h. $^-4 \times {}^-132 = {}^+528$

S–5. $(6, 3) \times (2, 6) = (30, 42)$

$(4, 1) \times (0, 4) = (4, 16)$

$(30, 42) \doteq (4, 16)$, because $30 + 16 = 42 + 4$.

S–7. a. $^-5 \times {}^+4$ can be represented by $(0, 5) \times (4, 0) = (0, 20)$, which represents $^-20$. Thus, the temperature decreases by $20°$.

b. $^-3 \times {}^-4$ can be represented by $(5, 8) \times (1, 5) = (45, 33)$, which represents $^+12$. So, at 11:00 AM there were 12 more gallons of water in the lake than at 3:00 PM.

c. $^+3 \times {}^-2$ can be represented by $(3, 0) \times (5, 7) = (15, 21)$, which represents $^-6$. Thus, the drive shaft turns counterclockwise at 6 rps. (In the above problems, you could choose completely different ordered-pair representatives.)

M–9. See Problem S–7 for suggestions.

APPENDIX C

EXERCISES 8–7

S–1. a. $+10 \div {}^-5 = {}^-2$ e. $+26 \div {}^-13 = {}^-2$
     b. ${}^-12 \div {}^+4 = {}^-3$ f. $+48 \div {}^+4 = {}^+12$
     c. ${}^-6 \div {}^-3 = {}^+2$ g. ${}^-63 \div {}^-21 = {}^+3$
     d. $+8 \div {}^+4 = {}^+2$ h. ${}^-16 \div {}^+1 = {}^-16$

∗S–3. a. $|3 - 9| \div |6 - 4| = 6 \div 2 = 3 = z.$
       Because $3 \leq 9$ and $6 > 4, p = 0.$ Then
       $(3, 9) \div (6, 4) = (0 \cdot 3, (1 - 0) \cdot 3) = (0, 3)$, which represents ${}^-3.$

     b. $|8 - 2| \div |3 - 2| = 6 \div 1 = 6 = z.$ Because $8 > 2$ and $3 > 2, p = 1.$
       Then $(8, 2) \div (3, 2) = (1 \cdot 6, (1 - 1) \cdot 6) = (6, 0)$, which represents ${}^+6.$

     c. $|6 - 0| \div |1 - 4| = 6 \div 3 = 2 = z.$ Because $6 > 0$ and $1 < 4, p = 0\cdot$
       Then $(6, 0) \div (1, 4) = (0 \cdot 2, (1 - 0) \cdot 2) = (0, 2)$, which represents ${}^-2.$

     d. $|6 - 6| \div |7 - 2| = 0 \div 5 = 0 = z.$ Because $6 \geq 6$ and $7 > 2, p = 1.$
       Then $(6, 6) \div (7, 2) = (1 \cdot 0, (1 - 1) \cdot 0) = (0, 0)$, which represents $0.$

     e. $|7 - 3| \div |8 - 5| = 4 \div 3$, which is not a whole number. So $+4 \div {}^+3$ is not defined.

E–5. a. $+10 \div {}^-5 = {}^-2$ if and only if $+10 = {}^-2 \times {}^-5.$ ${}^-2 \times {}^-5$ *does* equal $+10$; the answer to problem S–1 (a) is correct. Similarly,
     b. ${}^-12 \div {}^+4 = {}^-3$ if and only if ${}^-12 = {}^-3 \times {}^+4.$
     c. ${}^-6 \div {}^-3 = {}^+2$ if and only if ${}^-6 = {}^+2 \times {}^-3.$
     d. $+8 \div {}^+4 = {}^+2$ if and only if $+8 = {}^+2 \times {}^+4.$
     e. $+26 \div {}^-13 = {}^-2$ if and only if $+26 = {}^-2 \times {}^-13.$
     f. $+48 \div {}^+4 = {}^+12$ if and only if $+48 = {}^+12 \times {}^+4.$
     g. ${}^-63 \div {}^-21 = {}^+3$ if and only if ${}^-63 = {}^+3 \times {}^-21.$
     h. ${}^-16 \div {}^+1 = {}^-16$ if and only if ${}^-16 = {}^-16 \times {}^+1.$

[M–7.] a. $+8 \div {}^+2 = $ _____ (positive ÷ positive) ⎫ Choose one from
       b. ${}^-10 \div {}^+5 = $ _____ (negative ÷ positive) ⎪ each group, vary-
       c. ${}^-6 \div {}^-2 = $ _____ (negative ÷ negative) ⎬ ing the magni-
       d. $+9 \div {}^-3 = $ _____ (positive ÷ negative) ⎭ tudes of the numbers.

[M–9.] See text and Exercise sets 8–4, 8–5, 8–6, 8–7 for suggestions.

EXERCISES 8–8

S–1. $\{{}^-2, {}^-1\}$

S–3.   Some such equations are given below:

	Roots	Equation	p	q
a.	$\{1, 2\}$	$x^2 + {}^-3x + {}^+2 = 0$	$^-3$	$^+2$
	$\{2, 3\}$	$x^2 + {}^-5x + {}^+6 = 0$	$^-5$	$^+6$
	$\{3, 4\}$	$x^2 + {}^-7x + {}^+12 = 0$	$^-7$	$^+12$
b.	$\{^-1, {}^-2\}$	$x^2 + {}^+3x + {}^+2 = 0$	$^+3$	$^+2$
	$\{^-2, {}^-3\}$	$x^2 + {}^+5x + {}^+6 = 0$	$^+5$	$^+6$
	$\{^-3, {}^-4\}$	$x^2 + {}^+7x + {}^+12 = 0$	$^+7$	$^+12$
c.	$\{^+5, {}^-2\}$	$x^2 + {}^-3x + {}^-10 = 0$	$^-3$	$^-10$
	$\{^+4, {}^-3\}$	$x^2 + {}^-1x + {}^-12 = 0$	$^-1$	$^-12$
	$\{^-5, {}^+2\}$	$x^2 + {}^+3x + {}^-10 = 0$	$^+3$	$^-10$
	$\{^-4, {}^+3\}$	$x^2 + {}^+1x + {}^-12 = 0$	$^+1$	$^-12$

The regularities you might observe are:

1. If both roots are positive, then p is negative and q is positive.
2. If both roots are negative, then p and q are both positive.
3. If the roots are of opposite sign and the magnitude of the positive root is larger than that of the negative root, then p is negative and q is negative.
4. If the roots are of opposite sign and the magnitude of the negative root is larger than that of the positive root, then p is positive, and q is negative.

EXERCISES 8–8.1

S–1.   a.   yes
       b.   yes
       c.   yes
       d.   no
       See Sections 8–4, 8–5, 8–6, 8–7 for verification of these answers.

EXERCISES 8–8.2

S–1. $^+3 - {}^-4 = {}^+7$ and $^-4 - {}^+3 = {}^-7.$ $^-2 - {}^+7 = {}^-9$ and $^+7 - {}^-2 = {}^+9$

[S–3.]   For example,

$^+3 \div {}^-1 \neq {}^-1 \div {}^+3$ (commutative)

$^+48 \div ({}^-12 \div {}^-2) \neq ({}^+48 \div {}^-12) \div {}^-2$ (associative)

S–5.   a. $^+7 + {}^-5 = {}^-5 + {}^+7$ because
            $^+7 + {}^-5 = {}^+2$ and $^-5 + {}^+7 = {}^+2$
       b. $^-9 + {}^-2 = {}^-2 + {}^-9$ because
            $^-9 + {}^-2 = {}^-11$ and $^-2 + {}^-9 = {}^-11$
       c. $^-6 + {}^+4 = {}^+4 + {}^-6$ because
            $^-6 + {}^+4 = {}^-2$ and $^+4 + {}^-6 = {}^-2$

S–7. a. $^-7 \cdot {}^+5 = {}^-35$ and
$^+5 \cdot {}^-7 = {}^-35$

b. $^+8 \cdot {}^+9 = {}^+72$ and
$^+9 \cdot {}^+8 = {}^+72$

c. $^-4 \cdot {}^+12 = {}^-48$ and
$^+12 \cdot {}^-4 = {}^-48$

EXERCISES 8–8.3

S–1. a. $^-12 + {}^+0 = {}^-12$ c. $^+0 + {}^-6 = {}^-6$
b. $^+3 + {}^+0 = {}^+3$ d. $^+0 + {}^+0 = {}^+0$

S–3. a. $^+1 \cdot {}^-7 = {}^-7$ c. $^+8 \cdot {}^+1 = {}^+8$
b. $^+1 \cdot {}^+4 = {}^+4$ d. $^-6 \cdot {}^+1 = {}^-6$

[M–5.] For example, try exercises of the following type.

$$^+3 + {}^+0 = \underline{\hspace{2em}} \qquad \underline{\hspace{2em}} + {}^-3 = {}^-3$$

$$^+2 + {}^+0 = \underline{\hspace{2em}} \qquad {}^+6 + \underline{\hspace{2em}} = {}^+6$$

$$^+0 + {}^-5 = \underline{\hspace{2em}} \qquad {}^+0 + \underline{\hspace{2em}} = \underline{\hspace{2em}}$$

$$^+0 + {}^+3 = \underline{\hspace{2em}} \qquad\qquad\qquad \text{etc.}$$

$$^+4 + {}^+0 = \underline{\hspace{2em}}$$

EXERCISES 8–8.4

S–1. a. $^+4$ c. There is none.
b. $^-4$ d. $^-4$

S–3. a. $^+4, {}^-7, {}^+0, {}^+5, {}^-3$
b. $^-4, {}^+7, {}^+0, {}^-5, {}^+3$
c. They are additive inverses.

EXERCISES 8–8.5

S–1. a. $^+3 \cdot ({}^-6 + {}^+7) = {}^+3 \cdot {}^+1 = {}^+3$
$({}^+3 \cdot {}^-6) + ({}^+3 \cdot {}^+7) = {}^-18 + {}^+21 = {}^+3$
b. $^-5 \cdot ({}^+1 + {}^-4) = {}^-5 \cdot {}^-3 = {}^+15$
$({}^-5 \cdot {}^+1) + ({}^-5 \cdot {}^-4) = {}^-5 + {}^+20 = {}^+15$
c. $^-3 \cdot ({}^+6 + {}^-7) = {}^-3 \cdot {}^-1 = {}^+3$
$({}^-3 \cdot {}^+6) + ({}^-3 \cdot {}^-7) = {}^-18 + {}^+21 = {}^+3$

M–3. Because the student is drawing a conclusion about the entire set of integers based on one example, you might propose analogous arguments such as, "All students in the class are girls because Jane is a girl." This should help convince the student of the absurdity of his argument (unless the class happens to contain only girls).

EXERCISES 8–8.6

S–1.   Merely substitute $^-3$ for each occurrence of x in the general proof given in the text.

[E–3.]   a.   The assumptions or premises are:

(1) x is any integer
(2)   distributive property of integers
(3)   associative property of addition of integers
(4)   additive inverse property of integers
(5)   additive identity property of integers
(6)   additive property of equality (closure)
(7)   multiplicative property of equality (closure)

b.   Conclusion: For any integer x, $x \cdot 0 = 0$.

EXERCISES 8–8.7

S–1.   a. $^+3$ d. $^-14$
b. $^+2$ e. $^-135$
c. $^+1$

E–3.   The natural number 2 refers to the numerousity of sets. The rational number $\frac{2}{1}$ refers to the relative property of pairs of natural numbers. The integer $^+2$ refers to the more-less property of pairs of natural numbers.

EXERCISES 9–1

S–1.   a. $^+8 = {}^+8 \times {}^+1 = {}^+2 \times {}^+4 = {}^+8$
b. $^-42 = {}^-3 \times {}^+14 = {}^-6 \times {}^+7 = {}^-42$
c. $^+0 = {}^+0 \times {}^+4 = {}^+2 \times {}^+0 = {}^+0$
d. $^+40 = {}^-10 \times {}^-4 = {}^+2 \times {}^+20 = {}^+40$

S–3.   a. $^+24$ c. $^+35$
b. $^+51$ d. $^+12$ or $^-12$

S–5.   a.   yes   c.   yes
b.   no   d.   no

[S–7.]   Some possible representations are:

a. $\frac{^+1}{^+4}$, $\left(\frac{3}{4}, \frac{1}{2}\right)$ d. $\frac{^-2}{^+7}$, $\left(\frac{0}{7}, \frac{2}{7}\right)$

b. $\frac{^-1}{^+4}$, $\left(\frac{1}{2}, \frac{3}{4}\right)$ e. $\frac{^+2}{^+2}$, $\left(\frac{4}{3}, \frac{1}{3}\right)$

c. $\frac{^-19}{^-3}$, $\left(\frac{19}{3}, \frac{0}{3}\right)$ f. $\frac{^+0}{^-5}$, $\left(\frac{5}{2}, \frac{5}{2}\right)$

S–9. a. $\left(\frac{1}{4}, \frac{0}{4}\right) + \left(\frac{0}{7}, \frac{2}{7}\right) = \left(\frac{1}{4}, \frac{2}{7}\right) = \frac{-1}{28}$ because $\frac{2}{7} - \frac{1}{4} = \frac{1}{28}$

b. $\frac{+0}{1}$

c. $\frac{-15}{28}$

d. $\frac{+1}{4}$

S–11. a. $\left(\frac{1}{4}, \frac{0}{4}\right) \times \left(\frac{19}{3}, \frac{0}{3}\right) = \left(\frac{1}{4} \times \frac{19}{3} + \frac{0}{4} \times \frac{0}{3}, \frac{1}{4} \times \frac{0}{3} + \frac{0}{4} \times \frac{19}{3}\right)$

$$= \left(\frac{19}{12}, \frac{0}{12}\right) = \frac{+19}{12}$$

b. $\frac{-2}{28} = \frac{-1}{14}$

c. $\frac{+2}{28} = \frac{+1}{14}$

d. $\frac{+1}{4}$

∗E–13. Because $\frac{x}{y}$ and $\frac{x'}{y'}$ both represent R, we know that $x \cdot y' = y \cdot x'$.

Because $\frac{w}{z}$ and $\frac{w'}{z'}$ both represent R', we know that $w \cdot z' = z \cdot w'$.

$$R + R' = \frac{x}{y} + \frac{w}{z} = \frac{x \cdot z + y \cdot w}{y \cdot z} = \frac{(x' \cdot y' \cdot z' \cdot w')(x \cdot z + y \cdot w)}{(x' \cdot y' \cdot z' \cdot w') \cdot y \cdot z}$$

$$= \frac{(x \cdot y') \cdot (z \cdot w') \cdot (x' \cdot z') + (y \cdot x') \cdot (w \cdot z')(y' \cdot w')}{(y \cdot x') \cdot (z \cdot w')(y' \cdot z')}$$

$$= \frac{(x \cdot y')(z \cdot w')(x' \cdot z' + y' \cdot w')}{(x \cdot y') \cdot (z \cdot w')(y' \cdot z')}$$

$$= \frac{x' \cdot z' + y' \cdot w'}{y' \cdot z'} = \frac{x'}{y'} + \frac{w'}{z'}$$

Also, $R \cdot R' = \frac{x}{y} \cdot \frac{w}{z} = \frac{x \cdot w}{y \cdot z} = \frac{x \cdot w \cdot (y' \cdot z' \cdot x' \cdot w')}{y \cdot z \cdot (y' \cdot z' \cdot x' \cdot w')}$

$$= \frac{(x \cdot y')(w \cdot z')(x' \cdot w')}{(y \cdot x')(z \cdot w')(y' \cdot z')}$$

$$= \frac{(x \cdot y')(w \cdot z')(x' \cdot w')}{(x \cdot y')(w \cdot z')(y' \cdot z')}$$

$$= \frac{x' \cdot w'}{y' \cdot z'} = \frac{x'}{y'} \cdot \frac{w'}{z'}$$

S–15. a. $\frac{-4}{5}, \frac{+3}{7}, -1, 0$

b. If R' is the inverse of R, and R'', of R', then

$$R + R' = 0 \qquad\qquad \frac{+4}{5} + \frac{-4}{5} = 0$$

$$Q - R = Q + R' \qquad\qquad \frac{+2}{3} - \frac{+4}{5} = \frac{+2}{3} + \frac{-4}{5} = \frac{-2}{15}$$

$$R'' = R \qquad\qquad \left(\frac{+2}{5}\right)'' = \left(\frac{-2}{5}\right)' = \frac{+2}{5}$$

c. R' is the multiplicative inverse of R if $R \times R' = 1$. 0 has no multiplicative inverse; all other rational numbers have inverses.

The inverses of $\frac{+5}{4}$ and $\frac{-5}{4}$ are $\frac{+4}{5}$ and $\frac{-4}{5}$, respectively.

[S–17.] Using the definition involving non-negative rationals $\left(\text{e.g., } \dfrac{a}{b}\right)$

$$\text{let } R = \left\{\left(\frac{a}{b}, \frac{c}{d}\right)\right\}$$

$$\text{let } R' = \left\{\left(\frac{a'}{b'}, \frac{c'}{d'}\right)\right\}$$

$$R + R' = \left\{\left(\frac{a}{b} + \frac{a'}{b'}, \frac{c}{d} + \frac{c'}{d'}\right)\right\}$$

$$\qquad = \left\{\left(\frac{a'}{b'} + \frac{a}{b}, \frac{c'}{d'} + \frac{c}{d}\right)\right\} \qquad \text{(Addition of non-negative rationals is commutative.)}$$

$$\qquad = R' + R$$

and, using the definition involving integers (e.g., x, y),

let R be represented by $\dfrac{x}{y}$

let R' be represented by $\dfrac{x'}{y'}$

$R \cdot R'$ is represented by $\dfrac{x \cdot x'}{y \cdot y'} = \dfrac{x' \cdot x}{y' \cdot y}$ (Multiplication of integers is commutative.)

which represents $R' \cdot R$

S–19. a. additive identity property; additive and multiplicative inverse properties
b. multiplicative inverse property
c. additive identity and inverse properties
d. This is a field.

M–21. One possibility might be to paste labeled disks at appropriate positions on a number line to indicate equivalent fractions.

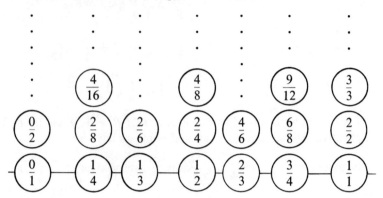

EXERCISES 9–2

S–1. a. 5, 5.0, 5.00, 5.000, 5.0000
      b. .4, .42, .428, .4285, .42857
      c. .4, .45, .450, .4500, .45000
    · d. .1, .12, .123, .1231, .12312

S–3. a, e, h converge to 0

  b, c, d, g converge to $\frac{1}{3}$

  f converges to $\frac{1}{33}$

*S–5. a. $\dfrac{1}{2 \times 10^n}$ $\qquad\qquad\qquad$ d. $\dfrac{10^n + 1}{3 \times 10^n}$

*S–7. $\dfrac{3n-1}{5n} = \dfrac{3}{5} - \dfrac{1}{5n} \to \dfrac{3}{5}$

  $\dfrac{3n+1}{5n} = \dfrac{3}{5} + \dfrac{1}{5n} \to \dfrac{3}{5}$

  Both converge to the same number; hence, they are equivalent. Also, the general term of the difference sequence is $\dfrac{2}{5n}$, which converges to 0.

[S–9.] $.55\overline{5}$ and $.255\overline{5}$, for example

S–11. $\dfrac{11}{7}$ because $\left(\dfrac{17}{7}, \dfrac{28}{14}, \dfrac{39}{21}, \ldots, \dfrac{11n+6}{7n}, \ldots \right) \to \dfrac{11}{7}$

S–13. Adding the corresponding distinguished decimal sequences term by term we obtain the sequence $.a_1, .a_1a_2, .a_1a_2a_3, \ldots$

[S–15.] Show $(1.000 \ldots) \times (a_1a_2 \ldots a_n \ldots) = a_1a_2 \ldots a_n \ldots$, by multiplying the corresponding distinguished decimal sequences term by term.

M–17.   Work out algorithms for converting repeating infinite decimals to fractional notation and for writing fractions as repeating infinite decimals. For example, suppose $F = .\overline{571428}571428 \ldots$.

Then

$$10^6 \times F = 571428.571428 \ldots$$
$$- F = \qquad -.571428 \ldots$$
$$\overline{(10^6 - 1)F = 571428}$$

or

$$F = \frac{571428}{999999} = \frac{4}{7}.$$

Conversely,

$$\begin{array}{r} .571428 \ldots \\ 7\overline{)4.000000} \end{array}$$

Discuss the correspondence between infinite decimals and points on the number line. Then let the students invent infinite decimals which do not repeat. They cannot represent rational numbers but can be located on the number line.

EXERCISES 9–3

S–1.   a. $(5, ^-1, 3, 8)$

  b. $\left(12, \frac{3}{2}, ^-15\right)$

  c. $(^-8, 9)$

  d. $\left(^-1, 1, \frac{1}{4}\right)$

  e. $(0, 0, 0)$

  f. $(^-8, ^-32, 0, ^+8)$

S–3.   The correspondence is one-to-one because $(a, 0) = (b, 0)$ if and only if $a = b$.

$$a + b \leftrightarrow (a + b, 0) = (a, 0) + (b, 0)$$

$$a \times b \leftrightarrow (a \times b, 0) = (a \times b + 0 \times 0, a \times 0 + 0 \times b)$$

$$= (a, 0) \times (b, 0)$$

S–5. $(0, 0)$ is the additive identity. $(a, b) + (0, 0) = (a + 0, b + 0) = (a, b)$.

  $(1, 0)$ is the multiplicative identity. $(a, b) \cdot (1, 0) = (a \cdot 1 - b \cdot 0, a \cdot 0 + b \cdot 1) = (a, b)$.

E–7.   Two n-dimensional vectors are equal if all pairs of corresponding components are equal.

  $(a, b, c) = (d, e, f)$ if $a = d$, $b = e$, and $c = f$.

Index

Index

Abelian group, 152
Absolute value, 300
Abstract system, 139
Addison-Wesley, *Elementary School Mathematics*, 373–374, 384–385, Appendix B *passim*
Addition, over complex numbers, 338
  over integers, 285
  over natural numbers, 179
  over positive rationals, 247
  over rationals, 316–317
  over reals, 331
  over vectors, 338
    See also Algorithm(s)
Additive identity, 182, 267, 309, 318, 339
Additive inverse, 310, 318, 339
Algebra, 341
Algebraic field, 318, 333, 339
Algebraic (mathematical) system, 137, 179
Algorithm(s), 218
  addition, 218–220, 249–251, 286–288
  division, 229, 261–263, 302–303
  "four-case," 286–288, 291–292, 296–298, 302–303
  multiplication, 226–227, 255–260, 295–298
  subtraction, 221–224, 253–254, 291–292
American Book Company, *Modern Mathematics Series*, 374–375, 384–385, Appendix B *passim*
Analytic geometry, 342
Angle, 350, 351, 352
  vertex of, 350
Antisymmetric property (of order relations), 88–90
Assertion, *see* Statement

Associative property, general definition (embodiments and systems), 148
  of addition, 181, 265–266, 307, 318, 339
  of multiplication, 190, 265–266, 307–308, 318, 339
Assumption, *see* Premise
Axiom(s), 51, 105–109, 149
  system, 51, 149
Axiomatic theories, 149
Axiomatization, 105–109
  a basic process ability, 6, 49–55

Binary operation, 93–96
  defined, 95
Binary relation, 79
Borrowing (subtraction algorithm), 222
Brace notation, 117
    See also Set notation

Calculus, 342
Canonical representation, 240, 282
Canonical set, 170
Cardinal number (cardinality), defined, 174
  vs. ordinality, 169
  set property, 169–170
Cartesian product (of sets), 188
Cauchy convergent sequence, 329
"Chant set," 172
Church, Alonzo, 61–62, 116
Clock arithmetic, 139–141
Closed curve, 347
Closure property, general definition (embodiments and systems), 148
  under addition, 181, 265, 285, 318
  under division, 265

451